Fundamental
Concepts in
the Design of
Experiments

Fundamental Concepts in the Design of Experiments

Third Edition

CHARLES R. HICKS

Professor of Statistics and Education

Purdue University

HOLT, RINEHART AND WINSTON

New York   Chicago   San Francisco
Philadelphia   Montreal   Toronto   London
Sydney   Tokyo   Mexico City
Rio de Janeiro   Madrid

Library of Congress Cataloging in Publication Data

Hicks, Charles Robert, 1920–
  Fundamental concepts in the design of experiments.

  Bibliography: p. 377
  Includes index.
  1. Experimental design. I. Title.
QA279.H53 1982   001.4'34   82-12062
ISBN 0-03-061706-5

CBS COLLEGE PUBLISHING
Holt, Rinehart and Winston
The Dryden Press
Saunders College Publishing

Contents

Preface to the Third Edition   xi
Preface to the Second Edition   xiii
Preface to the First Edition   xv

1
The Experiment, the Design, and the Analysis   1

1.1 INTRODUCTION TO EXPERIMENTAL DESIGN   1
1.2 THE EXPERIMENT   2
1.3 THE DESIGN   4
1.4 THE ANALYSIS   6
1.5 SUMMARY IN OUTLINE   6
1.6 EXAMPLES   7
PROBLEMS   14

2
Review of Statistical Inference   15

2.1 INTRODUCTION   15
2.2 ESTIMATION   17
2.3 TESTS OF HYPOTHESES   19
2.4 THE OPERATING CHARACTERISTIC CURVE   21
2.5 HOW LARGE A SAMPLE?   25
2.6 APPLICATION TO TESTS ON VARIANCES   25
2.7 APPLICATION TO TESTS ON MEANS   27
2.8 APPLICATION TO TESTS ON PROPORTIONS   30
PROBLEMS   32

3
Single-Factor Experiments with No Restrictions on Randomization   36

3.1 INTRODUCTION   36
3.2 ANALYSIS OF VARIANCE RATIONALE   38
3.3 AFTER ANOVA—WHAT?   46
3.4 TESTS ON MEANS   47
3.5 CONFIDENCE LIMITS ON MEANS   54

3.6 COMPONENTS OF VARIANCE 55
3.7 SUMMARY 59
PROBLEMS 61

4

Single-Factor Experiments—Randomized Block and
Latin Square Designs 66

4.1 INTRODUCTION 66
4.2 RANDOMIZED COMPLETE BLOCK DESIGN 67
4.3 ANOVA RATIONALE 72
4.4 MISSING VALUES 73
4.5 LATIN SQUARES 76
4.6 INTERPRETATIONS 78
4.7 GRAECO-LATIN SQUARES 79
4.8 EXTENSIONS 80
4.9 SUMMARY 80
PROBLEMS 80

5

Factorial Experiments 86

5.1 INTRODUCTION 86
5.2 FACTORIAL EXPERIMENTS 91
5.3 INTERPRETATIONS 94
5.4 COMPUTER PROGRAM 96
5.5 ANOVA RATIONALE 98
5.6 REMARKS 103
5.7 SUMMARY 104
PROBLEMS 105

6

2^f Factorial Experiments 112

6.1 INTRODUCTION 112
6.2 2^2 FACTORIAL 112
6.3 2^3FACTORIAL 120
6.4 2^f—REMARKS 123
6.5 SUMMARY 124
PROBLEMS 125

7

Qualitative and Quantitative Factors 129

7.1 INTRODUCTION 129
7.2 LINEAR REGRESSION 130
7.3 CURVILINEAR REGRESSION 137
7.4 ORTHOGONAL POLYNOMIALS 139
7.5 COMPUTER PROGRAM 144
7.6 QUALITATIVE AND QUANTITATIVE FACTORS 144
7.7 TWO FACTORS—ONE QUALITATIVE, ONE QUANTITATIVE 145
7.8 TWO FACTORS—BOTH QUANTITATIVE 152

7.9 SUMMARY 159
PROBLEMS 159

8
Multiple Regression 164

8.1 INTRODUCTION 164
8.2 THE MULTIPLE REGRESSION MODEL 165
8.3 COMPUTER PROGRAMS 175
8.4 REMARKS 176
PROBLEMS 177

9
3^f Factorial Experiments 187

9.1 INTRODUCTION 187
9.2 3^2 FACTORIAL 187
9.3 3^3 FACTORIAL 198
9.4 SUMMARY 207
PROBLEMS 207

10
Fixed, Random, and Mixed Models 210

10.1 INTRODUCTION 210
10.2 SINGLE-FACTOR MODELS 211
10.3 TWO-FACTOR MODELS 212
10.4 EMS RULES 214
10.5 EMS DERIVATIONS 218
10.6 THE PSEUDO-F TEST 221
10.7 REMARKS 223
PROBLEMS 224

11
Nested and Nested-Factorial Experiments 227

11.1 INTRODUCTION 227
11.2 NESTED EXPERIMENTS 228
11.3 ANOVA RATIONALE 232
11.4 NESTED-FACTORIAL EXPERIMENTS 233
11.5 COMPUTER PROGRAMS 238
11.6 REPEATED-MEASURES DESIGN AND NESTED-FACTORIAL
       EXPERIMENTS 239
11.7 SUMMARY 244
PROBLEMS 244

12
Experiments of Two or More Factors—Restrictions on Randomization 252

12.1 INTRODUCTION 252
12.2 FACTORIAL EXPERIMENT IN A RANDOMIZED BLOCK
       DESIGN 252

12.3   FACTORIAL EXPERIMENT IN A LATIN SQUARE DESIGN   258
12.4   REMARKS   259
12.5   SUMMARY   260
PROBLEMS   261

13

Factorial Experiment—Splot-Plot Design   265

13.1   INTRODUCTION   265
13.2   A SPLIT-SPLIT-PLOT DESIGN   273
13.3   SUMMARY   277
PROBLEMS   277

14

Factorial Experiment—Confounding in Blocks   281

14.1   INTRODUCTION   281
14.2   CONFOUNDING SYSTEMS   282
14.3   BLOCK CONFOUNDING—NO REPLICATION   285
14.4   BLOCK CONFOUNDING WITH REPLICATION   290
14.5   SUMMARY   299
PROBLEMS   300

15

Fractional Replication   303

15.1   INTRODUCTION   303
15.2   ALIASES   304
15.3   FRACTIONAL REPLICATIONS   306
15.4   SUMMARY   314
PROBLEMS   315

16

Miscellaneous Topics   318

16.1   INTRODUCTION   318
16.2   COVARIANCE ANALYSIS   318
16.3   RESPONSE-SURFACE EXPERIMENTATION   328
16.4   EVOLUTIONARY OPERATION (EVOP)   341
16.5   ANALYSIS OF ATTRIBUTE DATA   351
16.6   RANDOMIZED INCOMPLETE BLOCKS—RESTRICTION ON
       EXPERIMENTATION   355
16.7   YOUDEN SQUARES   360
PROBLEMS   362

17

Summary and Special Problems   366

Glossary of Terms   373
References   377

Statistical Tables　380

    TABLE A.　AREAS UNDER THE NORMAL CURVE　380
    TABLE B.　STUDENT'S *t* DISTRIBUTION　382
    TABLE C.　CHI SQUARE　383
    TABLE D.　*F* DISTRIBUTION　384
    TABLE E.　STUDENTIZED RANGES　390
    TABLE F.　COEFFICIENTS OF ORTHOGONAL POLYNOMIALS　392

Answers to Odd-Numbered Problems　393

Index　421

Preface to the Third Edition

One of the main reasons for a third edition has been the use of computers to simplify the analysis of data in an experiment. Computer programs follow most of the complex examples showing the input instructions to a computer, the printout from the computer, and the interpretation of the results.

The introduction has been completely revised to set the stage for experimental research while still maintaining the general theme of experiment, design, and analysis. Chapters 4 and 5 of the previous edition have been combined as they both represent restrictions on randomization, and a new chapter on multiple regression has been added. Some topics rarely taught in a single-semester course have been relegated to the chapter on miscellaneous topics.

Many illustrative examples have been replaced with examples that are more practical and better illustrate the technique. There are 60 percent more problems than in the second edition, resulting in more than 300 practical problems. Chapter 17 presents ten somewhat challenging problems for class discussion.

Two minor points: (1) Factorials have been introduced as 2^f and 3^f instead of as 2^n and 3^n, where f represents the number of factors in the experiment and n is reserved for the number of observations per treatment. (2) The use of single and double horizontal and vertical lines to show how the randomization has been restricted should help illustrate how data were collected.

I am indebted to McElrath and Associates, Inc. (MAI), Management Consultants, Minneapolis, Minnesota, for the opportunity to be exposed to many practical problems in various industries over the past 20 years. Many of the text examples have come from MAI clients. Several examples, especially in the field of education, have come from students' master's and doctoral research. I want to thank the following students for examples taken from their theses: Roger Bollenbacher, Robert Horn, James Lehman, Willard Lewallen, Charlotte Macy, Thomas Marshall, Larry Miller, Barbara O'Conner, Sallam Sallam, Nadia Sherief, Ruth Tanner, John Wetz, and Catherine Yeotis.

Last, and most significant, I wish to thank Purdue University for granting me sabbatical leave to complete this third revised edition.

Charles R. Hicks

Lafayette, Indiana
January, 1982

Preface to the Second Edition

At first glance the reader may not notice much change in this edition from the original. An attempt has been made to keep chapter and section headings the same. It cannot be considered as a major revision although there are many new features. The new material adds approximately 25% to the size of the book and the number of problems has been incrased by 70%.

Main features include a section on covariance analysis and analysis of attribute data in the 16th chapter, introduction of linear and curvilinear regression in Chapter 8, and an amplification of the concept of orthogonal contrasts and orthogonal polynomials. Some topics have been rearranged to appear earlier in the text and methods of testing means after analysis of variance have been changed and extended.

Several users of the original text have sent me helpful suggestions for this revision. Whenever a suggestion came from more than one person, I attempted to include it in this edition. Naturally, all suggestions were not followed and the responsibility for their adoption is mine alone. I am indebted to the following people who made detailed suggestions:

V. L. ANDERSON   *Purdue University*

O. A. BECKENDORF   *General Motors Institute*

H. C. CHARBOMEAU   *General Motors Institute*

R. I. FIELDS   *University of Louisville*

R. J. FOPMAN   *University of Cincinnati*

B. HARSHBARGER   *Virginia Polytechnic Inst.*

O. B. MOAN   *Arizona State University*

R. G. MORRIS   *General Motors Institute*

V. V. NAMILOV   *Moscow State University*

J. B. NEUHARDT   *The Ohio State University*

C. R. PETERSON   *General Motors Institute*

B. T. RHODES, JR.   *University of Houston*

H. W. STOKER   *The Florida State University*

B. J. TODD   *Clemson University*
L. A. WEAVER   *University of South Florida*
 I am also grateful to Purdue University for granting me sabbatical leave to complete this revised edition.

Charles R. Hicks

Lafayette, Indiana
March, 1973

Preface to the First Edition

It is the primary purpose of this book to present the fundamental concepts in the design of experiments using simple numerical problems, many from actual research work. These problems are selected to emphasize the basic philosophy of design. Another purpose is to present a logical sequence of designs that fit into a consistent outline; for every type of experiment, the distinctions among the experiment, the design, and the analysis are emphasized. Since this theme of experiment–design–analysis is to be highlighted throughout the text, the first chapter presents and explains this theme. The second chapter reviews statistical inference, and the remaining chapters follow a general outline, with an experiment–design–analysis summary at the end of each chapter.

The book is written for anyone engaged in experimental work who has a good background in statistical inference. It will be most profitable reading to those with a background in statistical methods including analysis of variance.

Some of the latest techniques have not been included. The reason for this is again the purpose of presenting the *fundamental* concepts in design to provide a background for reading journals in the field and understanding some of the more recent techniques. Although it is not assumed that every reader of this book is familiar with the calculus, some theory is presented where calculus is used. The reader should be able to grasp the underlying ideas of each chapter even if he must omit these sections.

This book evolved from a course in Design of Experiments given at Purdue University over the past four or five years. The author is indebted to Professors Gayle McElrath of the University of Minnesota and Clyde Kramer of Virginia Polytechnic Institute whose lectures, along with the author's, form the basis of this book. Credit is also due Mr. James Dick and Mr. Roland Rockwell, who took notes of the author's lectures.

I shall be forever indebted to the University of Florida for granting me a visiting professorship during the academic year 1962–1963 with time to work on this book, and to Mrs. Alvarez of their Industrial Engineering Department for typing the manuscript.

Charles R. Hicks

Lafayette, Indiana
November, 1963

*Fundamental
Concepts in
the Design of
Experiments*

1

The Experiment, the Design, and the Analysis

1.1 INTRODUCTION TO EXPERIMENTAL DESIGN

One of the most abused words in the English language is probably the word "research." Research means many things to many people. The high school student who visits the library to gather information on the Crusades is doing research. A man who checks the monthly sales of his company for the year 1980 is doing research. Many such studies do make use of bibliothecal or statistical tools or similar research tools but are a far cry from a scientist's conception of the term. Let us narrow the concept a bit and define *research* as "a systematic quest for undiscovered truth." (See Leedy [16].[1])

In this definition we note that the quest must be "systematic," using sound scientific procedures—not a haphazard, seat-of-the-pants approach—to solving a problem. The truth sought should be something that is not already known, which implies a thorough literature search to ascertain that the answer is not available in previous studies.

Not all studies are necessarily research, nor is all research experimental. A *true experiment* may be defined as a study in which certain independent variables are manipulated, their effect on one or more dependent variables is determined, and the levels of these independent variables are assigned at random to the experimental units in the study. Thus manipulation and randomization are essential for a true experiment from which one may be able to infer cause and effect. It is not always physically possible to assign some variables to the experimental units at random but it may be possible to run a quasi-experiment in which groups of units are assigned to various levels of the independent variable at random. For example, to study the effects of two methods of in-plant instruction on worker output, it may not be possible to assign workers at random to the two methods, but only to decide by a random flip of a coin which two of four intact classes will receive method I and which two will receive method II. In such quasi-experiments

---

[1] Numbers in brackets refer to the references at the end of the book.

the experimenter is obligated to demonstrate that the four classes were similar at the start of the study. Since they can be dissimilar in many different ways, this is no easy task. Randomizing all individuals to the four groups will assure a high probability of group equality on many variables.

There are other types of research that are not experimental. One of the most common is *ex-post-facto* research. The variables have already acted and the researcher only measures what has occurred. An example of some note would be the studies showing a high incidence of cancer in people who smoke heavily. There is no manipulation here as the researchers do not assign X number of cigarettes per day to some people and none to another group, and then wait to see whether or not cancer develops. Instead they study what they can find from people's historical records, and then attempt to make an inference. However, such inferences are dangerous, as the researchers must be able to rule out other competing hypotheses as to what may have caused cancer in the people studied.

In this category are survey research, which is also ex post facto, historical research, and developmental research. All of these may use statistical methods to analyze data collected but none are really experimental in nature.

An illustration familiar to statisticians is found in the techniques of regression and correlation. In a regression problem levels of the independent variable X are set and observations are made on the dependent variable Y, and the given level of X is assigned at random to each of the experimental units in the study. The purpose of such a study is to find an equation for predicting Y from X over the range of specified X's. In a correlation problem each experimental unit is measured on two variables, the n pairs (X, Y) are plotted to look for a relationship, and certain statistics are computed to indicate the strength of the relationship. In this case a strong correlation (high r) does not imply that X caused Y or that Y caused X. Hence regression analysis can often be used in a true experiment whereas correlational techniques are more likely to be used in ex-post-facto studies. This distinction between regression and correlation is often misunderstood.

Assuming that true experiments are our basic concern, what steps should be taken to make a meaningful study?

To attack this problem one may easily recognize three important phases of every project: the experimental or planning phase, the design phase, and the analysis phase. The theme of this book might be called: Experiment, Design, and Analysis.

1.2 THE EXPERIMENT

The experiment includes a statement of the problem to be solved. This sounds rather obvious, but in practice it often takes quite a while to get general agreement as to the statement of a problem. It is important to bring out all points of view to establish just what the experiment is intended to do. A careful statement of the problem goes a long way toward its solution.

Too often problems are tossed about in very general terms, a sort of "idea burping" (see [16]). As an example, note the difference in these two statements of a problem: What can Salk vaccine do for polio? Is there a difference in the percentage of first, second, and third grade children in the United States who contract polio within a year after being inoculated with Salk vaccine and those who are not inoculated? Much care must be taken to spell out the problem in terms that are well understood and that point the way in which the research might be conducted.

The statement of the problem must include reference to one or more criteria to be used in assessing the results of the study. The criterion is also called the dependent variable or the response variable Y; and there may be more than one Y of interest in a given study. The dependent variable may be qualitative, such as the number or percentage contracting polio, or quantitative, such as yield in grams of penicillin per kilogram of corn steep. Knowledge of this dependent variable is essential because the shape of its distribution often dictates what statistical tests can be used in the subsequent data analysis.

One must also ask: Is the criterion measurable, and if so, how accurately can it be measured? What instruments are necessary to measure it?

It is also necessary to define the independent variables or factors that may affect the dependent or response variable. Are these factors to be held constant, to be manipulated at certain specified levels, or to be averaged out by a process of randomization? Are levels of the factors to be set at certain fixed values, such as temperature at 70°F, 90°F, and 110°F, or are such levels to be chosen at random from among all possible levels? Are the factors to be varied quantitative (such as temperature) or qualitative (operators)? And how are the various factor levels to be combined?

If one factor A is set at three levels and a second factor B is set at two levels, can one set all six combinations as follows?

		Factor A		
		1	2	3
Factor B	1			
	2			

This is called a factorial experiment as all levels of A are combined with all levels of B. Sometimes one chooses three suppliers and then randomly selects two batches from each supplier to be analyzed chemically. Here the two batches are unique to a given supplier and the resulting arrangement is a nested or hierarchical arrangement.

	Supplier	
1	2	3
Batch	Batch	Batch
1 2	3 4	5 6

This involves six sets of data as did the above factorial, but now the factors are in a decidedly different arrangement.

All of these points should be considered as part of the experiment phase of any research.

1.3   THE DESIGN

Of primary importance, since the remainder of this book will be devoted to it, is the design phase of a project. Many times an experiment is agreed upon, data are collected, and conclusions are drawn with little or no consideration given to *how* the data were collected. First, how many observations are to be taken? Considerations of how large a difference is to be detected, how much variation is present, and what size risks are to be tolerated are all important in deciding on the size of the sample to be taken for a given experiment. Without this information the best alternative is to take as large a sample as possible. In practice this sample size is often quite arbitrary. However, as more and more tables become available, it should be possible to determine the sample size in a much more objective fashion (see Section 2.5).

Also of prime importance is the order in which the experiment is to be run, which should be random order. Once a decision has been made to control certain variables at specified levels, there are always a number of other variables that cannot be controlled. Randomization of the order of experimentation will tend to average out the effect of these uncontrolled variables.

For example, if an experimenter wishes to compare the average current flow through two types of computers and five of each type are to be tested, in what order are all ten to be tested? If the five of type I are tested, followed by the five of type II, and any general "drift" in line voltage occurs during the testing, it may appear that the current flow is greater on the first five (type I) than on the second five (type II), yet the real cause is the "drift" in line voltage. A random order for testing allows any time trends to average out. It is desirable to have the average current flow in the type I and type II computers equal, if the computer types do not differ in this respect. Randomization will help accomplish this. Randomization will also permit the experimenter to proceed as if the errors of measurement were independent, a common assumption in most statistical analyses.

What is meant by random order? Is the whole experiment to be completely randomized, with each observation made only after consulting a table of random numbers, or tossing dice, or flipping a coin? Or, possibly, once a temperature bath is prepared, is the randomization made only within this particular temperature and another randomization made at another temperature? In other words, what is the randomization procedure and how are the units arranged for testing? Once this step has been agreed upon, it is recommended that the experimenter keep a watchful eye on the experiment to see that it is actually conducted in the order prescribed.

Having agreed upon the experiment and the randomization procedure, a mathematical model can now be set up, which should describe the experiment. This model will show the response variable as a function of all factors to be studied and any restrictions imposed on the experiment as a result of the method of randomization.

Since the objective of the research project is to shed light on a stated problem, one should now express the problem in terms of a testable hypothesis or hypotheses. A research hypothesis states what the experimenter expects to find in the data. For example: "Inoculation with Salk vaccine will lower the incidence of polio." Or: "The four treatments will produce different average yields."

To test hypotheses, a statistician usually states the hypothesis in null form, since this is the only testable hypothesis. For example: "There will be no difference in the percentage of youngsters contracting polio between those inoculated and those not inoculated." Or: "The treatment means will have the same average yield." This null hypothesis form can usually be expressed in terms of the mathematical model set up in the last step of the design phase.

In this discussion of variables that influence the criterion either by being purposefully introduced into the study, representing restrictions on randomization, or being somewhat extraneous, one notes that there are basically three ways to handle independent variables in an experiment:

1. Rigidly controlled the variables remain fixed throughout the experiment, which, of course, means that any inferences drawn from the experiment are valid only for these fixed conditions. Here, for example, one might decide to study only children from the first three grades.
2. Manipulated or set at levels of interest. These would be the variables whose effect on Y are to be studied. They could be either qualitative or quantitative and their levels either fixed or random. To be effective, every effort should be made to include the extreme levels of such variables to provide opportunity to maximize their effect, if present.
3. Randomized. The order of experimentation is randomized to average out the effects of variables that cannot be controlled. Such averaging does not remove their effect completely as they still increase the variance of the observed data. Proper design of the experiment can also reduce or minimize the experimental error when such factors are anticipated.

A basic principle in such experimentation is referred to as the *max-min-con principle* (see Kerlinger [14]), where we seek to maximize the effect of the independent variables of interest, minimize the error variance, and control some of the variables at specified levels.

1.4 THE ANALYSIS

The final step, analysis, includes the procedure for data collection, data reduction, and the computation of certain test statistics to be used in making decisions about various aspects of an experiment. Analysis involves the computation of test statistics such as t, F, χ^2, and their corresponding decision rules for testing hypotheses about the mathematical model. Once the test statistics have been computed, decisions must be made. These decisions should be made in terms that are meaningful to the experimenter. They should not be couched in statistical jargon such as "the third-order $A \times B \times E$ interaction is significant at the 1 percent level," but instead should be expressed in graphical or tabular form, in order that they be clearly understood by the experimenter and by those persons who are to be "sold" by the experiment. The actual statistical tests should probably be included only in the appendix of a report on the experiment. These results should also be used as "feedback" to design a better experiment, once certain hypotheses seem tenable.

1.5 SUMMARY IN OUTLINE

  I. *Experiment*
    A. Statement of problem
    B. Choice of response or dependent variable
    C. Selection of factors to be varied
    D. Choice of levels of these factors
       1. Quantitative or qualitative
       2. Fixed or random
    E. How factor levels are to be combined

  II. *Design*
    A. Number of observations to be taken
    B. Order of experimentation
    C. Method of randomization to be used
    D. Mathematical model to describe the experiment
    E. Hypotheses to be tested

  III. *Analysis*
    A. Data collection and processing
    B. Computation of test statistics
    C. Interpretation of results for the experimenter

1.6 EXAMPLES

Example 1.1 Salk Vaccine Experiment. The experiment conducted in the United States in 1954 concerning the use of Salk vaccine in the control of polio has become a classic in its employment of good design principles. A more detailed description of the study can be found in *Statistics: A Guide to the Unknown* (see [23]).

Before the statement of the problem was agreed upon as given in Section 1.2, there was much discussion as to what population was to be sampled. It was decided that children in grades 1, 2, and 3 would be most appropriate, as it was among children of these ages that the disease was most prevalent.

Implied in the statement of the problem "Is there a difference in the percentage of first, second, and third grade children in the United States contracting polio within a year after being inoculated with Salk vaccine and those not inoculated?" is the concept of a criterion. The stated criterion is the percentage of children contracting polio. This dependent variable could be either the percent contracting the disease or the number contracting it. In discussing this variable it was noted that the overall percentage of children contracting polio was quite small—about eight or ten cases in 10,000 children. It was therefore necessary to take a fairly large sample in order to get any percentages large enough to discriminate between those who were inoculated and those who were not. The basic variable here is a qualitative one—a binomial variable—as children fall into just two categories: contracting polio or not contracting polio. One can safely assume an approximately normal distribution of this Y if the average number in the sample is expected to be four or more.

The main factor of interest in affecting Y is whether or not the child is inoculated, so it was planned to have an approximately equal number of children receive the vaccine and be given a placebo (a salt solution). This latter group would act as a control.

Other factors that might affect the contraction of polio included socio-economic status, grade level, doctors who make the diagnosis, and geographic area of residence. Grade level was considered and they agreed to sample from the first, second, and third grades. The other factors, or nuisance variables, were handled in the design procedures.

Since there is only one qualitative, fixed factor of interest—inoculated or not inoculated—there is no basic concern about combining factors. However, because three grade levels were to be considered, it was decided to stratify their sample by grades; that is, take approximately equal sample sizes of inoculated and noninoculated children from each grade level.

Much planning went into the design phase. At first it was proposed that children in grades 1 and 3 receive the vaccine and those in grade 2 be given the placebo treatment. This plan was later abandoned because polio is a contagious disease, and if children in a given grade get polio, others in that grade might contract it as well. Another early idea was to use children of parents who would consent to the inoculation as "experimental" and children of parents not consenting as a "control." This too was abandoned because parents of a higher socioeconomic status are more likely to consent than parents with a lower socioeconomic status and polio is a socially related disease—it affects more children from the upper echelons of society than from the lower!

The final decision was to do a complete randomization countrywide involving about 1 million randomly selected children from grades 1–3 in areas of the United States where polio had been quite prevalent in the past. Of these 1 million, more than 400,000 parents consented to have their children participate in the study. These children were then divided into two groups at random with group 1 getting the Salk vaccine and group 2 getting the placebo. To handle other concerns, the experiment was run as a "double-blind" experiment. The children did not know whether they were receiving the vaccine or the salt solution and the doctors who diagnosed the children did not know which children were inoculated and which were not.

The samples of approximately 200,000 each seemed adequate to permit the detection of differences in the percent contracting the disease, if it did show up.

A mathematical model could be written as

$$Y_{ij} = \mu + \tau_j + \varepsilon_{ij}$$

where Y_{ij} is equal to 0 or 1: 0 if no polio is diagnosed and 1 if polio is found. The subscipt i represents the child, a number from 1 to approximately 200,000, and j is 1 or 2: 1 if treated with Salk vaccine and 2 if not so treated. μ is a constant or general average of the 0's and 1's for the whole population. τ_j is the treatment effect: $j = 1$ if treated, $j = 2$ if not treated. ε_{ij} is a random error associated with a child receiving treatment j.

To see the hypothesis to be tested, note that

$$p_j = \frac{\sum_i^{n_j} Y_{ij}}{n_j}$$

is the proportion of children contracting polio who received treatment j. n_j is the number who received treatment j.

The statistical hypothesis to be tested is then

$$H_0: p_1' = p_2'$$

with alternative

$$H_1: p_1' < p_2'$$

where the sign $<$ indicates that we are interested in showing that the true proportion contracting polio will be less for treatment 1 (vaccine) than for treatment 2 (placebo). The primes on the p values are to show that the hypotheses are statements about population proportions whereas the samples will give p_1 and p_2—observed sample proportions. The data are shown in Table 1.1.

Table 1.1 Polio Experiment Results

Treatment	Sample Size	Number Contracting Polio	Proportion Contracting
Salk vaccine	$n_1 = 200{,}745$	56	$p_1 = 28 \times 10^{-5}$
Placebo	$n_2 = 201{,}229$	142	$p_2 = 71 \times 10^{-5}$
Totals	401,974	198	$\hat{p} = 49 \times 10^{-5}$

The test statistic for testing the hypothesis stated is

$$z = \frac{p_1 - p_2}{\sqrt{\hat{p}\hat{q}(1/n_1 + 1/n_2)}}$$

where $\hat{p}$ is the proportion contracting polio in the total sample of $n_1 + n_2$ children. Here $\hat{q} = 1 - \hat{p}$.

If one now assumes an α risk of 0.001—taking one chance in 1000 of rejecting H_0 when it is true—the rejection region for z is given by $z < -3.09$ (from the normal distribution table). Substituting the values above:

$$z = \frac{28 \times 10^{-5} - 71 \times 10^{-5}}{\sqrt{49 \times 10^{-5}(1 - 49 \times 10^{-5})[1/200{,}745 + 1/201{,}229]}}$$

$$z = -6.14$$

which gives a very, very strong indication that we should reject H_0 and conclude that the vaccine was indeed effective.

In fact, the probability of getting a z as low as or lower than -6.14 when there is really no difference in the two groups is less than one chance in a billion. These amazing results, of course, have been strongly substantiated by the almost complete eradication of polio in the United States.

Example 1.2 The following example [18] is presented to show the three phases of the design of an experiment. It is not assumed that the reader is familiar with the design principles or analysis techniques in this problem. The remainder of the book is devoted to a discussion of many such principles and techniques, including those used in this problem.

An experiment was to be designed to study the effect of several factors on the power requirements for cutting metal with ceramic tools. The metal was cut on a lathe and the y or vertical component of a dynamometer reading was recorded. As this y component is proportional to the horse-power requirements in making the cut, it was taken as the measured variable. The y component is measured in millimeters of deflection on a recording instrument. Some of the factors that might affect this deflection are tool types, angle of tool edge bevel, type of cut, depth of cut, feed rate, and spindle speed. After much discussion, it was agreed to hold depth of cut constant at 0.100 inch, feed rate constant at 0.012 in./min, and spindle speed constant at 1000 rpm. These levels were felt to represent typical operating conditions. The main objective of the study was to determine the effect of the other three factors (tool type, angle of edge bevel, and type of cut) on the power require-ments. As only two ceramic tool types were available, this factor was con-sidered at two levels. The angle of tool edge bevel was also set at two levels, 15° and 30°, representing the extremes for normal operation. The type of cut was either continuous or interrupted—again, two levels.

There are therefore two fixed levels for each of three factors, or eight experimental conditions (2^3), which may be set and which may affect the power requirements or y deflection on the dynamometer. This is called a 2^3 factorial experiment, since both levels of each of the three factors are to be combined with both levels of all other factors. The levels of two factors (type of tool and type of cut) are qualitative, whereas the angle of edge bevel (15° and 30°) is a quantitative factor.

The question of design for this experiment involves the number of tests to be made under each of the eight experimental conditions. After some preliminary discussion of expected variability under the same set of condi-tions and the costs of wrong decisions, it was decided to take four observa-tions under each of the eight conditions, making a total of 32 runs. The order in which these 32 units were to be put in a lathe and cut was to be completely randomized.

In order completely to randomize the 32 readings, the experimenter decided on the order of experimentation from the results of three coin toss-ings. A penny was used to represent the tool type T: heads for one type, tails for the other; a nickel represented the angle of bevel B: heads 30°, tails 15°; and a dime represented the type of cut C: heads interrupted, tails continuous.

Thus if the first set of tosses came up THT, it would mean that the first tool to be used in the lathe would be of tool type 1, bevel 2, and sub-jected to the continuous type of cut. A data layout is given in Figure 1.1 that shows each of the 32 experimental conditions. The numbers 1, 2, 3, 4, and 5 indicate the first five conditions to be run on the lathe, assuming the coins came up THT, TTH, HHT, HHT, HTH.

In this layout it should be noted that the same set of conditions may be repeated (for example, runs 3 and 4) before all eight conditions are run once.

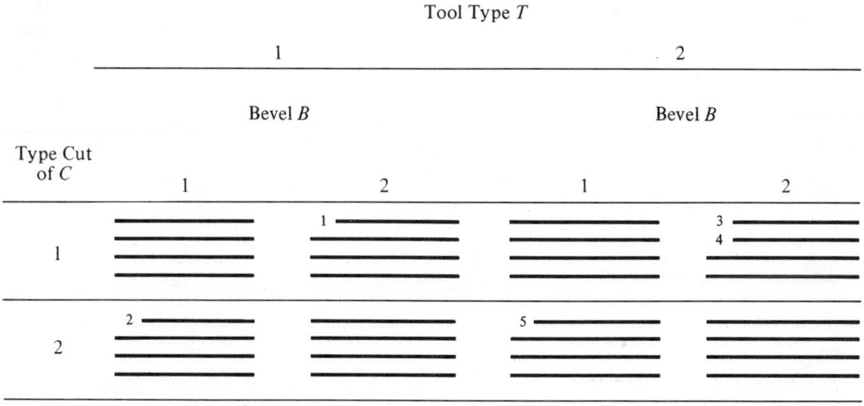

Tool Type T

Figure 1.1 Data layout of power consumption for ceramic tools.

The only restriction on complete randomization here is that once four repeated measures have occurred in the same cell, no more will be run using those same conditions.

The coin flipping continues until the order of all 32 runs has been decided upon. This is a 2^3 factorial experiment with four observations per cell, run in a completely randomized manner. Complete randomization ensures the averaging out of any effects that might be correlated with the time of the experiment. If the lathe-spindle speed should vary and all of the type 1 tools are run through first and the type 2 tools follow, this extraneous effect of lathe-spindle speed might appear as a difference between tool types if the speed were faster at first and slower near the end of the experiment.

The mathematical model for this experiment and design would be $Y_{ijkm} = \mu + T_i + B_j + TB_{ij} + C_k + TC_{ik} + BC_{jk} + TBC_{ijk} + \varepsilon_{m(ijk)}$, where Y_{ijkm} represents the measured variable, μ a common effect in all observations (the true mean of the population from which all the data came), T_i the tool type effect where $i = 1, 2$, B_j the angle of bevel where $j = 1, 2$, and C_k the type of cut where $k = 1, 2$. $\varepsilon_{m(ijk)}$ represents the random error in the experiment where $m = 1, 2, 3, 4$. The other terms stand for interactions between the main factors T, B, and C.

The analysis of this experiment consists of collecting 32 items of data in the spaces indicated in Figure 1.1 in a completely randomized manner. The results in millimeter deflection are given in Table 1.2.

This experiment and the mathematical model suggest a three-way analysis of variance (ANOVA), which yields the results in Table 1.3.

In testing the hypotheses that there is no type of tool effect, no bevel effect, no type of cut effect, and no interactions, the EMS column indicates that all observed mean squares are to be tested against the error mean

Table 1.2   Data for Power Requirement Example

	Tool Type T			
	1		2	
	Bevel Angle B		Bevel Angle B	
Type of Cut C	15°	30°	15°	30°
Continuous	29.0	28.5	28.0	29.5
	26.5	28.5	28.5	32.0
	30.5	30.0	28.0	29.0
	27.0	32.5	25.0	28.0
Interrupted	28.0	27.0	24.5	27.5
	25.0	29.0	25.0	28.0
	26.5	27.5	28.0	27.0
	26.5	27.5	26.0	26.0

Table 1.3   ANOVA for Power Requirement Example

Source of Variation	Degrees of Freedom (df)	Sum of Squares (SS)	Mean Square (MS)	Expected Mean Square (EMS)
Tool types T	1	2.82	2.82	$\sigma_e^2 + 16\phi_T$
Bevel B	1	20.32	20.32*	$\sigma_e^2 + 16\phi_B$
$T \times B$ interaction	1	0.20	0.20	$\sigma_e^2 + 8\phi_{TB}$
Type of cut C	1	31.01	31.01*	$\sigma_e^2 + 16\phi_C$
$T \times C$ interaction	1	0.01	0.01	$\sigma_e^2 + 8\phi_{TC}$
$B \times C$ interaction	1	0.94	0.94	$\sigma_e^2 + 8\phi_{BC}$
$T \times B \times C$ interaction	1	0.19	0.19	$\sigma_e^2 + 4\phi_{TBC}$
Error $\varepsilon_{m(ijk)}$	24	53.44	2.23	σ_e^2
Totals	31	108.93		

* One asterisk indicates significance at the 5 percent level; two, the 1 percent level; and three, the 0.1 percent level. This notation will be used throughout this book.

square of 2.23 with 24 degrees of freedom (df). The proper test statistic is the F statistic (Appendix, Table D) with 1 and 24 df. At the 5 percent significance level ($\alpha = 0.05$) the critical region of F is $F \geq 4.26$. Comparing each mean square with the error mean square indicates that only two hypotheses can be rejected: bevel has no effect on deflection and type of cut has no effect on deflection. None of the other hypotheses can be rejected, and it is concluded that only the angle of bevel and type of cut affect power consumption as measured by the y deflection on the dynamometer. Tool type appears to have little effect on the y deflection, and all interactions are negligible.

Table 1.4 Average y Deflections

Tool type T	1	28.1
	2	27.5
Bevel angle B	15°	27.0
	30°	28.6
Type of cut C	1	28.8
	2	26.8

Calculations on the original data of Table 1.2 show the average y deflections given in Table 1.4.

These averages seem to bear out the conclusions that bevel affects y deflection, with a 30° bevel requiring more power than the 15° bevel (note the difference in average y deflection of 1.6 mm), and that type of cut affects y deflection, with a continuous cut averaging 2.0 mm more deflection than an interrupted cut. The difference of 0.6 mm in average deflection due to tool type is not significant at the 5 percent level. Graphing all four B–C combinations in Figure 1.2 shows the meaning of no significant interaction.

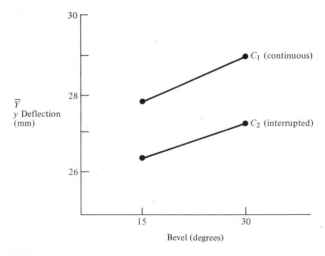

Figure 1.2 $B \times C$ interaction for power requirement example.

A brief examination of this graph indicates that the y deflection is increased by an increase in degree of bevel. The fact that the line for the continuous cut C_1 is above the line for the interrupted cut C_2 shows that the continuous cut requires more power. The fact that the lines are nearly parallel is characteristic of no interaction between two factors. Or it can easily be seen that an increase in the degree of bevel produced about the same average increase in y deflection regardless of which type of cut was made. This is another way to interpret the presence of no interaction.

The experiment here was a three-factor experiment with two levels for each factor. The design was a completely randomized design and the analysis was a three-way ANOVA with four observations per cell. From the results of this experiment the experimenter not only found that two factors (angle of bevel and type of cut) affect the power requirements, but also determined that within the range of the experiment it makes little difference which ceramic tool type is used and that there are no significant interactions among the three factors.

These are but two examples that might have been used to illustrate the three phases of the project: experiment, design, and analysis.

PROBLEMS

1.1 In your own area of interest, write out the statement of a problem that you believe is researchable.

1.2 For the problem stated in (1), list the dependent variable or variables of interest and how you propose to measure them.

1.3 For the problem stated in (1), list the independent variables that might affect your dependent variable(s) and explain how you would handle each variable.

1.4 Prepare a data layout for collecting data on your problem.

*Review of
Statistical Inference*

2.1 INTRODUCTION

In any experiment the experimenter is attempting to draw certain inferences or make a decision about some hypothesis or "hunch" concerning the situation being studied. Life consists of a series of decision-making situations. As you taste your morning cup of coffee, almost unconsciously you decide that it is better than, the same as, or worse than the coffee you have been drinking in the past. If, based on your idea of a "standard" cup of coffee, the new cup is sufficiently "bad," you may pour it out and make a new batch. If you believe it superior to your "standard," you may try to determine whether the brand is different than the one you are used to or the brewing is different, and so forth. In any event, it is the extreme differences that may cause you to decide to take action. Otherwise, you behave as if nothing has changed. In any decision you run the risk that the decision is wrong since it is based on a small amount of data.

In deciding whether or not to carry an umbrella on a given morning, you "collect" certain data: you tap the barometer, look at the sky, read the newspaper forecast, listen to the radio, and so on. After quickly assimilating all available data—including such predictions as "a 30 percent probability of rain today"—you make a decision. Somehow a compromise is made between the inconvenience of carrying the umbrella and the possibility of having to spend money for a cleaning bill for one's clothes.

In these instances, and in most everyday events, decisions are made in the light of uncertainty. *Statistics* may be defined as a tool for decision making in the light of uncertainty. Uncertainty does not imply no knowledge, but only that the exact outcome is not completely predictable. If ten coins are tossed, one knows that the number of heads will be some integer between 0 and 10. However, each specific integer has a certain chance of occurring and various results may be predicted in terms of their chance or probability of occurrence. When the results of an experiment that are observed could have occurred only 5 times in 100 by chance alone, most experimenters consider that this is a rare event and will state that the results are statistically significant at the 5 percent significance level. In such cases the hypothesis being tested is usually rejected as untenable. When statistical methods are used in experimentation, one can assess the magnitude of the risks taken in making a particular decision.

Statistical inference refers to the process of inferring something about a population from a sample drawn from that population. The population consists of all possible values of some random variable Y, where Y represents the response or dependent variable being measured. The response Y may represent tensile strength, weight, score, reaction time, or whatever criterion is being used to evaluate the experimental results. Characteristics of the population of this random variable are called *parameters*. In general θ will be used to designate any given population parameter. The average or expected value of the random variable is designated as $E(Y) = \mu$. If the probability function defining the random variable is known,

$$E(Y) = \sum Y_i p(Y_i)$$

where $p(Y_i)$ is a discrete probability function, and

$$E(Y) = \int Y f(Y) \, dY$$

where $f(Y)$ is a continuous probability density function.

The long-range average of squared deviations from the mean of a population is called the *population variance* σ_Y^2. Or

$$E[(Y - \mu_Y)^2] = \sigma_Y^2$$

The square root of this parameter is called *population standard deviation* σ_Y. Quantities computed from the sample values drawn from the population are called *sample statistics*, or simply, statistics. Examples include the sample mean

$$\bar{Y} = \sum_i^n Y_i/n$$

where n is the number of observations in the sample, and the sample variance

$$s^2 = \sum_{i=1}^n (Y_i - \bar{Y})^2/(n - 1)$$

Italic letters will be used to designate sample statistics. The symbol u will be used to designate a general statistic corresponding to the population parameter θ.

Most statistical theory is based on the assumption that samples drawn are *random samples*, that is, that each member of the population has an equal chance of being included in the sample and that the pattern of variation in the population is not changed by this deletion of the n members for the sample.

The notion of statistical inference may be divided into two parts: (1) estimation and (2) tests of hypotheses.

2.2   ESTIMATION

The objective of statistical estimation is to make an estimate of a population parameter based on a sample statistic drawn from this population. Two types of estimates are usually needed, point estimates and interval estimates.

A *point estimate* is a single statistic used to estimate a parameter. For example, the sample mean $\bar{Y}$ is a point estimate of the population mean μ. Point estimates are usually expected to have certain desirable characteristics. They should be unbiased, consistent, and have minimum variance.

An *unbiased statistic* is one whose expected or average value taken over an infinite number of similar samples equals the population parameter being estimated. Symbolically, $E(u) = \theta$, where $E(\cdot)$ means the expected value of the statistic in $(\cdot)$. The sample mean is an unbiased statistic, since it can be proved that

$$E(\bar{Y}) = \mu_Y \qquad \text{(or simply, } \mu)$$

Likewise, the sample variance as defined above is unbiased, since

$$E(s^2) = E\left[\sum_i (Y_i - \bar{Y})^2/(n - 1)\right] = \sigma_Y^2 \qquad \text{(or } \sigma^2)$$

Note that the sum of squares $\sum_{i=1}^n (Y_i - \bar{Y})^2$ must be divided by $n - 1$ and not by n if s^2 is to be unbiased. The sample standard deviation

$$s = \sqrt{\sum_{i=1}^n (Y_i - \bar{Y})^2/(n - 1)}$$

is not unbiased, since it can be shown that

$$E(s) \neq \sigma$$

This somewhat subtle point is proved in Burr (see [7], pp. 101–104).

A few basic theorems involving expected values that will be useful later in this book include

$$E(k) = k \quad \text{(where } k \text{ is a constant)}$$
$$E(kY) = kE(Y)$$
$$E(Y_1 + Y_2) = E(Y_1) + E(Y_2)$$
$$E(Y - \mu_Y)^2 = E(Y^2) - \mu_Y^2 = \sigma_y^2$$

The statistic $\sum_i (Y_i - \bar{Y})^2$ is used so frequently that it is often referred to as the *sum of squares*, or SS_Y. It is actually the sum of the squares of deviations of the sample values from their sample mean.

From the above it is easily seen that

$$E\left[\sum_i (Y_i - \bar{Y})^2\right] = (n - 1)\sigma_Y^2$$

or

$$E(SS_Y) = (n - 1)\sigma_Y^2$$

where $n - 1$ represents the degrees of freedom associated with the sum of squares on the left. In general, for any statistic u,

$$E\left[\sum_i (u_i - \bar{u})^2\right] = (\text{df on } SS_u)\sigma_u^2$$

or

$$E(SS_u) = (\text{df on } SS_u)\sigma_u^2$$

When a sum of squares is divided by its degrees of freedom, it indicates an averaging of the squared deviations from the mean, and this statistic, SS/df, is called a *mean square*, or MS. Hence

$$E(MS) = E(SS/df) = \sigma^2$$

of the variable or statistic being studied.

A *consistent statistic* is one whose value comes closer and closer to the parameter as the sample size is increased. Symbolically,

$$\lim_{n \to \infty} \Pr(|u_n - \theta| < \varepsilon) \to 1$$

which expresses the notion that as n increases, the probability approaches certainty (or 1) that the statistic u_n, which depends on n, will be within a distance ε, however small, of the true parameter θ.

Minimum variance applies where two or more statistics are being compared. If u_1 and u_2 are two estimates of the same parameter θ, the estimate having the smaller standard deviation is called the minimum-variance estimate. This is shown diagrammatically in Figure 2.1. In Figure 2.1, $\sigma_{u_1} < \sigma_{u_2}$, and u_1 is then said to be the *minimum-variance estimate* as variance is the square of the standard deviation.

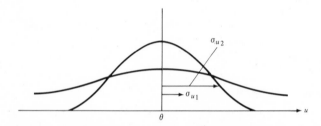

Figure 2.1   Minimum-variance estimates.

An *interval estimate* consists of the interval between two values of a sample statistic that is asserted to include the parameter in question. The band of values between these two limits is called a *confidence* interval for the parameter, since its width may be determined from the degree of confidence that is assumed when we say that the parameter lies in this band. Confidence limits (the end points of the confidence interval) are determined from the observed sample statistic, the sample size, and the degree of confidence desired. A 95 percent confidence interval on μ is given by

$$\bar{Y} \pm 1.96 \frac{\sigma}{\sqrt{n}}$$

where 1.96 is taken from normal distribution values (see Appendix, Table A), n is the sample size, and σ is the population standard deviation. If 100 sample means based on n observations each are computed, and 100 confidence intervals are set up using the above formula, we expect that about 95 of the 100 intervals will include μ. If only one interval is set up based on one sample of n, as is usually the case, we can state that we have 95 percent confidence that this interval includes μ. If σ is unknown, the Student's t distribution (Appendix, Table B) is used, and confidence intervals are given by

$$\bar{Y} \pm t_{1 - \alpha/2} \frac{s}{\sqrt{n}}$$

where s is the sample standard deviation and t has $n - 1$ df. $100(1 - \alpha)$ percent gives the degree of confidence desired.

2.3   TESTS OF HYPOTHESES

A *statistical hypothesis* is an assumption about the population being sampled. It usually consists of assigning a value to one or more parameters of the population. For example, it may be hypothesized that the average number of miles per gallon obtained with a certain carburetor is 19.5. This is expressed as $H_0: \mu = 19.5$ mi/gal. The basis for the assignment of this value to μ usually rests on past experience with similar carburetors. Another example would be to hypothesize that the variance in weight of filled vials for the week is 40 grams^2 or $H_0: \sigma^2 = 40$ g^2. When such hypotheses are to be tested, the other parameters of the population are either assumed or estimated from data taken on a random sample from this population.

A *test of a hypothesis* is simply a rule by which a hypothesis is either accepted or rejected. Such a rule is usually based on sample statistics, called *test statistics*, when they are used to test hypotheses. For example, the rule might be to reject $H_0: \mu = 19.5$ mi/gal if a sample of 25 carburetors averaged

18.0 mi/gal ($\bar{Y}$) or less when tested. The *critical region* of a test statistic consists of all values of the test statistic where the decision is made to reject H_0. In the example above, the critical region for the test statistic $\bar{Y}$ is where $\bar{Y} \leq 18.0$ mi/gal.

Since hypothesis testing is based on observed sample statistics computed on n observations, the decision is always subject to possible errors. If the hypothesis is really true and it is rejected by the sample, a *type I error* is committed. The probability of a type I error is designated as α. If the hypothesis is accepted when it is not true, that is, some alternative hypothesis is true, a *type II error* has been made and its probability is designated as β. These α and β error probabilities are often referred to as the risks of making incorrect decisions, and one of the objectives in hypothesis testing is to design a test whose α and β risks are both small. In most such test procedures α is set at some predetermined level, and the decision rule is then formulated in such a way as to minimize the other risk, β. In quality control work, α is the producer's risk and β the consumer's risk.

To review hypothesis testing, a series of steps can be taken that will apply to most types of hypotheses and test statistics. To help clarify these steps and to illustrate the procedure, a simple example is given parallel to the steps.

Steps in Hypothesis Testing	*Examples*
1. Set up the hypothesis and its alternative.	1. $H_0: \mu = 19.5$ mi/gal. $H_1: \mu < 19.5$ mi/gal.
2. Set the significance level of the test α and the sample size n.	2. $\alpha = 0.05, n = 25$.
3. Choose a test statistic to test H_0 noting any assumptions necessary when applying this statistic.	3. Test statistic: $\bar{Y}$ or standardized $\bar{Y}$ $$Z = \frac{\bar{Y} - \mu}{\sigma/\sqrt{n}}$$ (Assume σ known and $=2$.)
4. Determine the sampling distribution of this test statistic when H_0 is true.	4. $\bar{Y}$ is normally distributed with mean μ and standard deviation $\sigma/\sqrt{n}$. Or Z is $N(0, 1)$.
5. Set up a critical region on this test statistic where H_0 will be rejected in $(100)\alpha$ percent of the samples when H_0 is true.	5. 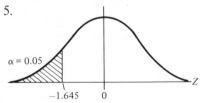

Critical region: $Z \leq -1.645$

6. Choose a random sample of n observations, compute the test statistic, and make a decision on H_0.

6. If $n = 25$ and $\bar{Y} = 18.9$ mi/gal,

$$Z = \frac{18.9 - 19.5}{2/\sqrt{25}} = -1.5.$$

As $-1.5 > -1.645$, do not reject H_0.

In the example above a one-sided or one-tailed test was used. This is dictated by the alternative hypothesis, since we only wish to reject H_0 when low values of $\bar{Y}$ are observed. The size of the significance level α is often set in an arbitrary fashion such as 0.05 or 0.01. It should reflect the seriousness of rejecting many carburetors when they are really satisfactory, or when the actual mean of the lot (population) is 19.5 mi/gal or better. In using the normal variate Z, σ is assumed known; a different test statistic would be used if σ were unknown, namely, Student's t. The critical region may also be expressed in terms of $\bar{Y}$ using the critical Z value of -1.645:

$$-1.645 = \frac{\bar{Y}_c - 19.5}{2/\sqrt{25}}$$

or $\bar{Y}_c = 18.8$, and the decision rule can be expressed as: Reject H_0 if $\bar{Y} \leq 18.8$. Here $\bar{Y} = 18.9$ and the hypothesis is not rejected.

The procedure outlined may be used to test many different hypotheses. The nature of the problem will indicate what test statistic is to be used, and proper tables can be found to set up the required critical region. Well known are tests such as those on a single mean, two means with various assumptions about the corresponding variances, one variance, and two variances. These tests are reviewed in Sections 2.6–2.8.

2.4   THE OPERATING CHARACTERISTIC CURVE

In the example given above no mention was made of the type II error of size β. β is the probability of accepting the original hypothesis H_0 when it is not true or when some alternative hypothesis, H_1, is true. Now H_1 in this example states that $\mu < 19.5$ mi/gal, and there are many possible means that satisfy this alternative hypothesis. Thus β is a function of the μ that is less than the hypothesized μ value. To see how β varies with μ, let us consider several possible μ values and sketch the distributions of sample means $\bar{Y}$ that these various μ values would generate. Also note the critical region that already has been established as $\bar{Y} = 18.8$ based on H_0.

From Figure 2.2 it can be seen that as one considers μ farther and farther to the left of 19.5, the β error decreases. The size of any β error can be determined for any given μ from Figure 2.2 and a normal distribution table. For example, if μ is assumed to be 19.0, the critical mean 18.8 is standardized

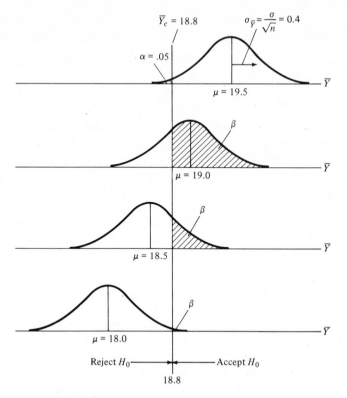

Figure 2.2   The effect of the mean on the β error.

with respect to this assumed mean and

$$Z = \frac{18.8 - 19.0}{2/\sqrt{25}} = \frac{-0.2}{0.4} = -0.5$$

indicating that the rejection point is one half of a standard deviation below 19.0. For a normal distribution (see Appendix, Table A), a $Z = -0.5$ has an area above it of $1 - 0.3085$ or 0.6915. Thus the β error for $\mu = 19.0$ is 0.69, or 69 percent of the samples would still accept the original hypothesis that $\mu = 19.5$ when it has really dropped to 19.0.

If one considers the mean μ at 18.0 as in the bottom sketch of Figure 2.2

$$Z = \frac{18.8 - 18.0}{0.4} = 2.00$$

and the β error is only 0.0228.

If β is plotted for various values of μ, the resulting curve is called the *operating characteristic* (or OC) *curve* for this test. Table 2.1 summarizes the computations for several assumed values of μ.

Table 2.1 Data for Operating Characteristic Curve

μ	$Z = \dfrac{18.8 - \mu}{2/\sqrt{25}}$	β	$1 - \beta$
18.0	+2.00	0.02	0.98
18.3	+1.25	0.11	0.89
18.5	+0.75	0.23	0.77
19.0	-0.50	0.69	0.31
19.2	-1.00	0.84	0.16
19.5	-1.75	0.96	0.04*

* (approximately = α)

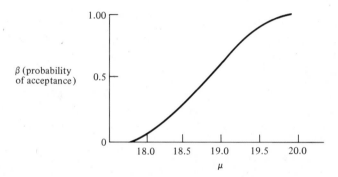

Figure 2.3 Operating characteristic curve.

The plotted OC curve is then as shown in Figure 2.3.

Some people prefer to plot $1 - \beta$ versus μ. This curve is called the *power curve* of the test as $1 - \beta$ is called the power of the test against the alternative μ.

It should be noted that when μ is quite a distance from the hypothesized value of 19.5, the test does a fairly good job of detecting this shift; that is, if $\mu = 18.5$, the probability of rejecting 19.5 is 0.77, which is fairly high. On the other hand, if μ has shifted only slightly from 19.5, say 19.2, the probability of detection is only 0.16. The power of a test may be increased by increasing the sample size or increasing the risk α.

2.5 HOW LARGE A SAMPLE?

The question of how large a sample to take from a population for making a test is one often asked of a statistician. This question can be answered provided the experimenter can answer each of the following questions.

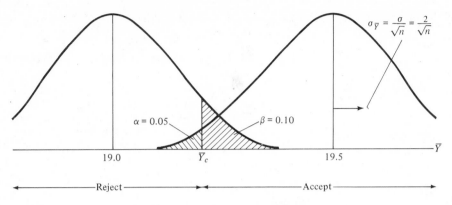

Figure 2.4   Determining sample size.

1. How large a shift (from 19.5 to 19.0) in a parameter do you wish to detect?
2. How much variability is present in the population? (Based on past experience, $\sigma = 2$ mi/gal.)
3. What size risks are you willing to take? ($\alpha = 0.05$ and $\beta = 0.10$.)

If numerical values can be at least estimated in answering the above questions, the sample size may be determined.

Set up two sampling distributions of $\bar{Y}$, one when H_0 is true, $\mu = 19.5$, and the other when the alternative to be detected is true, $\mu = {}^{.}19.0$ (Figure 2.4). Indicate by $\bar{Y}_c$ a value between the two μ's that will become a critical point, rejecting H_0 for observed values of $\bar{Y}$ below it and accepting H_0 for $\bar{Y}$ values above it. Indicate the α and β risks on the diagram. Set up two simultaneous equations, standardizing $\bar{Y}_c$ first with respect to a μ of 19.5 (α equation) and second with respect to a μ of 19.0 (β equation). Solve these equations for n and $\bar{Y}_c$:

α equation: $\quad \dfrac{\bar{Y}_c - 19.5}{2/\sqrt{n}} = -1.645 \quad$ (based on $\alpha = 0.05$)

β equation: $\quad \dfrac{\bar{Y}_c - 19.0}{2/\sqrt{n}} = +1.282 \quad$ (based on $\beta = 0.10$)

Subtracting the second equation from the first and multiplying both sides of each equation by $2/\sqrt{n}$ gives

$$-0.5 = -2.927\left(\frac{2}{\sqrt{n}}\right)$$

$$\sqrt{n} = \frac{5.854}{0.5} = 11.71$$

$$n = (11.71)^2 = 136.9 \text{ or } 137$$

Keeping $\alpha = 0.05$ gives

$$\bar{Y}_c = 19.5 - 1.645\left(\frac{2}{\sqrt{137}}\right) = 19.22 \text{ mi/gal}$$

The decision rule is then: Choose a random sample of 137 carburetors, and if the mean mi/gal of these is less than 19.22, reject H_0; otherwise, accept H_0.

Excellent tables are available for determining n for several tests of hypotheses, such as those in Davies [9, pp. 606–615] or Owen [20, pp. 19, 23, 36, 41, ff.].

2.6   APPLICATION TO TESTS ON VARIANCES

To review how the concepts of confidence limits and hypothesis testing applies to variance tests consider the following.

I. Tests on a single variance.
$H_0: \sigma^2 = \sigma_0^2$
$H_1: \sigma^2 \neq \sigma_0^2$ [or $\sigma^2 > \sigma_0^2$ or $\sigma^2 < \sigma_0^2$]
Choose α and n.
Test statistic is

$$\frac{(n-1)s^2}{\sigma_0^2}$$

which follows the chi-square distribution with $(n - 1)$ degrees of freedom, if one can assume that Y is normally distributed.
If $\chi^2 \geq \chi_{\alpha/2}^2$ or $\leq \chi_{1-\alpha/2}^2$, reject H_0.
[If $\chi^2 \geq \chi_{\alpha}^2$, reject, or if $\chi^2 \leq \chi_{1-\alpha}^2$, reject.]
To set confidence limits on σ^2 one computes

$$\frac{(n-1)s^2}{\chi_{\alpha/2}^2} \leq \sigma^2 \leq \frac{(n-1)s^2}{\chi_{1-\alpha/2}^2}$$

using the chi-square table with $(n - 1)$ df to give $100(1 - \alpha)$ percent confidence limits.

As an example consider that one might wish to determine whether a process standard deviation is greater than a specified value of 5 ounces. If a sample of ten weights is found to have an average of 35.2 ounces and a standard deviation of 7 ounces, what can we conclude about the standard deviation of this process?

We hypothesize $H_0: \sigma^2 = 25$ oz^2 and take as alternative $H_1: \sigma^2 > 25$ oz^2. Selecting $\alpha = 0.05$ and $n = 10$ gives $n - 1 = 9$ df. One should reject H_0 if

$$\chi^2 \geq \chi_{0.05}^2(9 \text{ df}) = 16.919$$

based on Table C of the Appendix.

From the sample data

$$\chi^2 = \frac{(n-1)s^2}{\sigma_0^2} = \frac{9(49)}{25} = 17.64$$

so we reject the hypothesis and conclude that our process standard deviation is greater than 5 ounces.

We might wish to set 90 percent confidence limits on σ^2 (or σ) based on our observed variance of 49:

$$\frac{9(49)}{\chi_{0.05}^2} \leq \sigma^2 \leq \frac{9(49)}{\chi_{0.95}^2}$$

$$\frac{9(49)}{16.919} \leq \sigma^2 \leq \frac{9(49)}{3.325}$$

$$26.07 \leq \sigma^2 \leq 132.63$$

or

$$5.1 \leq \sigma^2 \leq 11.5 \text{ ounces}$$

and we feel 90 percent confident that the process standard deviation is between 5.1 and 11.5. We note that this band does not include the hypothesized value of 5 ounces. We also note that this is a two-sided confidence band whereas the hypothesis test was one-sided. By making the α for confidence limits twice the α in the test of H_0, the results are comparable.

II. Tests on two independent variances.

As another example of a test of a hypothesis, consider testing whether or not the variances of two normal populations are equal. This example is included here as the test statistic involved will have many applications in later chapters.

In accordance with the steps outlined in Section 2.3, we have the following.

$H_0: \sigma_1^2 = \sigma_2^2 \qquad H_1: \sigma_1^2 > \sigma_2^2$.
$\alpha = 0.05$.
Test statistic $F = s_1^2/s_2^2$ (often called the variance ratio, where s_1^2 is based on $n_1 - 1$ df and s_2^2 is based on $n_2 - 1$ df).
If the two samples are independently chosen from normal populations and H_0 is true, the F statistic follows a skewed distribution, formed as the ratio of two independent chi-square distributions. A table giving a few percentiles of this F distribution appears as Appendix Table D. This table is entered with $n_1 - 1$ df for the numerator and $n_2 - 1$ df for the denominator.

The critical region is set at $F \geq F_{0.95}$ in this example ($\alpha = 0.05$) with the degrees of freedom dependent on the size of the two samples.

If the first sample results are

$$n_1 = 8 \qquad s_1^2 = 156$$

and the second are

$$n_2 = 10 \qquad s_2^2 = 100$$

then

$$F = \frac{156}{100} = 1.56$$

and the critical region is $F \geq 3.29$ for 7 and 9 df. Hence the hypothesis is not rejected.

In this example only a one-sided test has been considered as one is usually concerned with showing that one variance is greater or less than another. It is customary then to place the larger observed variance in the numerator of F and call it s_1^2.

2.7   APPLICATION TO TESTS ON MEANS

I. Tests on a single mean.
$H_0: \mu = \mu_0$
$H_1: \mu \neq \mu_0$ [or $\mu > \mu_0$ or $\mu < \mu_0$]
Choose n and α.
Test statistic when:

A. σ is known.

$$Z = \frac{\bar{Y} - \mu_0}{\sigma\sqrt{n}}$$

Z is $N(0, 1)$
Reject H_0 if $|Z| \geq Z_{1-\alpha/2}$ [or $Z \geq Z_{1-\alpha}$, or $Z \leq Z_\alpha$].

The problem of the carburetors in Section 2.3 illustrates this test.

B. σ is unknown, normal population.

$$t = \frac{\bar{Y} - \mu_0}{s/\sqrt{n}}$$

t follows a t distribution (Appendix, Table B) with $n - 1$ degrees of freedom.
Reject H_0 if $|t| \geq t_{1-\alpha/2}$ [or $t \geq t_{1-\alpha}$ or $t < -t_{1-\alpha}$].

Consider a sample of six cylinder blocks whose cope hardness values are 70, 75, 60, 75, 65, and 80. Is there evidence here that the average cope hardness has changed from its specified value of 75?

$$H_0: \mu = 75$$
$$H_1: \mu \neq 75$$

If $\alpha = 0.05$, $n = 6$, one should reject H_0 if $|t| \geq t_{0.975}$ with 5 df or if $|t| \geq 2.57$. From the given sample of six: $\bar{Y} = 70.8$, $s = 7.4$. So,

$$t = \frac{70.8 - 75}{7.4/\sqrt{6}} = -1.39$$

Since $|t| = 1.39 < 2.57$, one cannot reject H_0. We might have used our data to set 95 percent confidence limits on μ. From Section 2.2 $100(1 - \alpha)$ percent confidence limits are

$$\bar{Y} \pm t_{1-\alpha/2}s/\sqrt{n}$$
$$70.8 \pm 2.57(7.4)/\sqrt{6}$$
$$70.8 \pm 7.8$$

or from 63.0 to 78.6, which includes our hypothesized value of 75.

II. Tests on two means.
    A. If independent samples:

        1. If variances are known: σ_1^2 and σ_2^2.
            $H_0: \mu_1 = \mu_2$
            $H_1: \mu_1 \neq \mu_2$ [or $\mu_1 > \mu_2$]
            Choose α, n_1, and n_2.
            Test statistic is

$$Z = \frac{\bar{Y}_1 - \bar{Y}_2}{\sqrt{\dfrac{\sigma_1^2}{n_1} + \dfrac{\sigma_2^2}{n_2}}}.$$

            Reject H_0 if $|Z| \geq Z_{1-\alpha/2}$ [or $Z \geq Z_{1-\alpha}$].

        2. If variances are unknown but equal and populations are normal: Test statistic is

$$t = \frac{\bar{Y}_1 - \bar{Y}_2}{\sqrt{\dfrac{(n_1 - 1)s_1^2 + (n_2 - 1)s_2^2}{n_1 + n_2 - 2}\left(\dfrac{1}{n_1} + \dfrac{1}{n_2}\right)}}$$

            with $n_1 + n_2 - 2$ df. One rejects if $|t| \geq t_{1-\alpha/2}$ [or $t \geq t_{1-\alpha}$].

        3. If variances are unknown and unequal: If a preliminary F test shows the variances to be unequal, no test on means may be necessary as the two samples are so heterogeneous with respect to their variances. If, however, a test on means is desired, some tests are given in statistics books. This situation is often referred to as the Behrens–Fisher problem. One method is to take as the test

statistic

$$t' = \frac{\bar{Y}_1 - \bar{Y}_2}{\sqrt{\dfrac{s_1^2}{n_1} + \dfrac{s_2^2}{n_2}}}$$

and check its significance using a special formula for its degrees of freedom:

$$df = \frac{\left(\dfrac{s_1^2}{n_1} + \dfrac{s_2^2}{n_2}\right)^2}{\dfrac{\left(\dfrac{s_1^2}{n_1}\right)^2}{n_1 + 1} + \dfrac{\left(\dfrac{s_2^2}{n_2}\right)^2}{n_2 + 1}} - 2$$

B. If dependent or correlated samples:
Here one often has the same sample "before and after" some treatment has been applied. The usual procedure is to take differences between the first and second observations on the same piece, person, or whatever, and test the hypothesis that the mean difference μ_D is zero. This reduces the problem to a test on a single mean:

$$H_0: \mu_D = 0$$
$$H_1: \mu_D \neq 0 \quad (\mu_D > 0)$$

n differences, α
Test statistic is

$$t = \frac{\bar{d}}{s_d/\sqrt{n}} \quad \text{with } n - 1 \text{ df}$$

Reject if $|t| \geq t_{1-\alpha/2}$ [or $t \geq t_{1-\alpha}$].
Choose a random sample of n, take two sets of measurements, one on each unit in the sample, compute the differences. Using these differences compute $\bar{d}$ and s_d, then t, and render a decision on H_0.

Some examples illustrating the foregoing tests are given in the following.

Example 1   If two samples randomly selected from two independent normal populations give

$$n_2 = 4, \quad \bar{Y}_2 = 12.0, \quad s_2^2 = 2.0, \quad n_1 = 9, \quad \bar{Y}_1 = 16.0 \quad s_1^2 = 5.0$$

is there enough evidence to claim that the mean of population 1 is greater than the mean of population 2?

$$H_0: \mu_1 = \mu_2$$
$$H_1: \mu_1 > \mu_2$$
$$\sigma_1^2 = \sigma_2^2 \quad (\text{note } F = 2.5 < F_{8,3}(0.975))$$

Reject if

$$t > t_{0.95,11 \text{ df}} = 1.796$$

$$t = \frac{16.0 - 12.0}{\sqrt{\left[\frac{8(5) + 3(2)}{11}\right]\left[\frac{1}{9} + \frac{1}{4}\right]}} = 3.25$$

Hence reject H_0 and claim μ_1 is greater than μ_2.

Example 2   A group of ten students was pretested before instruction and posttested after six weeks of instruction with the following achievement scores.

Student	Before	After	Differences $= d$
1	14	17	3
2	12	16	4
3	20	21	1
4	8	10	2
5	11	10	−1
6	15	14	−1
7	17	20	3
8	18	22	4
9	9	14	5
10	7	12	5

Is there evidence of an improvement in achievement over this six-week period?

$$H_0: \mu_D = 0$$
$$H_1: \mu_D > 0$$

Above

$$\bar{d} = 2.5, \quad s_d = 2.22$$

$$t = \frac{2.5}{2.22/\sqrt{10}} = 3.56$$

Since t is $> t_{9,0.95} = 1.833$, we reject H_0 and conclude that there has been an improvement.

2.8   APPLICATION TO TESTS ON PROPORTIONS

I. Tests on a single proportion p.
$$H_0: p = p_0$$
$$H_1: p \neq p_0 \text{ [or } p > p_0 \text{ or } p < p_0\text{]}$$
Choose α and n (n had best be large enough so that $np \geq 4$), then the

test statistic is

$$Z = \frac{\hat{p} - p_0}{\sqrt{\dfrac{p_0 q_0}{n}}}$$

where $q_0 = 1 - p_0$ and $\hat{p}$ is the observed proportion in a sample of n. Here Z is $N(0, 1)$ and one rejects if $|Z| > Z_{1-\alpha/2}$ [or $Z > Z_{1-\alpha}$ or $Z < -Z_{1-\alpha}$].

As an example consider whether or not a class of 25 students has an excessive number of left-handed students if we find ten left-handed students and the national average is 20 percent.

$H_0: p = 0.20$
$H_1: p > 0.20$
$n = 25$, $\alpha = 0.05$
Reject H_0 if $Z \geq 1.65$. Here

$$Z = \frac{10/25 - 0.20}{\sqrt{(0.2)(0.8)/25}} = 2.5$$

so we reject H_0 and conclude that our class does have an excessive number of left-handers.

II. Tests on two independent proportions.
$H_0: p_1 = p_2$
$H_1: p_1 \neq p_2$ [or $p_1 > p_2$]
Choose α, n_1, and n_2.
Test statistic is

$$Z = \frac{\hat{p}_1 - \hat{p}_2}{\sqrt{\hat{p}\hat{q}\left(\dfrac{1}{n_1} + \dfrac{1}{n_2}\right)}}$$

where $\hat{p}$ is the proportion of defectives in both samples together:

$$\hat{p} = \frac{n_1 \hat{p}_1 + n_2 \hat{p}_2}{n_1 + n_2}$$

Reject H_0 if $|Z| \geq Z_{1-\alpha/2}$ [or $Z \geq Z_{1-\alpha}$]

If the number of defectives found in a sample of 100 drawn from a production process was 12 on Monday and a sample of 200 from the same process showed 16 defectives on Tuesday, has the process improved?

$H_0: p_1 = p_2$
$H_1: p_1 > p_2$ where p_1 is proportion on Monday.
$\alpha = 0.05$, $n_1 = 100$, $n_2 = 200$.

Reject if $Z \geq 1.64$.
Here

$$Z = \frac{12/100 - 16/200}{\sqrt{(28/300)(272/300)[1/100 + 1/200]}}$$

$$Z = 1.12$$

so one cannot say that the process has improved significantly.

PROBLEMS

2.1 For the following data on tensile strength in psi (pounds per square inch), deter-
     mine the mean and variance of this sample:

Tensile Strength (psi)	Frequency
18,461	2
18,466	12
18,471	15
18,476	10
18,481	8
18,486	3
Total	50

2.2 Using the results of Problem 2.1, test the hypothesis that the mean tensile strength
     of the population sampled is 18,470 psi (assume that $\sigma^2 = 40$, $\alpha = 0.05$).

2.3 Plot the operating characteristic curve for Problem 2.2.

2.4 Determine how large a sample would be needed to detect a 10-psi increase in the
     mean tensile strength of Problem 2.2 if $\alpha = 0.05$, $\beta = 0.02$, and $\sigma^2 = 40$.

2.5 Repeat Problem 2.4 for the detection of a 10-psi shift in the mean in either direction.

2.6 For a sample mean of 124 g (grams) based on 16 observations from a population
     whose variance is 25 g^2 set up 90 percent confidence limits on the mean of the
     population.

2.7 For a sample variance of 62 based on 12 observations test the hypothesis that the
     population variance is 40. Use a one-sided test and a 1 percent significance level.

2.8 For Problem 2.7 set up 95 percent confidence limits (two-sided) on σ^2.

2.9 Two samples are taken, one from each of two machines. For this process

$$n_1 = 8, \qquad \bar{Y}_1 = 42.2 \text{ g}, \qquad s_1^2 = 10 \text{ g}^2$$
$$n_2 = 15, \qquad \bar{Y}_2 = 44.5 \text{ g}, \qquad s_2^2 = 18 \text{ g}^2$$

Test these results for a significant difference in variances.

2.10 Test the results in Problem 2.9 for a significant difference in means.

2.11 Reflection light box readings before and after dichromating the interior of a metal cone were

Test Number	1	2	3	4	5	6	7	8
Before	6.5	6.0	7.0	6.8	6.5	6.8	6.2	6.5
After	4.4	4.2	5.0	5.0	4.8	4.6	5.2	4.9

Test for a significant difference in mean light-box readings.

2.12 To test the hypothesis that the defective fraction p of a process is 0.20, a sample of 100 pieces was drawn at random.

1. Use an $\alpha = 0.05$ and set up a critical region for the number of observed defectives to test this hypothesis against a two-sided alternative.

2. If the process now shifts to a 0.10 fraction defective, find the probability of an error of the second kind for this alternative.

3. Without repeating the work, would you expect the β error to be the same if the process shifted to 0.30? Explain.

2.13 Given the following sample data on a random variable, Y: 12, 8, 14, 20, 26, 26, 20, 21, 18, 24, 30, 21, 18, 16, 10, 20. Assuming $\sigma = 7$, and $\alpha = 0.05$, test each of the following hypotheses:

1. $H_0: \mu = 12$
   $H_1: \mu > 12$
2. $H_0: \mu = 16$
   $H_1: \mu \neq 16$
3. $H_0: \mu = 18$
   $H_1: \mu > 18$

2.14 For test 1 in Problem 2.13 evaluate the power of the test with respect to several alternative hypotheses from $\mu = 13$ to $\mu = 24$. Plot the power curve.

2.15 Repeat the results of Problem 2.14 with $\alpha = 0.01$. Plot on the same graph and compare the two power curves.

2.16 Given the following data on the current flow in amperes through a cereal forming machine: 8.2, 8.3, 8.2, 8.6, 8.0, 8.4, 8.5, 7.3. Test the hypothesis of the standard deviation $\sigma = 0.35$ ampere versus the alternative that $\sigma > 0.35$ ampere.

2.17 For the data in Problem 2.16 test the hypothesis that the true mean current flow is $\mu = 7.0$.

2.18 The percent moisture content in a puffed cereal where samples are taken from two different "guns" showed

Gun I: 3.6, 3.8, 3.6, 3.3, 3.7, 3.4
Gun II: 3.7, 3.9, 4.2, 4.2, 4.9, 3.6, 3.5, 4.0

Test the hypothesis of equal variances and equal means. Use any assumptions you believe appropriate.

2.19 Pretest data for experimental and control groups on course content in a special vocational–industrial course indicated:

$$\text{Experimental:} \quad \bar{Y}_1 = 9.333, s_1 = 4.945, n_1 = 12$$
$$\text{Control:} \quad \bar{Y}_2 = 8.375, s_2 = 1.187, n_2 = 8$$

Make any statistical tests that the data might suggest. Comment on the results.

2.20 The WISC performance scores of ten boys and eight girls were recorded as follows:

Boys: 101, 86, 72, 129, 99, 118, 104, 125, 90, 107

Girls: 97, 107, 94, 101, 90, 108, 108, 86

Test whether or not there is a significant difference in the means of boys and girls. Take $\alpha = 0.10$.

2.21 An occupational therapist conducted a study to evaluate the relative merits of two prosthetic devices designed to facilitate manual dexterity. The therapist assigned 21 patients with identical handicaps to wear one or the other of the two devices while performing a certain task. Eleven patients wore device A and ten wore device B. The researcher recorded the time each patient required to perform a certain task, with the following results:

$$\bar{y}_A = 65 \text{ seconds} \quad s_A^2 = 81$$
$$\bar{y}_B = 75 \text{ seconds} \quad s_B^2 = 64$$

Do these data provide sufficient evidence to indicate that device A is more effective than device B?

2.22 An anthropologist believes that the proportion of individuals in two populations with double occipital hair whorls is the same. To see whether there is any reason to doubt this hypothesis, the anthropologist takes independent random samples from each of the two populations and determines the number in each sample with this characteristic. Results showed:

Population	n	Number with Characteristic
1	100	23
2	120	32

Is there reason to doubt this hypothesis?

2.23 A psychologist randomly selected ten wives and their husbands from among the residents of an urban area, and asked them to complete a questionnaire designed to measure their level of satisfaction with the community in which they lived. Do the following results indicate that the husbands are better satisfied with the community than are their wives?

					Scores					
Wife	33	57	32	54	52	34	60	40	59	39
Husband	44	60	55	68	40	48	57	49	47	52

2.24 Measurements were made at each of two plants on the tensile strength of samples of the same type of steel. The data are given below in thousands of psi. Can it be stated that the steels made in the two plants have the same mean tensile strength?

Plant 1	Plant 2
207	219
195	228
209	222
218	197
196	225
217	184
	218
	211

2.25 The standard medical treatment for a certain ailment has proved to be about 85 percent effective. One hundred patients were given an alternative treatment with the result that 78 showed marked improvement. Is the new treatment actually inferior or could the difference noted be reasonably attributed to chance?

Single-Factor Experiments
with No Restrictions
on Randomization

3.1 INTRODUCTION

In this and several subsequent chapters single-factor experiments will be considered. In this chapter no restrictions will be placed on the randomization so that the design will be completely randomized. Many of the techniques of analysis for a completely randomized single-factor experiment can be applied with little alteration to more complex experiments.

For example, the single factor could be steel manufacturers, where the main interest of an analyst centers on the effect of several different manufacturers on the hardness of steel purchased from them. It could be temperature in an instance where the experimenter is concerned about the effect of temperature on penicillin yield. Whenever only one factor is varied, whether the levels be quantitative or qualitative, fixed or random, the experiment is referred to as a *single-factor experiment*, and the symbol τ_j will be used to indicate the effect of the jth level of the factor. τ_j suggests that the general factor may be thought of as a "treatment" effect.

If the order of experimentation applied to the several levels of the factor is completely random, so that any material to which the treatments might be applied is considered approximately homogeneous, the design is called a *completely randomized design*. The number of observations for each level of the treatment or factor will be determined from cost considerations and the power of the test. The model then becomes

$$Y_{ij} = \mu + \tau_j + \varepsilon_{ij}$$

where Y_{ij} represents the ith observation ($i = 1, 2, \ldots, n_j$) on the jth treatment ($j = 1, 2, \ldots, k$ levels). For example, Y_{23} represents the second observation using level 3 of the factor. μ is a common effect for the whole experiment, τ_j represents the effect of the jth treatment, and ε_{ij} represents the random error present in the ith observation on the jth treatment.

The error term ε_{ij} is usually considered a normally and independently distributed (NID) random effect whose mean value is zero and whose variance is the same for all treatments or levels. This is expressed as: ε_{ij}'s are NID $(0, \sigma_e^2)$ where σ_e^2 is the common variance within all treatments. μ is always a fixed parameter, and $\tau_1, \tau_2, \ldots, \tau_j, \ldots, \tau_k$ are considered to be fixed parameters if the levels of treatments are fixed. It is also assumed that

$$\sum_{j=1}^{k} \tau_j = 0$$

If the k levels of treatments are chosen at random, the τ_j's are assumed NID $(0, \sigma_\tau^2)$. Whether the levels are fixed or random depends upon how these levels are chosen in a given experiment.

The analysis of a single factor completely randomized experiment usually consists of a one-way analysis of variance test where the hypothesis $H_0: \tau_j = 0$ for all j, is tested. If this hypothesis is true, then no treatment effects exist and each observation Y_{ij} is made up of its population mean μ and a random error ε_{ij}. After an analysis of variance (ANOVA), many other tests may be made, and some of these will be shown in the example that follows.

Example 3.1 In the manufacture of clothing a wear-testing machine is used to measure the resistance to abrasion of different fabrics. The dependent variable Y gives the loss of weight of the material in grams after a specified number of cycles and the problem is to determine whether or not there is a difference in the average weight loss among four competing fabrics. Here there is but one factor of interest: the fabric at levels A, B, C, and D. These are qualitative and fixed levels of this single factor. It was agreed to test four samples of each fabric and completely randomize the order of testing of the 16 samples. The mathematical model for this example then is

$$Y_{ij} = \mu + \tau_j + \varepsilon_{ij}$$

with $i = 1, 2, \ldots, 4$ and $j = 1, 2, \ldots, 4$ since there are four samples for each of four treatments (fabrics). The data are shown in Table 3.1.

Table 3.1 Fabric Wear Resistance Data

		Fabric	
A	B	C	D
1.93	2.55	2.40	2.33
2.38	2.72	2.68	2.40
2.20	2.75	2.31	2.28
2.25	2.70	2.28	2.25

An analysis of variance performed on these data gave the results in Table 3.2.

Table 3.2 Fabric Data ANOVA

Source of Variation	df	SS	MS
Between Fabrics: τ_j	3	0.5201	0.1734
Within fabrics or error: ε_{ij}	12	0.2438	0.0203
Totals	15	0.7639	

To test $H_0: \tau_j = 0$ for all $j = 1, 2, 3$, and 4, the test statistic is

$$F_{3,12} = \frac{0.1734}{0.0203} = 8.54$$

which is significant at the 1 percent significance level since Appendix Table D shows 5.95 as the 1 percent F ($F_{0.99}$) for 3 and 12 degrees of freedom. We can then reject the hypothesis and claim that there is a considerable difference in average wear resistance among the four fabrics.

3.2 ANALYSIS OF VARIANCE RATIONALE

To review the basis for the F test in a one-way analysis of variance, k populations, each representing one level of treatment, can be considered with observations as shown in Table 3.3.

Table 3.3 Population Layout for One-Way ANOVA

			Treatment		
	1	2	$\cdots$ j	$\cdots$	k
	Y_{11}	Y_{12}	Y_{1j}		Y_{1k}
	Y_{21}	Y_{22}	Y_{2j}		Y_{2k}
	Y_{31}	Y_{32}	—		—
	—	—	—		—
	Y_{i1}	Y_{i2}	Y_{ij}		Y_{ik}
Population means	$\mu_{.1}$	$\mu_{.2}$	$\cdots$ $\mu_{.j}$ $\cdots$		$\mu_{.k}$

Here the use of the "dot notation" indicates a summing over all observations in the population. Since each observation could be returned to the population and measured, there could be an infinite number of observations taken on each population, so the average or $E(Y_{i1}) = \mu_{.1}$, and so on. μ will represent the average Y_{ij} over all populations, or $E(Y_{ij}) = \mu$. In the model τ_j, the treatment effect can also be indicated by $\mu_{.j} - \mu$, and then the model is either

$$Y_{ij} = \mu + \tau_j + \varepsilon_{ij}$$

or

$$Y_{ij} \equiv \mu + (\mu_{.j} - \mu) + (Y_{ij} - \mu_{.j})$$

This last expression is seen to be an identity true for all values of Y_{ij}. Expressed another way,

$$Y_{ij} - \mu \equiv (\mu_{.j} - \mu) + (Y_{ij} - \mu_{.j}) \tag{3.1}$$

Since these means are unknown, random samples are drawn from each population and estimates can be made of the treatment means and the grand mean. If n_j observations are taken for each treatment where the numbers need not be equal, a sample layout would be as shown in Table 3.4.

Table 3.4 Sample Layout for One-Way ANOVA

	Treatment					
	1	2	$\cdots$	j	$\cdots$	k
	Y_{11}	Y_{12}	$\cdots$	Y_{1j}	$\cdots$	Y_{1k}
	Y_{21}	Y_{22}	$\cdots$	Y_{2j}	$\cdots$	Y_{2k}
	$\vdots$	$\vdots$		$\vdots$		$\vdots$
	Y_{i1}	Y_{i2}	$\cdots$	Y_{ij}		Y_{ik}
	$\vdots$			$\vdots$		$\vdots$
	$Y_{n_1 1}$	$\vdots$		$Y_{n_j j}$		
		$Y_{n_2 2}$				$Y_{n_k k}$
Totals	$T_{.1}$	$T_{.2}$	$\cdots$	$T_{.j}$	$\cdots$	$T_{.k}$ $T_{..}$
Number	n_1	n_2	$\cdots$	n_j	$\cdots$	n_k N
Means	$\bar{Y}_{.1}$	$\bar{Y}_{.2}$	$\cdots$	$\bar{Y}_{.j}$	$\cdots$	$\bar{Y}_{.k}$ $\bar{Y}_{..}$

Here $T_{.j}$ represents the total of the observations taken under treatment j, n_j represents the number of observations taken for treatment j, and $\bar{Y}_{.j}$ is the observed mean for treatment j. $T_{..}$ represents the grand total of all observations taken where

$$T_{..} = \sum_{j=1}^{k} \sum_{i=1}^{n_j} Y_{ij} = \sum_{j=1}^{k} T_{.j}$$

and

$$N = \sum_{j=1}^{k} n_j$$

and $\bar{Y}_{..}$ is the mean of all N observations. Note too that

$$\bar{Y}_{..} = \sum_{j=1}^{k} n_j \bar{Y}_{.j}/N$$

If these sample statistics are substituted for their corresponding population parameters in Equation (3.1), we get a sample equation (also an identity) of the form

$$Y_{ij} - \bar{Y}_{..} \equiv (\bar{Y}_{.j} - \bar{Y}_{..}) + (Y_{ij} - \bar{Y}_{.j}) \tag{3.2}$$

This equation states that the deviation of any observation from the grand mean can be broken into two parts: the deviation of the observation from its own treatment mean plus the deviation of the treatment mean from the grand mean.

If both sides of Equation (3.2) are squared and then added over both i and j, we have

$$\sum_{j=1}^{k} \sum_{i=1}^{n_j} (Y_{ij} - \bar{Y}_{..})^2 = \sum_{j=1}^{k} \sum_{i=1}^{n_j} (\bar{Y}_{.j} - \bar{Y}_{..})^2 + \sum_{j=1}^{k} \sum_{i=1}^{n_j} (Y_{ij} - \bar{Y}_{.j})^2$$

$$+ 2 \sum_{j=1}^{k} \sum_{i=1}^{n_j} (\bar{Y}_{.j} - \bar{Y}_{..})(Y_{ij} - \bar{Y}_{.j}) \tag{3.3}$$

Examining the last expression on the right, we find that

$$\sum_{j=1}^{k} \sum_{i=1}^{n_j} (\bar{Y}_{.j} - \bar{Y}_{..})(Y_{ij} - \bar{Y}_{.j}) = \sum_{j=1}^{k} (\bar{Y}_{.j} - \bar{Y}_{..}) \left[\sum_{i=1}^{n_j} (Y_{ij} - \bar{Y}_{.j}) \right]$$

The term in brackets is seen to equal zero, as the sum of the deviations about the mean within a given treatment equals zero. Hence

$$\sum_{j=1}^{k} \sum_{i=1}^{n_j} (Y_{ij} - \bar{Y}_{..})^2 = \sum_{j=1}^{k} \sum_{i=1}^{n_j} (\bar{Y}_{.j} - \bar{Y}_{..})^2 + \sum_{j=1}^{k} \sum_{i=1}^{n_j} (Y_{ij} - \bar{Y}_{.j})^2 \tag{3.4}$$

This may be referred to as "the fundamental equation of analysis of variance," and it expresses the idea that the total sum of squares of deviations from the grand mean is equal to the sum of squares of deviations between treatment means and the grand mean plus the sum of squares of deviations within treatments. In Chapter 2 an unbiased estimate of population variance was determined by dividing the sum of squares $\sum_{i=1}^{n} (Y_i - \bar{Y})^2$ by the corresponding number of degrees of freedom, $n - 1$. If the hypothesis being tested in analysis of variance is true, namely, that $\tau_j = 0$ for all j, or that there is no treatment effect, then $\mu_{.1} = \mu_{.2} = \cdots = \mu_{.j} \cdots = \mu_{.k}$ and all there is in the model is the population mean μ and random error ε_{ij}. Then, any one of the three terms in Equation (3.4) may be used to give an unbiased estimate of this common population variance. For example, dividing the left-hand term by its degrees of freedom, $N - 1$ will yield an unbiased estimate of population variance σ^2. Within the jth treatment, $\sum_i (Y_{ij} - \bar{Y}_{.j})^2$ divided by $n_j - 1$ df would yield an unbiased estimate of the variance within the jth treatment. If the variances within the k treatments are really all alike, their estimates may be pooled to give $\sum_{j=1}^{k}\sum_{i=1}^{n_j} (Y_{ij} - \bar{Y}_{.j})^2$ with degrees of freedom

$$\sum_{j}^{k} (n_j - 1) = N - k$$

which will give another estimate of the population variance. Still another estimate can be made by first estimating the variance $\sigma_{\bar{Y}}^2$ between means drawn from a common population with $\sigma_{\bar{Y}}^2 = n_j\sigma_{\bar{Y}_{.j}}^2$. An unbiased estimate for $\sigma_{\bar{Y}}^2$ is given by $\sum_{j=1}^{k} (\bar{Y}_{.j} - \bar{Y}_{..})^2/(k - 1)$ so that an unbiased estimate of

$$\sigma_{\bar{Y}}^2 = \sum_{j=1}^{k} n_j(\bar{Y}_{.j} - \bar{Y}_{..})^2/(k - 1)$$

which is found by summing the first term on the right-hand side of Equation (3.4) over i and dividing by $k - 1$ df. Thus there are three unbiased estimates of σ^2 possible from the data in a one-way ANOVA if the hypothesis is true. Now all three are not independent since the sum of squares is additive in Equation (3.4). However, it can be shown that if each of the terms (sums of squares) on the right of Equation (3.4) is divided by its proper degrees of freedom, it will yield two independent chi-square distributed unbiased estimates of σ^2 when H_0 is true. If two such independent unbiased estimates of the same variance are compared, their ratio can be shown to be distributed as F with $k - 1$, $N - k$ df. If, then, H_0 is true, the test of the hypothesis can be made using a critical region of the F distribution with the observed F at $k - 1$ and $N - k$ df given by

$$F_{k-1, N-k} = \frac{\sum_{j=1}^{k} n_j(\bar{Y}_{.j} - \bar{Y}_{..})^2/(k - 1)}{\sum_{j=1}^{k} \sum_{i=1}^{n_j} (Y_{ij} - \bar{Y}_{.j})^2/(N - k)} \tag{3.5}$$

The critical region is usually taken as the upper tail of the F distribution, rejecting H_0 if $F \geq F_{1-\alpha}$ where α is the area above $F_{1-\alpha}$. In this F ratio the sum of squares between treatments is always put into the numerator, and then a significant F will indicate that the differences between means has something in it besides the estimate of variance. It probably indicates that there is a real difference in treatment means $(\mu_{.1}, \mu_{.2}, \ldots)$ and that H_0 should be rejected. These unbiased estimates of population variance, sums of squares divided by df, are also referred to as *mean squares*.

The actual computing of sums of squares indicated in Equation (3.5) is much easier if they are first expanded and rewritten in terms of treatment totals. These computing formulas are given in Table 3.5. Applying the

Table 3.5 One-Way ANOVA

Source	df	SS	MS
Between treatments τ_j	$k-1$	$\sum_{j=1}^{k} n_j (\bar{Y}_{.j} - \bar{Y}_{..})^2$ $= \sum_{j=1}^{k} \frac{T_{.j}^2}{n_j} - \frac{T_{..}^2}{N}$	$SS_{\text{treatment}}/(k-1)$
Within treatments or error ε_{ij}	$N-k$	$\sum_{j=1}^{k} \sum_{i=1}^{n_j} (Y_{ij} - \bar{Y}_{.j})^2$ $= \sum_{j=1}^{k} \sum_{i=1}^{n_j} Y_{ij}^2 - \sum_{j=1}^{k} \frac{T_{.j}^2}{n_j}$	$SS_{\text{error}}/(N-k)$
Totals	$N-1$	$\sum_{j=1}^{k} \sum_{i=1}^{n_j} (Y_{ij} - \bar{Y}_{..})^2$ $= \sum_{j=1}^{k} \sum_{i=1}^{n_j} Y_{ij}^2 - \frac{T_{..}^2}{N}$	

formulas in Table 3.5 to the problem data in Table 3.1, Table 3.6 shows treatment totals $T_{.j}$'s and the sums of squares of the Y_{ij}'s for each treatment along with the results added across all treatments.

Table 3.6   Fabric Wear Resistance Data

	Fabric				
	A	B	C	D	
	1.93	2.55	2.40	2.33	
	2.38	2.72	2.68	2.40	
	2.20	2.75	2.31	2.28	
	2.25	2.70	2.28	2.25	
$T_{.j}$:	8.76	10.72	9.67	9.26	$T_{..} = 38.41$
n_j:	4	4	4	4	$N = 16$
$\sum_{i=1}^{n_j} Y_{ij}^2$:	19.2918	28.7534	23.4769	21.4498	$\sum_{j}^{k} \sum_{i}^{n_j} Y_{ij}^2 = 92.9719$

The sums of squares can be computed quite easily from this table. The total sum of squares states, "Square each observation, add over all observations, and subtract the correction term." The last is the grand total squared and divided by the total number of observations

$$SS_{total} = \sum_{j=1}^{k} \sum_{i=1}^{n_j} Y_{ij}^2 - \frac{T_{..}^2}{N} = 92.9719 - \frac{(38.41)^2}{16} = 0.7639$$

The sum of squares between treatments is found by totaling n_j observations for each treatment, squaring this total, dividing by the number of observations, adding for all treatments, and then subtracting the correction term:

$$SS_{treatment} = \sum_{j=1}^{k} \frac{T_{.j}^2}{n_j} - \frac{T_{..}^2}{N}$$

$$= \frac{(8.76)^2}{4} + \frac{(10.72)^2}{4} + \frac{(9.67)^2}{4} + \frac{(9.26)^2}{4} - \frac{(38.41)^2}{16} = 0.5201$$

The sum of squares for error is then determined by subtraction:

$$SS_{error} = SS_{total} - SS_{treatment} = 0.7639 - 0.5201 = 0.2438$$

These results are displayed as in Table 3.2, and the F test is run on the $H_0: \tau_j = 0$ as shown before.

The computation of these sums of squares can be simplified in two ways. One could code the data by subtracting a convenient constant such as 2.00 and then multiplying all Y's by 100 to eliminate the decimals. The subtraction of a constant will not affect the sums of squares but multiplying by 100 will multiply all SS terms by $(100)^2$ or 10,000. However, since the test statistic F is the ratio of two mean squares, the F will be unaffected by this coding scheme.

Since the advent of computers, the original data can be fed into a computer program for a one-way ANOVA and the computer will print out the ANOVA table. To illustrate this procedure the programs used in this text will be SPSS (Statistical Package for the Social Sciences) programs, although the input and output formats are quite similar for most available commercial programs. Table 3.7 shows the input for Example 3.1 in an SPSS one-way ANOVA.

The problem is given an identifying number PROB 7. Line 1 is a code to identify the programmer. Lines 2 and 3 indicate that an SPSS program is desired. Line 5 identifies the problem again. Line 6 shows the variables Y and X herein labeled WL, FAB for wear loss and fabric. Line 7 shows the number of spaces necessary for the data: 4.0 for WL or Y—including the decimal point, and 1.0 for the fabric identified by level 1, 2, 3, or 4. Lines 9 through 14 indicate what statistics are to be computed on the data. The ONEWAY ANOVA is Y by X or WL by FAB with levels 1 through 4 of the independent variable. Lines 10–13 will be described later as statistics of interest to the experimenter. After indicating a desire for all statistics, line 14, the input data are typed in as shown. (If card input is used, each step, 1–32,

Table 3.7   SPSS Input for Example 3.1

```
+ + +CREATE PROB7
        1. 000 = 77509,I13
        2. 000 = COMMON (SPSS)
        3. 000 = SPSS
        4. 000 = #EOR
        5. 000 = RUN NAME              STAT 502 PROB7
        6. 000 = VARIABLE LIST         WL,FAB
        7. 000 = INPUT FORMAT          (F4.0, 1X, F1.0)
        8. 000 = N OF CASES            16
        9. 000 = ONEWAY                WL BY FAB (1,4)/
       10. 000 =                       CONTRAST = 1 0 0 −1/
       11. 000 =                       CONTRAST = 0 1 −1 0/
       12. 000 =                       CONTRAST = 1 −1 −1 1/
       13. 000 =                       RANGES = SNK
       14. 000 = STATISTICS            ALL
       15. 000 = READ INPUT DATA
       16. 000 = 1.93 1
       17. 000 = 2.38 1
       18. 000 = 2.20 1
       19. 000 = 2.25 1
       20. 000 = 2.55 2
       21. 000 = 2.72 2
       22. 000 = 2.75 2
       23. 000 = 2.70 2
       24. 000 = 2.40 3
       25. 000 = 2.68 3
       26. 000 = 2.31 3
       27. 000 = 2.28 3
       28. 000 = 2.33 4
       29. 000 = 2.40 4
       30. 000 = 2.28 4
       31. 000 = 2.25 4
       32. 000 = FINISH
       33. 000 = #S
```

---

is punched on a separate card.) The program is terminated on line 32 and it is sent to the computer. In a very short time, a printout is received as in Table 3.8.

These results are the same as shown in Table 3.2 with an additional item F PROB giving the probability of this F (8.534) or higher occurring by chance. So one need not look up a significant F value in the F table.

In addition to the ANOVA table, the information given in Table 3.9 is also printed.

Table 3.8 ANOVA for Example 3.1 by SPSS

VARIABLE WL BY FAB

ANALYSIS OF VARIANCE

SOURCE	D.F.	SUM OF SQUARES	MEAN SQUARES	F RATIO	F PROB.
BETWEEN GROUPS	3	.5201	.1734	8.534	.0026
WITHIN GROUPS	12	.2438	.0203		
TOTAL	15	.7639			

Table 3.9 Group Data on Example 3.1

GROUP	COUNT	MEAN	STANDARD DEVIATION	STANDARD ERROR	MINIMUM	MAXIMUM	95 PCT CONF INT FOR MEAN
GRP 1	4	2.1900	.1892	.0946	1.9300	2.3800	1.8889 TO 2.4911
GRP 2	4	2.6800	.0891	.0445	2.5500	2.7500	2.5383 TO 2.8217
GRP 3	4	2.4175	.1823	.0911	2.2800	2.6800	2.1275 TO 2.7075
GRP 4	4	2.3150	.0656	.0328	2.2500	2.4000	2.2107 TO 2.4193
TOTAL	16	2.4006			1.9300	2.7500	

This table shows the mean and standard deviation of each group—or treatment level—1 through 4 and its corresponding standard error ($s_{\bar{Y}} = s_Y/\sqrt{n}$) and 95 percent confidence limits on its true group mean.

3.3   AFTER ANOVA—WHAT?

After concluding, as in the problem above, that there is a difference in treatment means, we must ask more questions such as: Which treatment is the best? Does the mean wear resistance of fabric A differ from that of C? Does the mean of A and B together differ from that of C and D? To answer questions following an analysis of variance, one must consider the assumptions made when the experiment was planned originally. The decision tree of Figure 3.1 may be helpful in outlining some tests after ANOVA.

In the next sections we examine some of the tests indicated in Figure 3.1.

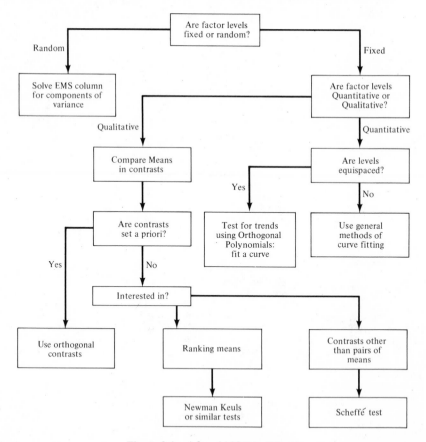

Figure 3.1   After ANOVA, What?

3.4   TESTS ON MEANS

In considering qualitative levels of treatments as in Example 3.1, any comparison of means will depend upon whether the decision on what means to compare was made before the experiment was run or after the results were examined.

Tests on Means Set Prior to Experimentation—Orthogonal Contrasts

If the above decision is made prior to the running of an experiment, such comparisons can usually be set without disturbing the risk α of the original ANOVA. This means that the contrasts must be chosen with care, and the number of such contrasts should not exceed the number of degrees of freedom between the treatment means. The method usually used here is called the *method of orthogonal contrasts*.

In comparing treatments 1 and 2 it should seem quite logical to examine the difference in their means or their totals since all treatment totals are based on the same sample size n. One such comparison might be $T_{.1} - T_{.2}$. If, on the other hand, one wishes to compare the first and second treatment with the third, the third treatment total should be weighted by a factor of 2 since it is based on only four observations and the treatment totals of 1 and 2 together represent eight observations. Such a comparison would then be $T_{.1} + T_{.2} - 2T_{.3}$. Note that for each of the two comparisons or contrasts, the sum of the coefficients of the treatment totals always adds to zero:

$$T_{.1} - T_{.2} \text{ has coefficients } +1 - 1 = 0$$

$$T_1 + T_{.2} - 2T_{.3} \text{ has coefficients } +1 + 1 - 2 = 0$$

Hence a *contrast* C_m is defined for any linear combination of treatment totals as follows:

$$C_m = c_{1m}T_{.1} + c_{2m}T_{.2} + \cdots + c_{jm}T_{.j} + \cdots + c_{km}T_{.k}$$

where

$$c_{1m} + c_{2m} + \cdots + c_{jm} + \cdots + c_{km} = 0$$

or, more compactly,

$$C_m = \sum_{j=1}^{k} c_{jm}T_{.j}$$

where

$$\sum_{j=1}^{k} c_{jm} = 0$$

In dealing with several contrasts it is highly desirable to have the contrasts independent of each other. When the contrasts are independent—one having no projection on any of the others—they are said to be orthogonal contrasts, and independent tests of hypotheses can be made by comparing the mean square of each such contrast with the mean square of the error term in the experiment. Each contrast carries one degree of freedom.

It may be recalled that when two straight lines, say

$$c_{11}X + c_{21}Y = b_1$$
$$c_{12}X + c_{22}Y = b_2$$

are perpendicular, or orthogonal, to each other, the slope of one line is the negative reciprocal of the slope of the other line. Changing the two lines above into slope-intercept form,

$$Y = -c_{11}/c_{21}X + b_1/c_{21}$$
$$Y = -c_{12}/c_{22}X + b_2/c_{22}$$

To be orthogonal,

$$-c_{11}/c_{21} = -(-c_{22}/c_{12})$$

or

$$c_{11}c_{12} = -c_{21}c_{22}$$

or

$$c_{11}c_{12} + c_{21}c_{22} = 0$$

The sum of the products of the corresponding coefficients (on X and on Y) must therefore add to zero. This idea may be extended to a general definition.

Two contrasts C_m and C_q are said to be *orthogonal contrasts*, provided

$$\sum_{j=1}^{k} c_{jm}c_{jq} = 0$$

for equal n's.

The sums of squares for a contrast are given by

$$SS_{C_m} = \frac{C_m^2}{n \sum_{j=1}^{k} c_{jm}^2}$$

To apply this procedure to the problem above, three orthogonal contrasts may be set up since there are 3 df between treatments. One such set of three might be

$$C_1 = T_{.1} \qquad\qquad - T_{.4}$$
$$C_2 = \qquad\quad T_{.2} - T_{.3}$$
$$C_3 = T_{.1} - T_{.2} - T_{.3} + T_{.4}$$

C_1 is a contrast to compare the mean of the first treatment with the fourth, C_2 compares the second treatment with the third, and C_3 compares the average of treatments 1 and 4 with the average of 2 and 3. The coefficients of the $T_{.j}$'s for the three contrasts are given in Table 3.10.

Table 3.10 Orthogonal Coefficients

	$T_{.1}$	$T_{.2}$	$T_{.3}$	$T_{.4}$
C_1	+1	0	0	−1
C_2	0	+1	−1	0
C_3	+1	−1	−1	+1

It can be seen from Table 3.10 that the sum of coefficients c_{jm} adds up to zero for each contrast, and that the sum of products of coefficients of each pair of contrasts is also zero.

For the example given in Section 3.2, using the data in Table 3.6, the contrasts are

$$C_1 = +1(8.76) + 0(10.72) + 0(9.67) - 1(9.26) = -0.50$$
$$C_2 = 0(8.76) + 1(10.72) - 1(9.67) + 0(9.26) = 1.05$$
$$C_3 = +1(8.76) - 1(10.72) - 1(9.67) + 1(9.26) = -2.37$$

The corresponding sums of squares are

$$\text{SS}_{C_1} = \frac{(-0.50)^2}{4(2)} = 0.0312$$

$$\text{SS}_{C_2} = \frac{(1.05)^2}{4(2)} = 0.1378$$

$$\text{SS}_{C_3} = \frac{(-2.37)^2}{4(4)} = 0.3511$$

$$\text{Total} = \overline{0.5201}$$

Note that these three orthogonal sums of squares add up to the treatment sums of squares. Each contrast has 1 df, so one may test hypotheses implied by the contrasts as follows:

$$H_{0_1}: \tau_1 = \tau_4 \quad \text{or} \quad \mu_1 = \mu_4 \qquad F_{1,12} = \frac{0.0312/1}{0.0203} = 1.54$$

$$H_{0_2}: \tau_2 = \tau_3 \quad \text{or} \quad \mu_2 = \mu_3 \qquad F_{1,12} = \frac{0.1378/1}{0.0203} = 6.76$$

$$H_{0_3}: \tau_1 + \tau_4 = \tau_2 + \tau_3 \qquad F_{1,12} = \frac{0.3511/1}{0.0203} = 17.30$$

or

$$\mu_{1+4} = \mu_{2+3}$$

Since the F for 1 and 12 df is 4.75 at the 5 percent significance level and 9.33 at the 1 percent significance level, one can reject H_{0_2} at the 5 percent level and conclude that the mean wear resistance of fabric 2 (or B) differs from that of fabric 3 (or C). At the 1 percent level one rejects H_{0_3} and concludes that the mean wear resistances of fabrics 1 and 4 (A and D) differ from that of 2 and 3 (B and C).

If a computer program is used to run the ANOVA, it can be programmed to compute any desired contrasts. In Table 3.7 lines 10, 11, and 12 indicate the three contrasts desired. The printout is as shown in Table 3.11.

Table 3.11   Orthogonal Contrasts for Example 3.1

| | CONTRAST COEFFICIENT MATRIX | | | |
	GRP 1	GRP 2	GRP 3	GRP 4
CONTRAST 1	1.0	0	0	-1.0
CONTRAST 2	0	1.0	-1.0	0
CONTRAST 3	1.0	-1.0	-1.0	1.0

| | POOLED VARIANCE ESTIMATE | | | |
	VALUE	S. ERROR	T VALUE	D.F.	T PROB.
CONTRAST 1	$-.1250$	.1008	-1.2403	12.0	.239
CONTRAST 2	.2625	.1008	2.6046	12.0	.023
CONTRAST 3	$-.5925$	.1425	-4.1570	12.0	.001

At first glance these results seem different from the calculated values above; however, the "value" of the contrast given in Table 3.11 is based on means instead of totals, so multiplying each by $n_j = 4$ yields the contrasts of -0.5, 1.05, and -2.37. Then the tests are run as t tests instead of F tests. In the case of 1 df it can be shown that $t_{df}^2 = F_{1,df}$. Here, for example, for the second contrast $t = 2.6$, which squares into 6.76 as seen above.

As this method of orthogonal contrasts is used quite often in experimental design work, the definitions and formulas are given below for the case of unequal numbers of observations per treatment.

C_m is a *contrast* if

$$\sum_{j=1}^{k} n_j c_{jm} = 0$$

and C_m and C_q are *orthogonal contrasts* if

$$\sum_{j=1}^{k} n_j c_{jm} c_{jq} = 0$$

The *sum of squares for such contrasts* is given by

$$SS_{C_m} = \frac{C_m^2}{\sum_{j=1}^{k} n_j c_{jm}^2}$$

Tests on Means After Experimentation

If the decision on what comparisons to make is withheld until after the data are examined, comparisons may still be made, but the α level is altered because such decisions are not taken at random but are based on observed results. Several methods have been introduced to handle such situations [17], but only the Newman–Keuls range test [15] and Scheffé's test [21] are described here.

Range Test   After the data have been compiled the following steps are taken.

1. Arrange the k means in order from low to high.
2. Enter the ANOVA table and take the error mean square with its degrees of freedom.
3. Obtain the standard error of the mean for each treatment

$$s_{\bar{Y}_{.j}} = \sqrt{\frac{\text{error mean square}}{\text{number of observations in } \bar{Y}_{.j}}}$$

   where the error mean square is the one used as the denominator in the F test on means $\bar{Y}_{.j}$.
4. Enter a Studentized range table (Appendix, Table E) of significant ranges at the α level desired, using $n_2 =$ degrees of freedom for error mean square and $p = 2, 3, \ldots, k$, and list these $k - 1$ ranges.
5. Multiply these ranges by $s_{\bar{Y}_{.j}}$ to form a group of $k - 1$ least significant ranges.
6. Test the observed ranges between means, beginning with largest versus smallest, which is compared with the least significant range for $p = k$; then test largest versus second smallest with the least significant range for $p = k - 1$; and so on. Continue this for second largest versus smallest, and so forth, until all $k(k - 1)/2$ possible pairs have been tested. The sole exception to this rule is that no difference between two means can be declared significant if the two means concerned are both contained in a subset with a nonsignificant range.

   To see how this works, consider the data of Example 3.1 as given in Table 3.6. Following the steps given above:

1. $k = 4$ means are   2.19   2.32   2.42   2.68
   for treatments: A   D   C   B

2. From Table 3.2 the error mean square is 0.0203 with 12 df.
3. Standard error of a mean is

$$s_{\bar{Y}_{.j}} = \sqrt{\frac{0.0203}{4}} = 0.0712$$

4. From the Appendix, Table E, at the 5 percent level the significant ranges are, for $n_2 = 12$,

$$
\begin{array}{cccc}
p: & 2 & 3 & 4 \\
\text{Ranges:} & 3.08 & 3.77 & 4.20
\end{array}
$$

5. Multiplying by the standard error of 0.0712, the least significant ranges (LSR) are

$$
\begin{array}{cccc}
p: & 2 & 3 & 4 \\
\text{LSR:} & 0.22 & 0.27 & 0.30
\end{array}
$$

6. Largest versus smallest: B versus A, $0.49 > 0.30$.
   Largest versus next smallest: B versus D, $0.36 > 0.27$.
   Largest versus next largest: B versus C, $0.26 > 0.22$.
   Second largest versus smallest: C versus A, $0.23 < 0.27$.
   Second largest versus next largest: C versus D, $0.10 < 0.22$.
   Third largest versus smallest: D versus A, $0.13 < 0.22$.

From the results in step 6 one sees that B differs significantly from A, D, and C, but A, D, and C do not differ significantly from each other. These results are shown in Figure 3.2, where any means underscored by the same line are not significantly different.

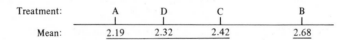

Figure 3.2   Newman-Keuls test means.

Scheffé's Test   Since the Newman–Keuls test is restricted to comparing pairs of means and it is often desirable to examine other contrasts which represent combinations of treatments, many experimenters prefer a test devised by Scheffé [21]. Scheffé's method uses the concept of contrasts presented earlier but the contrasts need not be orthogonal. In fact, any and all conceivable contrasts may be tested for significance. Since comparing means in pairs is a special case of contrasts, Scheffé's scheme is more general than the Newman–Keuls. However, since the Scheffé method must be valid for such a large set of possible contrasts, it requires larger observed differences to be significant than some of the other schemes. To see how Scheffé's method applies, the following steps are taken after the data have been compiled.

1. Set up all contrasts of interest to the experimenter and compute their numerical values.
2. Determine the significant F statistic for the ANOVA just performed based on α and degrees of freedom $k - 1$, $N - k$.
3. Compute $A = \sqrt{(k - 1)F}$, using the F from step 2.
4. Compute the standard error of each contrast to be tested. This standard error is given by

$$s_{C_m} = \sqrt{(\text{error mean square}) \sum_j n_j c_{jm}^2}$$

5. If a contrast C_m is numerically larger than A times s_{C_m}, it is declared significant. Or, if $|C_m| > As_{C_m}$, reject the hypothesis that the true contrast among means is zero.

Applying this technique to the coded data of Tables 3.1, 3.2, and 3.6, proceed as follows.

1. Consider two contrasts:

$$C_1 = T_{.1} - T_{.2} \qquad\qquad = 8.76 - 10.72 \qquad\qquad = -1.96$$
$$C_2 = 3T_{.1} - T_{.2} - T_{.3} - T_{.4} = 26.28 - 10.72 - 9.67 - 9.26 = -3.37$$

(Note that these are not orthogonal. The first compares treatment A with B and the second compares the mean of treatment A with the mean of the other three treatments.)
2. Since $\alpha = 0.05$, $k - 1 = 3$, $N - k = 12$, $F = 3.49$.
3. $A = \sqrt{3(3.49)} = 3.24$.
4. $s_{C_1} = \sqrt{0.0203[4(1)^2 + 4(-1)^2]} = 0.40$ and $As_{C_1} = 1.30$.
   $s_{C_2} = \sqrt{0.0203[4(3)^2 + 4(-1)^2 + 4(-1)^2 + 4(-1)^2]} = 0.99$ and $As_{C_2} = 3.21$.
5. Since $|C_1| = 1.96$ is > 1.30, this contrast is significant.
   Since $|C_2| = 3.37$ is > 3.21, this contrast is also significant.

If the computer is used, it can be instructed to carry out the Newman–Keuls test, called SNK in the SPSS program. Item 13 in Table 3.7 asks for this test. The printout is given in Table 3.12.

This table determines subsets among the means and it is seen that groups 1, 4, and 3 (fabrics A, D, and C) form one group and group 2 (fabric B) forms another, and that the subgroups do not overlap.

Table 3.12   Newman–Keuls Printout for Example 3.1

MULTIPLE RANGE TEST

STUDENT-NEWMAN–KEULS PROCEDURE

RANGES FOR THE .050 LEVEL—

          3.08   3.77   4.20

THE RANGES ABOVE ARE TABULAR VALUES.

THE VALUE ACTUALLY COMPARED WITH MEAN(J)–MEAN(I) IS . .

.1008 — RANGE — SQRT(1/N(I) + 1/N(J))

HOMOGENEOUS SUBSETS (SUBSETS OF GROUPS, WHOSE HIGHEST AND LOWEST MEANS DO NOT DIFFER BY MORE THAN THE SHORTEST SIGNFICANT RANGE FOR A SUBSET OF THAT SIZE)

SUBSET 1

GROUP	GRP 1	GRP 4	GRP 3
MEAN	2.1900	2.3150	2.4175

SUBSET 2

GROUP	GRP 2
MEAN	2.6800

3.5   CONFIDENCE LIMITS ON MEANS

After an analysis of variance it is often desirable to set confidence limits on a treatment mean. The $100(1 - \alpha)$ percent confidence limits on $\mu_{.j}$ are given by

$$\overline{Y}_{.j} \pm t_{1-\alpha/2} \sqrt{\frac{\text{mean square used to test treatment mean square}}{n_j}} \quad (3.6)$$

where the mean square used to test treatment mean square is the error mean square in a one-way ANOVA but may be a different mean square in more complex analyses. The degrees of freedom used with the Student's t statistic are the degrees of freedom that correspond to the mean square used for testing the treatment mean.

In Example 3.1, 95 percent confidence limits on the mean wear resistance for fabric B, μ_2 would be

$$\overline{Y}_{.2} \pm t_{.975} \sqrt{\frac{0.0203}{4}}$$

with 12 df on t. By substituting and taking the t value from Appendix Table B, the limits are $2.68 \pm 2.18(0.0712)$ or 2.68 ± 0.16 or from 2.52 to 2.84.

3.6   COMPONENTS OF VARIANCE

In Example 3.1 the levels of the factor were considered as fixed, since only four fabrics were available and a decision was desired on the effect of these four fabrics only. If, however, the levels of the factor are random (such as operators, days, or samples where the levels in the experiment might have been chosen at random from a large number of possible levels), the model is called a *random model*, and inferences are to be extended to all levels of the population (of which the observed four levels are random samples). In a random model, the experimenter is not usually interested in testing hypotheses, setting confidence limits, or making contrasts in means, but in estimating components of variance. How much of the variance in the experiment might be considered as due to true differences in treatment means, and how much might be due to random error about these means?

Example 3.2   To see how to analyze and interpret a component of variance or random model, consider the following problem.

A company supplies a customer with several hundred batches of a raw material every year. The customer is interested in a high yield from the raw material in terms of percent usable chemical. He usually makes three sample determinations of yield from each batch in order to control the quality of the incoming material. He expects and gets some variation between determinations on a given batch, but he suspects that there may be significant batch-to-batch variation as well.

To check this, he selects five batches at random from several batches available and runs three yield determinations per batch. His 15 yield determinations are completely randomized. The mathematical model is again

$$Y_{ij} = \mu + \tau_j + \varepsilon_{i(j)}$$

except that in this experiment the k levels of the treatment (batches) are chosen at random, rather than being fixed levels. The data are shown in Table 3.13.

Table 3.13   Chemical Yield by Batch Data

		Batch		
1	2	3	4	5
74	68	75	72	79
76	71	77	74	81
75	72	77	73	79

An ANOVA performed on these data gave the results in Table 3.14.

The F test gives $F_{4,10} = 36.94/1.80 = 20.5$, which is highly significant. Since these batches are but a random sample of batches, we may be interested

Table 3.14   Chemical Yield ANOVA

Source	df	SS	MS	EMS
Between batches τ_j	4	147.74	36.94	$\sigma_\varepsilon^2 + n\sigma_\tau^2$
Error $\varepsilon_{i(j)}$	10	17.99	1.80	σ_ε^2
Totals	14	165.73		

in how much of the variance in the experiment might be attributed to batch differences and how much to random error. To help answer these questions, another column has been added to the ANOVA table (Table 3.14). This is the expected mean square (EMS) column. It can be derived by inserting the mathematical model into the operational formulas for mean squares in Table 3.5 and then computing the expected values of these mean squares. The results will depend upon whether a fixed or random model is used as will be discussed in greater length in the next part of this section.

The interpretation of Table 3.14 then is that the error variance σ_ε^2 is best estimated as 1.80, its corresponding mean square. Also, the mean square 36.94 is an estimate of $\sigma_\varepsilon^2 + 3\sigma_\tau^2$. If the numerical mean squares are set equal to the variance components that they are estimating,

$$s_\varepsilon^2 = 1.80 \qquad s_\varepsilon^2 + 3s_\tau^2 = 36.94$$

where s^2 is used since these are estimates of the corresponding σ^2's.

Solving these expressions,

$$s_\tau^2 = \frac{36.94 - 1.80}{3} = 11.71$$

The total variance can be estimated as

$$s_{total}^2 = s_\tau^2 + s_\varepsilon^2 = 11.71 + 1.80 = 13.51$$

and 11.71/13.51 or 86.7 percent of the total variance is attributable to batch differences and only 1.80/13.51 or 13.3 percent is attributable to errors within batches.

It is interesting to note that the above estimate of total variance of 13.51 would give a standard deviation of 3.68. One might therefore expect all the data in such a small experiment ($N = 15$) to fall within four standard deviations or within $4(3.68) = 14.72$. The actual range is 13 (high, 81—low, 68). This breakdown of total variance into its two components then seems quite reasonable.

Some readers may wonder why the total sum of squares is never divided by its degrees of freedom to estimate the total variance. If the methods of the next section are used to determine the EMS for the total, it is found to be a biased estimate of the total variance. In fact, a derivation will show that the

EMS for the total in a one-way ANOVA is equal to

$$\text{EMS}_{\text{total}} = \frac{n(k-1)}{N-1}\sigma_\tau^2 + \sigma_\varepsilon^2$$

Since the coefficient of σ_τ^2 is always less than one, this EMS will be less than the total variance given above as $\sigma_\tau^2 + \sigma_\varepsilon^2$.

Expected Mean Square Derivation

For a single-factor experiment

$$Y_{ij} = \mu + \tau_j + \varepsilon_{ij} \tag{3.7}$$

From Table 3.5 the sum of squares for treatments with equal n's is

$$\text{SS}_{\text{treatment}} = \sum_{j=1}^{k} n(\bar{Y}_{.j} - \bar{Y}_{..})^2$$

From model Equation (3.7)

$$\bar{Y}_{.j} = \sum_{i=1}^{n} Y_{ij}/n = \sum_{i=1}^{n} (\mu + \tau_j + \varepsilon_{ij})/n$$

$$= \frac{n\mu}{n} + \frac{n\tau_j}{n} + \sum_{i=1}^{n} \varepsilon_{ij}/n$$

$$= \mu + \tau_j + \sum_{i=1}^{n} \varepsilon_{ij}/n \tag{3.8}$$

Also,

$$\bar{Y}_{..} = \sum_{j=1}^{k}\sum_{i=1}^{n} Y_{ij}/nk = \sum_{j}^{k}\sum_{i}^{n} (\mu + \tau_j + \varepsilon_{ij})/nk$$

$$= \frac{nk\mu}{nk} + n\sum_{j}^{k} \tau_j/nk + \sum_{j}^{k}\sum_{i}^{n} \varepsilon_{ij}/nk$$

$$= \mu + \sum_{j}^{k} \tau_j/k + \sum_{j}^{k}\sum_{i}^{n} \varepsilon_{ij}/nk \tag{3.9}$$

Subtracting Equation (3.9) from Equation (3.8) gives

$$\bar{Y}_{.j} - \bar{Y}_{..} = \tau_j - \sum_{j=1}^{k} \tau_j/k + \sum_{i=1}^{n} \varepsilon_{ij}/n - \sum_{i}^{n}\sum_{j}^{k} \varepsilon_{ij}/nk$$

Squaring gives

$$(\bar{Y}_{.j} - \bar{Y}_{..})^2 = \left(\tau_j - \sum_{j=1}^{k} \tau_j/k\right)^2 + \frac{1}{n^2}\left(\sum_{i}^{n} \varepsilon_{ij} - \sum_{i}^{n}\sum_{j}^{k} \varepsilon_{ij}/k\right)^2$$

$$(+ \text{ cross products})$$

Multiplying by n and summing over j gives

$$
SS_{treatment} = \sum_{j=1}^{k} n(\bar{Y}_{.j} - \bar{Y}_{..})^2
$$

$$
= n \sum_{j}^{k} \left(\tau_j - \sum_{j}^{k} \tau_j/k \right)^2
$$

$$
+ \frac{n}{n^2} \sum_{j}^{k} \left(\sum_{i}^{n} \varepsilon_{ij} - \sum_{i}^{n}\sum_{j}^{k} \varepsilon_{ij}/k \right)^2 + n \sum_{j} (\text{cross product})
$$

The expected value operator may now be applied to this $SS_{treatment}$:

$$
E(SS_{treatment}) = nE\left[\sum_{j}^{k} \left(\tau_j - \sum_{j}^{k} \tau_j/k \right)^2 \right]
$$

$$
+ \frac{1}{n} E\left[\sum_{j}^{k} \left(\sum_{i}^{n} \varepsilon_{ij} - \sum_{i}^{n}\sum_{j}^{k} \varepsilon_{ij}/k \right)^2 \right]
$$

as it can be shown that the expected value of the cross-product term equals zero.

If the treatment levels are fixed,

$$
\sum_{j=1}^{k} \tau_j = \sum_{j=1}^{k} (\mu_{.j} - \mu) = 0
$$

and the $E(SS_{treatment})$ becomes

$$
E(SS_{treatment}) = n \sum_{j=1}^{k} \tau_j^2 + \frac{1}{n}(nk - n)\sigma_\varepsilon^2
$$

since errors are random and $\sum_{j=1}^{k} \tau_j^2$ is a constant. The $E(MS_{treatment}) = E[SS_{treatment}/(k-1)]$, so

$$
E(MS_{treatment}) = \left[n \sum_{j=1}^{k} \tau_j^2/(k-1) \right] + \frac{n(k-1)}{n(k-1)} \sigma_\varepsilon^2
$$

If, on the other hand, treatment levels are random and their variance is σ_τ^2

$$
E(MS_{treatment}) = \frac{n(k-1)\sigma_\tau^2}{(k-1)} + \sigma_\varepsilon^2
$$

the EMS term corresponding to the treatments is then either

$$
\sigma_\varepsilon^2 + n\left[\sum_{j} \tau_j^2/(k-1) \right]
$$

or

$$
\sigma_\varepsilon^2 + n\sigma_\tau^2
$$

depending upon whether treatments are fixed or random. When fixed, we shall designate $\sum_j \tau_j^2/(k - 1)$ as ϕ_τ, so the fixed treatment EMS is $\sigma_\varepsilon^2 + n\phi_\tau$.

For the error mean square,

$$SS_{error} = \sum_{j=1}^{k} \sum_{i=1}^{n} (Y_{ij} - \bar{Y}_{.j})^2$$

Subtracting Equation (3.8) from Equation (3.7) gives

$$Y_{ij} - \bar{Y}_{.j} = \varepsilon_{ij} - \sum_{i=1}^{n} \varepsilon_{ij}/n$$

Squaring and adding gives

$$\sum_{j}^{k} \sum_{i}^{n} (Y_{ij} - \bar{Y}_{.j})^2 = \sum_{j}^{k} \sum_{i}^{n} \left(\varepsilon_{ij} - \sum_{i} \varepsilon_{ij}/n \right)^2$$

Taking the expected value, we have

$$E(SS_{error}) = E\left[\sum_{j}^{k} \sum_{i}^{n} \left(\varepsilon_{ij} - \sum_{i} \varepsilon_{ij}/n \right)^2 \right]$$

$$= \sum_{j}^{k} E\left[\sum_{i} \left(\varepsilon_{ij} - \sum_{i} \varepsilon_{ij}/n \right)^2 \right]$$

$$= \sum_{j}^{k} (n - 1)\sigma_\varepsilon^2$$

$$= k(n - 1)\sigma_\varepsilon^2$$

and

$$E(MS_{error}) = E[SS_{error}/k(n - 1)] = \sigma_\varepsilon^2$$

as shown in Table 3.14.

3.7 SUMMARY

In this chapter consideration has been given to

Experiment	Design	Analysis
I. Single factor	Completely randomized	One-way ANOVA

In applying the ANOVA techniques, certain assumptions should be kept in mind:

1. The process is in control, that is, it is repeatable.
2. The population distribution being sampled is normal.
3. The variance of the errors within all k levels of the factor are homogeneous.

Many texts [4], [9] discuss these assumptions and what may be done if they are not met in practice.

Several studies have shown that lack of normality in the dependent variable Y does not seriously affect the analysis when the number of observations per treatment is the same for all treatments. To check for homogeneous variances within treatments, several tests have been suggested. One quick check is to examine the ranges of the observations within each treatment. For Example 3.1 these are 0.45, 0.20, 0.40, and 0.15. If the average range $\bar{R}$ is multiplied by a factor D_4, which is found in most quality control texts, and all ranges are less than $D_4\bar{R}$, it is quite safe to assume homogeneous variances. Here $\bar{R} = 0.30$ and $D_4\bar{R} = 0.68$ based on samples of four. Since all four ranges are well below 0.68, homogeneity of variance may be reasonably assured. Values of D_4 for a few selected sample sizes are given in Table 3.15.

Table 3.15 D_4 Values for Sample Size n

If $n =$	2	3	4	5	6	7	8	9	10
$D_4 =$	3.267	2.575	2.282	2.115	2.004	1.924	1.864	1.816	1.777

If a more precise test is desired, our computer program includes three tests: Cochran's C test, Bartlett-Box F test, and a maximum/minimum variance test. Two of these show probabilities of obtaining the observed statistic or a larger value when the variances are homogeneous. If the probabilities are well above 0.05, homogeneity is a safe assumption. Table 3.16 shows the printout for Example 3.1.

Table 3.16   Homogeneity of Variances Tests

COCHRANS C = MAX. VARIANCE/SUM(VARIANCES) =	.4406,	P = .557 (APPROX.)
BARTLETT-BOX F =	1.265,	P = .287
MAXIMUM VARIANCE/MINIMUM VARIANCE =	8.326	

With some measured variables it may be that the means of the treatments and the variances within the treatments are related in some way. For example, a Poisson variable has its mean and variance equal. It can be shown mathematically that a necessary and sufficient condition for normality of a variable is that means and variances drawn from that population are independent of each other. If it is known or observed that sample means and sample variances are not independent but are related to each other, it may be possible to "break" this correlation by a suitable transformation of the original variable Y. Table 3.17 gives several suggested transformations to try if certain specified relationships hold.

Table 3.17 Transformations in ANOVA

If	is proportional to	Transform Y_{ij} to
s_j^2	$\bar{Y}_{.j}$	$\sqrt{Y_{ij}}$ (Poisson case)
s_j^2	$\bar{Y}_{.j}(1 - \bar{Y}_{.j})$	arc sin $\sqrt{Y_{ij}}$ (binomial case)
s_j	$\bar{Y}_{.j}$	$\log Y_{ij}$ or $\log(Y_{ij} + 1)$
s_j	$\bar{Y}_{.j}^2$	$1/Y_{ij}$

Computer programs can be instructed simply to change the original variable Y_{ij} to some transformed variable as shown above and carry out the ANOVA on the transformed variable.

PROBLEMS

3.1 Assuming a completely randomized design, do a one-way analysis of variance on the following data in order to familiarize yourself with the technique.

	Factor A Level				
	1	2	3	4	5
Measurement	8	4	1	4	10
	6	−2	2	6	8
	7	0	0	5	7
	5	−2	−1	5	4
	8	3	−3	4	9

3.2 The cathode warm-up time in seconds was determined for three different tube types using eight observations on each type of tube. The order of experimentation was completely randomized. The results were

	Tube Type					
	A		B		C	
Warm-up time	19	20	20	40	16	19
(seconds)	23	20	20	24	15	17
	26	18	32	22	18	19
	18	35	27	18	26	18

Do an analysis of variance on these data and test the hypothesis that the three tube types require the same average warm-up time.

3.3 For Problem 3.2 set up orthogonal contrasts between the tube types and test your contrasts for significance.

3.4 Set up 95 percent confidence limits for the average warm-up time for tube type C in Problem 3.2.

3.5 Use the Newman–Keuls range method to test for differences between tube types.

3.6 The following data are on the pressure in a torsion spring for several settings of the angle between the legs of the spring in a free position.

	Angle of Legs of Spring (degrees)				
	67	71	75	79	83
Pressure (psi)	83	84	86	89	90
	85	85	87	90	92
		85	87	90	
		86	87	91	
		86	88		
		87	88		
			88		
			88		
			88		
			89		
			90		

Assuming a completely randomized design, complete a one-way analysis of variance for this experiment and state your conclusion concerning the effect of angle on the pressure in the spring.

3.7 Set up orthogonal contrasts for the angles in Problem 3.6.

3.8 Show that the expanded forms for the sums of squares in Table 3.5 are correct.

3.9 Assume that the levels of factor A in Problem 3.1 were chosen at random and determine the proportion of variance attributable to differences in level means and the proportion due to error.

3.10 Verify the results given in Table 3.14 based on the data of Table 3.13.

3.11 Set up two or more contrasts for the data of Problem 3.2 and test their significance by the Scheffé method.

3.12 Since the Scheffé method is not restricted to equal sample sizes, set up several contrasts for Problem 3.6 and test for significance by the Scheffé method.

3.13 It is suspected that the environmental temperature in which batteries are activated affects their activated life. Thirty homogeneous batteries were tested, six at each of five temperatures, and the data shown below were obtained. Analyze and interpret the data.

	Temperature (°C)				
	0	25	50	75	100
Activated life	55	60	70	72	65
(seconds)	55	61	72	72	66
	57	60	73	72	60
	54	60	68	70	64
	54	60	77	68	65
	56	60	77	69	65

3.14 A highway research engineer wishes to determine the effect of four types of sub-grade soil on the moisture content in the top soil. He takes five samples of each type of subgrade soil and the total sum of squares is computed as 280, whereas the sum of squares among the four types of subgrade soil is 120.

1. Set up an analysis-of-variance table for these results.

2. Set up a mathematical model to describe this problem, define each term in the model, and state the assumptions made on each term.

3. Set up a test of the hypothesis that the four types of subgrade soil have the same effect on moisture content in the top soil.

4. Set up a set of orthogonal contrasts for this problem.

5. Explain briefly how to set up a test on means after the analysis of variance for these data.

6. Set up an expression for 90 percent confidence limits on the mean of type 2 subgrade soil. Insert all numerical values that are known.

3.15 Data collected on the effect of four fixed types of television tube coating on the conductivity of the tubes is given as

Coating			
I	II	III	IV
56	64	45	42
55	61	46	39
62	50	45	45
59	55	39	43
60	56	43	41

Do an analysis of variance on these data and test the hypothesis that the four coatings yield the same average conductivity.

3.16 For Problem 3.15 set up orthogonal contrasts between tube coatings and test them for significance.

3.17 Use a Newman–Keuls range test on the data of Problem 3.15 and discuss any significant results that you find.

3.18 Set up contrasts and test using Scheffé's method to answer the following questions: Do the means of coating I and II differ? The means of II and IV? The mean of I and II versus the mean of III and IV?

3.19 Explain why confidence limits on a treatment mean as given in Table 3.9 might differ from the method suggested in Section 3.5. Which method would you recommend under what conditions?

3.20 To determine the effect of several teaching methods on student achievement, 30 students were assigned to five treatment groups with six students per group. The treatments given for the semester are as follows:

Treatment	Description
1	Current textbook
2	Textbook A with teacher
3	Textbook A with machine
4	Textbook B with teacher
5	Textbook B with machine

At the end of a semester of instruction, achievement scores were recorded and some of the statistics were found to be

	Treatment				
	1	2	3	4	5
Totals	120	600	720	240	420

Source	df	SS	MS	F	$F_{0.95}$
Between treatments		340			
Error					
Total		465			

a. Complete the ANOVA table.
b. Write the mathematical model assumed here and state the hypothesis to be tested.
c. Test the hypothesis and state your conclusion.

3.21 Set up one set of orthogonal contrasts that might seem reasonable from the treatment descriptions in Problem 3.20.

3.22 Test whether or not the mean achievement under treatment 1 differs significantly from the mean achievement under treatment 5 in a Newman–Keuls sense for Problem 3.20.

3.23 Determine the standard error of the contrast $4T_{.1} - T_{.2} - T_{.3} - T_{.4} - T_{.5}$ based on Problem 3.20 data.

3.24 Three fertilizers are tried on 27 plots of land in a random fashion such that each fertilizer is applied to nine plots. The total yield for each fertilizer type is given by

	Type		
	1	2	3
$T_{.j}$	240	320	180

and the ANOVA table is

Source	df	SS	MS
Between fertilizers	2	622	311
Error	24	1440	60
Total	26	2062	

a. Set up one set of orthogonal contrasts that might be used on these data.
b. For your first contrast, determine the sum of squares due to this contrast.
c. For your second contrast, find its standard error.
d. If one wished to compare types 1 and 2 and also the average of 1 and 3 versus 2, which method would you recommend and why?

3.25 Birth weights in pounds of five Poland China pigs were recorded from each of six randomly chosen litters. The ANOVA printout showed:

Source	df	SS	MS
Between litters	5	6.0	1.20
Within litters	24	9.6	0.40

Determine what percentage of the total variance in this experiment can be attributed to litter to litter differences.

4

Single-Factor Experiments—Randomized Block and Latin Square Designs

4.1 INTRODUCTION

Consider the problem of determining whether or not different brands of tires exhibit different amounts of tread loss after 20,000 miles of driving. A fleet manager wishes to consider four brands that are available and make some decision about which brand might show the least amount of tread wear after 20,000 miles. The brands to be considered are *A*, *B*, *C*, and *D*, and although driving conditions might be simulated in the laboratory, he wants to try these four brands under actual driving conditions. The variable to be measured is the difference in maximum tread thickness on a tire between the time it is mounted on the wheel of a car and after it has completed 20,000 miles on this car. The measured variable Y_{ij} is this difference in thickness in mils (0.001 inch), and the only factor of interest is brands, say, τ_j where $j = 1, 2, 3,$ and 4.

Since the tires must be tried on cars and since some measure of error is necessary, more than one tire of each brand must be used and a set of four of each brand would seem quite practical. This means 16 tires, four each of four different brands, and a reasonable experiment would involve at least four cars. Designating the cars as I, II, III, and IV, one might put brand *A*'s four tires on car I, brand *B*'s on car II, and so on, with a design as shown in Table 4.1.

Table 4.1 Design 1 for Tire Brand Test

	Car			
	I	II	III	IV
Brand	*A*	*B*	*C*	*D*
distribution	*A*	*B*	*C*	*D*
	A	*B*	*C*	*D*
	A	*B*	*C*	*D*

66

Table 4.2   Design 2 for Tire Brand Test

	Car			
	I	II	III	IV
Brand distribution	$C(12)$	$A(14)$	$D(10)$	$A(13)$
and loss in thickness	$A(17)$	$A(13)$	$C(11)$	$D(9)$
	$D(13)$	$B(14)$	$B(14)$	$B(8)$
	$D(11)$	$C(12)$	$B(13)$	$C(9)$

One look at this design shows its fallibility, since averages for brands are also averages for cars. If the cars travel over different terrains, using different drivers, any apparent brand differences are also car differences. This design is called completely confounded, since we cannot distinguish between brands and cars in the analysis.

A second attempt at design might be to try a completely randomized design, as given in Chapter 3. Assigning the 16 tires to the four cars in a completely random manner might give results as in Table 4.2. In Table 4.2 the loss in thickness is given for each of the 16 tires. The purpose of complete randomization here is to average out any car differences that might affect the results. The model would be

$$Y_{ij} = \mu + \tau_j + \varepsilon_{ij}$$

with

$$j = 1, 2, 3, 4 \qquad i = 1, 2, 3, 4$$

An analysis of variance on these data gives the results in Table 4.3.

The F test shows $F_{3,12} = 2.43$, and the 5 percent critical region for F is $F_{3,12} \geq 3.49$ (Appendix, Table D), so there is no reason to reject the hypothesis of equal average tread loss among the four brands.

Table 4.3   ANOVA for Design 2

Source	df	SS	MS
Brands	3	30.69	10.2
Error	12	50.25	4.2
Totals	15	80.94	

4.2 RANDOMIZED COMPLETE BLOCK DESIGN

A more careful examination of design 2 in Table 4.2 will reveal some glaring disadvantages of the completely randomized design in this problem. One thing to be noted is that brand A is never used on car III or brand B on car I. Also, any variation within brand A may reflect variation between cars I, II, and IV. Thus the random error may not be merely an experimental error but may include variation between cars. Since the chief objective of experimental design is to reduce the experimental error, a better design might be one in which car variation is removed from error variation. Although the completely randomized design averaged out the car effects, it did not eliminate the variance among cars. A design that requires that each brand be used once on each car is a *randomized complete block design*, given in Table 4.4.

Table 4.4   Design 3: Randomized Block Design for Tire Brand Test

	Car			
	I	II	III	IV
Brand distribution	$B(14)$	$D(11)$	$A(13)$	$C(9)$
and loss in thickness	$C(12)$	$C(12)$	$B(13)$	$D(9)$
	$A(17)$	$B(14)$	$D(11)$	$B(8)$
	$D(13)$	$A(14)$	$C(10)$	$A(13)$

In this design the order in which the four brands are placed on a car is random and each car gets one tire of each brand. In this way better comparisons can be made between brands since they are all driven over approximately the same terrain, and so on. This provides a more homogeneous environment in which to test the four brands. In general, these groupings for homogeneity are called *blocks* and randomization is now restricted within blocks. This design also allows the car (block) variation to be independently assessed and removed from the error term. The model for this design is

$$Y_{ij} = \mu + \beta_i + \tau_j + \varepsilon_{ij} \qquad (4.1)$$

where β_i now represents the block effect (car effect) in the example above.

The analysis of this model is a two-way analysis of variance, since the block effect may now also be isolated. A slight rearrangement of the data in Table 4.4 gives Table 4.5.

The total sum of squares is computed as in Chapter 3

$$SS_{total} = \sum_{j}^{4} \sum_{i}^{4} Y_{ij}^2 - \frac{T_{..}^2}{N}$$

$$= 2409 - \frac{(193)^2}{16} = 80.94$$

Table 4.5 Randomized Block Design Data for Tire Brand Test

Car	Brand				$T_{i.}$
	A	B	C	D	
I	17	14	12	13	56
II	14	14	12	11	51
III	13	13	10	11	47
IV	13	8	9	9	39
$T_{.j}$	57	49	43	44	$T_{..} = 193$
$\sum_i Y_{ij}^2$	823	625	469	492	$\sum_j \sum_i Y_{ij}^2 = 2409$

The brand (treatment) sum of squares is computed as usual

$$SS_{brands} = \sum_j \frac{T_{.j}^2}{n} - \frac{T_{..}^2}{N}$$

$$= \frac{(57)^2 + (49)^2 + (43)^2 + (44)^2}{4} - \frac{(193)^2}{16} = 30.69$$

Since the car (block) effect is similar to the brand effect but totaled across the rows of Table 4.5, the car sum of squares is computed exactly like the brand sum of squares, using row totals $T_{i.}$ instead of column totals. Calling the number of treatments (brands) in general k, then

$$SS_{car} = \sum_{i=1}^{n} \frac{T_{i.}^2}{k} - \frac{T_{..}^2}{N}$$

$$= \frac{(56)^2 + (51)^2 + (47)^2 + (39)^2}{4} - \frac{(193)^2}{16} = 38.69$$

The error sum of squares is now the remainder after subtracting both brand and car sum of squares from the total sum of squares

$$SS_{error} = SS_{total} - SS_{brand} - SS_{car}$$
$$= 80.94 - 30.69 - 38.69 = 11.56$$

Table 4.6 is an ANOVA table for these data.

To test the hypothesis, $H_0: \mu_{.1} = \mu_{.2} = \mu_{.3} = \mu_{.4}$, the ratio is

$$F_{3,9} = \frac{10.2}{1.3} = 7.8$$

Table 4.6   ANOVA for Randomized Block
Design of Tire Brand Test

Source	df	SS	MS	EMS
Brands	3	30.69	10.2	$\sigma_\varepsilon^2 + 4\phi_\tau$
Cars	3	38.69	12.9	$\sigma_\varepsilon^2 + 4\sigma_\beta^2$
Error	9	11.56	1.3	σ_ε^2
Totals	15	80.94		

which is significantly larger than the corresponding critical F even at the 1 percent level (Appendix, Table D). The hypothesis of equal brand means is thus rejected. It is to be noted that this hypothesis could not be rejected using a completely randomized design. The randomized block design that allows for removal of the block (car) effect definitely reduced the error variance estimate from 4.2 to 1.3.

It is also possible, if desired, to test the hypothesis that the average tread loss of all four cars is the same. $H_1: \mu_{1.} = \mu_{2.} = \mu_{3.} = \mu_{4.}$, and $F_{3,9} = 12.9/1.3 = 9.9$, which is also significant at the 1 percent level (Appendix, Table D). Here this hypothesis is rejected and a car-to-car variation is detected.

Even though an effect due to cars (blocks) has been isolated, the main objective is still to test brand differences. Thus it is still a single-factor experiment, the blocks representing only a restriction on complete randomization due to the environment in which the experiment was conducted. Other examples include testing differences in materials that are fed into several different machines, testing differences in fertilizers that must be spread on several different plots of ground, testing the effect of different teaching methods on several pupils. In these examples the blocks are machines, plots, and pupils, respectively, and the levels of the factors of interest can be randomized within each block.

When data are presented in tabular form there is usually no way to determine how the data were collected. Was the randomization complete over all N observations or was the experiment run in blocks with randomization restricted to within the blocks? To help in signifying the design of the experiment, it is suggested that in the case of a completely randomized design, no horizontal or vertical lines be drawn on the data, as in Table 4.2. When randomization has been restricted, either vertical or horizontal lines as shown in Tables 4.4 and 4.5, respectively, can be used to indicate this restriction on the randomization. As more complex designs are presented, this scheme may require some double lines for one restriction and single lines for a second restriction, and so on. It is hoped, however, that such a scheme will help the observer in noting just how the randomization has been restricted.

Since blocking an experiment is a very useful procedure to reduce the experimental error, a second example may be helpful as it illustrates the situation in which the same experimental units (animals, people, parts, etc.) are measured before the experiment and then again at the conclusion of the experimental treatment. These before–after, pretest–posttest designs can be treated very well as randomized block designs where the experimental units are the blocks. Interest is primarily in the effect of the treatment and the block effects can be removed to reduce the experimental error. Some refer to these designs as repeated-measures designs because data are repeated on the same units a second time. The following example will illustrate the procedure.

Example 4.1 A study on a physical strength measurement in pounds on seven subjects before and after a specified training period gave the results shown in Table 4.7.

Table 4.7 Pretest and Posttest Strength Measures

Subjects	Pretest	Posttest
1	100	115
2	110	125
3	90	105
4	110	130
5	125	140
6	130	140
7	105	125

This problem could be handled by the methods of Chapter 2 using a t test on differences where d = posttest measure − pretest measure. Here $\bar{d} = 15.71, s_d = 3.45$, and $t = 12.05$, which is highly significant ($p = 0.00002$) with 6 df.

If it is considered as a single-factor experiment at two levels (pre- and post-) with seven blocks making up a randomized block design, the data and ANOVA appear as shown in Table 4.8.

Here the large F indicates a strong treatment effect. In fact, the F reported is equal to t^2 given by the method of differences as $t = 12.05$ when squared $(12.05)^2 = 145.2$.

Table 4.8   Strength Measure Data and Analysis

Subjects	Pretest	Posttest	$T_{i.}$
1	100	115	215
2	110	125	235
3	90	105	195
4	110	130	240
5	125	140	265
6	130	140	270
7	105	125	230
Totals $T_{.j}$	770	880	$T_{..} = 1650$
$\sum\limits_{i} Y_{ij}^2$:	85,850	111,600	$\sum\limits_{j}\sum\limits_{i} Y_{ij}^2 = 197,450$

Source	df	SS	MS	F	Probability
Tests (treatments)	1	864.29	864.29	145	0.00002
Subjects (blocks)	6	2085.71	347.62		
Error	6	35.71	5.95		
Totals	13	2985.71			

4.3   ANOVA RATIONALE

For this randomized complete block design, the model is

$$Y_{ij} = \mu + \beta_i + \tau_j + \varepsilon_{ij} \tag{4.2}$$

or

$$Y_{ij} = \mu + (\mu_{i.} - \mu) + (\mu_{.j} - \mu) + (Y_{ij} - \mu_{i.} - \mu_{.j} + \mu) \tag{4.3}$$

where $\mu_{i.}$ represents the true mean of block i. The last term can be obtained by subtracting the treatment and block deviations from the overall deviation as follows:

$$(Y_{ij} - \mu) - (\mu_{i.} - \mu) - (\mu_{.j} - \mu) \equiv Y_{ij} - \mu_{i.} - \mu_{.j} + \mu$$

Best estimates of the parameters in Equation (4.3) give the sample model (after moving $\bar{Y}_{..}$ to the left of the equation):

$$Y_{ij} - \bar{Y}_{..} = (\bar{Y}_{i.} - \bar{Y}_{..}) + (\bar{Y}_{.j} - \bar{Y}_{..}) + (\bar{Y}_{ij} - \bar{Y}_{i.} - \bar{Y}_{.j} + \bar{Y}_{..})$$

Squaring both sides and adding with $i = 1, 2, \ldots, n, j = 1, 2, \ldots, k$,

$$\sum_{i=1}^{n} \sum_{j=1}^{k} (Y_{ij} - \bar{Y}_{..})^2 = \sum_{i}^{n} \sum_{j}^{k} (\bar{Y}_{i.} - \bar{Y}_{..})^2 + \sum_{i}^{n} \sum_{j}^{k} (\bar{Y}_{.j} - \bar{Y}_{..})^2$$

$$+ \sum_{i}^{n} \sum_{j}^{k} (Y_{ij} - \bar{Y}_{i.} - \bar{Y}_{.j} + \bar{Y}_{..})^2$$

$$+ \text{3 cross products} \tag{4.4}$$

A little algebraic work on the sums of the cross products will show that they all reduce to zero and the remaining equation becomes the fundamental equation of a two-way analysis of variance. The equation states that

$$SS_{total} = SS_{block} + SS_{treatment} + SS_{error}$$

one sum of squares for each variable term in the model [Equation (4.2)]. Each sum of squares has associated with it its degrees of freedom, and dividing any sum of squares by its degrees of freedom will yield an unbiased estimate of population variance σ_ε^2 if the hypotheses under test are true.

The breakdown of degrees of freedom here is

$$\begin{array}{cccc} \text{total} & \text{blocks} & \text{treatments} & \text{error} \\ (nk - 1) = & (n - 1) + & (k - 1) & + (n - 1)(k - 1) \end{array}$$

The error degrees are derived from the remainder

$$(nk - 1) - (n - 1) - (k - 1) = nk - n - k + 1 = (n - 1)(k - 1)$$

It can be shown that each sum of squares on the right of Equation (4.4) when divided by its degrees of freedom provides mean squares that are independently chi-square distributed, so that the ratio of any two of them is distributed as F.

The sums of squares formulas given in Equation (4.4) are usually expanded and rewritten to give formulas that are easier to apply. These are shown in Table 4.9. They are the formulas applied to the data of Table 4.5 where $nk = N$.

4.4 MISSING VALUES

Occasionally in a randomized block design an observation is lost. A vial may break, an animal may die, or a tire may disintegrate, so that there occurs one or more missing observations in the data. For a single-factor completely randomized design this presents no problem, since the analysis of variance can be run with unequal n_j's. But for a two-way analysis this means a loss of orthogonality, since for some blocks the $\sum_j \tau_j$ no longer equals zero, and for some treatment the $\sum_i \beta_i$ no longer equals zero. When

Table 4.9 ANOVA for Randomized Block Design

Source	df	SS	MS
Between blocks β_i	$n - 1$	$\displaystyle\sum_i^n \frac{T_{i.}^2}{k} - \frac{T_{..}^2}{nk}$	$SS_{block}/(n - 1)$
Between treatments τ_j	$k - 1$	$\displaystyle\sum_j^k \frac{T_{.j}^2}{n} - \frac{T_{..}^2}{nk}$	$SS_{treatment}/(k - 1)$
Error ε_{ij}	$(n - 1)(k - 1)$	$\displaystyle\sum_i^n \sum_j^k Y_{ij}^2 - \sum_{i=1}^n \frac{T_{i.}^2}{k}$ $\displaystyle - \sum_{j=1}^k \frac{T_{.j}^2}{n} + \frac{T_{..}^2}{nk}$	$SS_{error}/(n - 1)(k - 1)$
Totals	$nk - 1$	$\displaystyle\sum_i^n \sum_j^k Y_{ij}^2 - \frac{T_{..}^2}{nk}$	

blocks and treatments are orthogonal, the block totals are added over all treatments, and vice versa. If one or more observations are missing, the usual procedure is to replace the value with one that makes the sum of the squares of the errors a minimum.

In the tire brand test example suppose that the brand C tire on car III blew out and was ruined before completing the 20,000 miles. The resulting data after coding by subtracting 13 mils appear in Table 4.10, where y is inserted in place of this missing value.

Table 4.10 Missing Value Example

Car	A	B	C	D	$T_{i.}$
		Brand			
I	4	1	-1	0	4
II	1	1	-1	-2	-1
III	0	0	y	-2	$y - 2$
IV	0	-5	-4	-4	-13
$T_{.j}$	5	-3	$y - 6$	-8	$y - 12 = T_{..}$

Now,

$$SS_{error} = SS_{total} - SS_{treatment} - SS_{block}$$

$$= \sum_i \sum_j Y_{ij}^2 - \sum_j \frac{T_{.j}^2}{n} - \sum_i \frac{T_{i.}^2}{k} + \frac{T_{..}^2}{nk}$$

For this example,

$$SS_{error} = 4^2 + 1^2 + \cdots + y^2 + \cdots + (-4)^2$$
$$- \frac{(5)^2 + (-3)^2 + (y-6)^2 + (-8)^2}{4}$$
$$- \frac{(4)^2 + (-1)^2 + (y-2)^2 + (-13)^2}{4} + \frac{(y-12)^2}{16}$$

To find the y value that will minimize this expression, it is differentiated with respect to y and set equal to zero. As all constant terms have their derivatives zero,

$$\frac{d(SS_{error})}{dy} = 2y - \frac{2(y-6)}{4} - \frac{2(y-2)}{4} + \frac{2(y-12)}{16} = 0$$

Solving gives

$$16y - 4y + 24 - 4y + 8 + y - 12 = 0$$
$$9y = -20$$
$$y = -\frac{20}{9} = -2.2 \quad \text{or } 10.8 \text{ mils}$$

If this value is now used in the y position, the resulting ANOVA table is as shown in Table 4.11. This is an approximate ANOVA as the sums of squares are slightly biased. The resulting ANOVA is not too different from before, but the degrees of freedom for the error term are reduced by one, since there are only 15 actual observations, and y is determined from these 15 readings.

This procedure can be used on any reasonable number of missing values by differentiating the error sum of squares partially with respect to each such missing value and setting it equal to zero, giving as many equations as unknown values to be solved for these missing values.

Table 4.11 ANOVA for Tire Brand Test Example Adjusted for a Missing Value (approximate)

Source	df	SS	MS
Brands τ_j	3	28.7	9.5
Cars β_i	3	38.3	12.7
Error ε_{ij}	8	11.2	1.4
Totals	14	78.2	

In general, for one missing value y_{ij} it can be shown that

$$y_{ij} = \frac{nT'_{i.} + kT'_{.j} - T'_{..}}{(n-1)(k-1)} \tag{4.5}$$

where the primed totals $T'_{i.}$, $T'_{.j}$, and $T'_{..}$ are the totals indicated without the missing value y. In our example,

$$y_{ij} = y_{33} = \frac{4(-2) + 4(-6) - (-12)}{(3)(3)}$$

$$y_{33} = \frac{-20}{9} = -2.2 \text{ (as before)}$$

4.5 LATIN SQUARES

The reader may have wondered about a possible position effect in the problem on testing tire brands. Experience shows that rear tires get different wear than front tires and even different sides of the same car may show different amounts of tread wear. In the randomized block design, the four brands were randomized onto the four wheels of each car with no regard for position. The effect of position on wear could be balanced out by rotating the tires every 5000 miles giving each brand 5000 miles on each wheel. However, if this is not feasible, the positions can impose another restriction on the randomization in such a way that each brand is not only used once on each car but also only once in each of the four possible positions: left front, left rear, right front, and right rear.

A design in which each treatment appears once and only once in each row (position) and once and only once in each column (cars) is called a *Latin square design*. Interest is still centered on one factor, treatments, but two restrictions are placed on the randomization. An example of one such 4 × 4 Latin square is shown in Table 4.12.

Such a design is only possible when the number of levels of both restrictions equals the number of treatment levels. In other words, it must be a

Table 4.12   4 × 4 Latin Square Design

Position	I	II	III	IV
		Car		
1	C	D	A	B
2	B	C	D	A
3	A	B	C	D
4	D	A	B	C

square. It is not true that all randomization is lost in this design, as the particular Latin square to be used on a given problem may be chosen at random from several possible Latin squares of the required size. Tables of such squares are found in Fisher and Yates [11].

The analysis of the data in a Latin square design is a simple extension of previous analyses where the data are now added in a third direction— positions. If the data of Table 4.5 were imposed on the Latin square of Table 4.12, the results could be those shown in Table 4.13.

Table 4.13   Latin Square Design Data on Tire Wear

Position	Car				
	I	II	III	IV	$T_{..k}$
1	C 12	D 11	A 13	B 8	44
2	B 14	C 12	D 11	A 13	50
3	A 17	B 14	C 10	D 9	50
4	D 13	A 14	B 13	C 9	49
$T_{i..}$	56	51	47	39	193

Treatment totals for A, B, C, and D brands are 57, 49, 43, and 44 as before, where the model is now

$$Y_{ijk} = \mu + \beta_i + \tau_j + \gamma_k + \varepsilon_{ijk}$$

and γ_k represents the positions effect. Since the only new totals are for positions, a position sum of squares can be computed as

$$SS_{position} = \frac{(44)^2 + (50)^2 + (50)^2 + (49)^2}{4} - \frac{(193)^2}{16} = 6.19$$

and

$$SS_{error} = SS_{total} - SS_{brand} - SS_{car} - SS_{position}$$
$$= 80.94 - 30.69 - 38.69 - 6.19 = 5.37$$

Table 4.14 is the ANOVA table for these data.

Once again another restriction placed on the randomization has further reduced the experimental error, although the position effect is not significant at the 5 percent level. This further reduction of error variance is attained at the expense of degrees of freedom, since now the estimate of σ_ε^2 is based on only 6 df instead of 9 df as in the randomized block design. This means less precision in estimating this error variance. But the added restrictions should be made if the environmental conditions suggest them. After discovering

Table 4.14   Latin Square ANOVA

Source	df	SS	MS	EMS
Brands τ_j	3	30.69	10.2	$\sigma_\varepsilon^2 + 4\phi_\tau$
Cars β_i	3	38.69	12.9	$\sigma_\varepsilon^2 + 4\sigma_\beta^2$
Positions γ_k	3	6.19	2.1	$\sigma_\varepsilon^2 + 4\phi_\gamma$
Error ε_{ijk}	6	5.37	0.9	σ_ε^2
Totals	15	80.94		

that position had no significant effect, some investigators might "pool" the position sum of squares with the error sum of squares and obtain a more precise estimate of σ_ε^2, namely, 1.3, as given in Table 4.6. However, there is a danger in "pooling," as it means "accepting" a hypothesis of no position effect, and the investigator has no idea about the possible error involved in "accepting" a hypothesis. Naturally, if the degrees of freedom on the error term are reduced much below that of Table 4.14, there will have to be some pooling to get a reasonable yardstick for assessing other effects.

4.6   INTERPRETATIONS

After any ANOVA one usually wishes to investigate the single factor further for a more detailed interpretation of the effect of the treatments. Here, after the Latin square ANOVA of Table 4.14, we are probably concerned about which brand to buy. Using a Newman–Keuls approach, the results are as follows:

1. Means: 10.75   11.00   12.25   14.25
   For brands: *C*   *D*   *B*   *A*

2. $s_e^2 = 0.9$ with 6 df.

3. $s_{\bar{Y}_{.j.}} = \sqrt{\dfrac{0.9}{4}} = 0.47$

4. For $p =$ 2   3   4
   Tabled ranges: 3.46   4.34   4.90

5. LSRs: 1.63,   2.04,   2.30.

6. Testing averages:

$$A\text{–}C = 3.50 > 2.30*$$
$$A\text{–}D = 3.25 > 2.04*$$
$$A\text{–}B = 2.00 > 1.63*$$
$$B\text{–}C = 1.50 < 2.04$$

Thus only A differs from the other brands in tread wear. A has the largest tread wear and so is the poorest brand to buy. One could recommend brand B, C, or D, whichever is the cheapest or whichever is "best" according to some criterion other than tread wear.

4.7 GRAECO-LATIN SQUARES

In some experiments still another restriction may be imposed on the randomization. The design may be a Graeco-Latin square such as Table 4.15 exhibits.

Table 4.15 Graeco-Latin Square Design

Position	\multicolumn{4}{c}{Car}
	I	II	III	IV
1	$A\alpha$	$B\beta$	$C\gamma$	$D\delta$
2	$B\gamma$	$A\delta$	$D\alpha$	$C\beta$
3	$C\delta$	$D\gamma$	$A\beta$	$B\alpha$
4	$D\beta$	$C\alpha$	$B\delta$	$A\gamma$

In this design the third restriction is at levels α, β, γ, δ, and not only do these each appear once and only once in each row and each column, but they appear once and only once with each level of treatment A, B, C, or D. The model for this would be

$$Y_{ijkm} = \mu + \beta_i + \tau_j + \gamma_k + \omega_m + \varepsilon_{ijkm}$$

where ω_m is the effect of the latest restriction with levels α, β, γ, and δ. An outline of the analysis appears in Table 4.16.

Table 4.16 Graeco-Latin ANOVA Outline

Source	df
β_i	3
τ_j	3
γ_k	3
ω_m	3
ε_{ijkm}	3
Total	15

Such a design may not be very practical, as only 3 df are left for the error variance.

4.8 EXTENSIONS

In some situations it is not possible to place all treatments in every block of an experiment. This leads to a design known as an *incomplete block design*. Sometimes a whole row or column of a Latin square is missing and the resulting incomplete Latin square is called a *Youden square*. These two special designs are discussed in Chapter 16.

4.9 SUMMARY

Experiment	Design	Analysis
I. Single factor	1. Completely randomized $Y_{ij} = \mu + \tau_j + \varepsilon_{ij}$	1. One-way ANOVA
	2. Randomized block $Y_{ij} = \mu + \tau_j + \beta_i + \varepsilon_{ij}$ a. Complete b. Incomplete, balanced c. Incomplete, general	2. a. Two-way ANOVA b. Special ANOVA c. Regression method
	3. Latin square $Y_{ijk} = \mu + \beta_i + \tau_j$ $+ \gamma_k + \varepsilon_{ijk}$ a. Complete b. Incomplete, Youden square	3. a. Three-way ANOVA b. Special ANOVA (like 2b)
	4. Graeco-Latin square $Y_{ijkm} = \mu + \beta_i + \tau_j$ $+ \gamma_k + \omega_m + \varepsilon_{ijkm}$	4. Four-way ANOVA

It might be emphasized once again that, so far, interest has been centered on a single factor (treatments) for the discussion in Chapters 3 and 4. The special designs simply represent restrictions on the randomization. In the next chapter, two or more factors are considered.

PROBLEMS

4.1 The effects of four types of graphite coaters on light box readings are to be studied. As these readings might differ from day to day, observations are to be taken on each of the four types every day for three days. The order of testing of the four types on any given day can be randomized. The results are

Day	Graphite Coater Type			
	M	A	K	L
1	4.0	4.8	5.0	4.6
2	4.8	5.0	5.2	4.6
3	4.0	4.8	5.6	5.0

Analyze these data as a randomized block design and state your conclusions.

4.2 Set up orthogonal contrasts among coater types and analyze for Problem 4.1.

4.3 Use the Newman–Keuls range test to compare four coater-type means for Problem 4.1.

4.4 If the reading on type K for the second day was missing, what missing value should be inserted and what is the analysis now?

4.5 Set up at least four contrasts in Problem 4.1 and use the Scheffé method to test the significance of these.

4.6 In a research study at Purdue University on metal-removal rate, five electrode shapes A, B, C, D, and E were studied. The removal was accomplished by an electric discharge between the electrode and the material being cut. For this experiment five holes were cut in five workpieces, and the order of electrodes was arranged so that only one electrode shape was used in the same position on each of the five workpieces. Thus, the design was a Latin square design, with workpieces (strips) and positions on the strip as restrictions on the randomization. Several variables were studied, one of which was the Rockwell hardness of metal where each hole was to be cut. The results were as follows:

Strip	Position				
	1	2	3	4	5
I	A(64)	B(61)	C(62)	D(62)	E(62)
II	B(62)	C(62)	D(63)	E(62)	A(63)
III	C(61)	D(62)	E(63)	A(63)	B(63)
IV	D(63)	E(64)	A(63)	B(63)	C(63)
V	E(62)	A(61)	B(63)	C(63)	D(62)

Analyze these data and test for an electrode effect, position effect, and strip effect on Rockwell hardness.

4.7   The times in hours necessary to cut the holes in Problem 4.6 were recorded as follows:

	Position				
Strip	1	2	3	4	5
I	$A(3.5)$	$B(2.1)$	$C(2.5)$	$D(3.5)$	$E(2.4)$
II	$E(2.6)$	$A(3.3)$	$B(2.1)$	$C(2.5)$	$D(2.7)$
III	$D(2.9)$	$E(2.6)$	$A(3.5)$	$B(2.7)$	$C(2.9)$
IV	$C(2.5)$	$D(2.9)$	$E(3.0)$	$A(3.3)$	$B(2.3)$
V	$B(2.1)$	$C(2.3)$	$D(3.7)$	$E(3.2)$	$A(3.5)$

Analyze these data for the effect of electrodes, strips, and positions on time.

4.8   Analyze the electrode effect further in Problem 4.7 and make some statement as to which electrodes are best if the shortest cutting time is the most desirable factor.

4.9   For an $m \times m$ Latin square, prove that a missing value may be estimated by

$$Y_{ijk} = \frac{m(T'_{i..} + T'_{.j.} + T'_{..k}) - 2T'_{...}}{(m - 1)(m - 2)}$$

where the primes indicate totals for row, treatment, and column containing the missing value and $T'_{...}$ is the grand total of all actual observations.

4.10 A composite measure of screen quality was made on screens using four lacquer concentrations, four standing times, four acryloid concentrations (A, B, C, D), and four acetone concentrations (α, β, Γ, Δ). A Graeco-Latin square design was used with data recorded as follows:

	Lacquer Concentration			
Standing Time	$\frac{1}{2}$	1	$1\frac{1}{2}$	2
30	$C\beta(16)$	$B\Gamma(12)$	$D\Delta(17)$	$A\alpha(11)$
20	$B\alpha(15)$	$C\Delta(14)$	$A\Gamma(15)$	$D\beta(14)$
10	$A\Delta(12)$	$D\alpha(6)$	$B\beta(14)$	$C\Gamma(13)$
5	$D\Gamma(9)$	$A\beta(9)$	$C\alpha(8)$	$B\Delta(9)$

Do a complete analysis of these data.

4.11 Explain why a three-level Graeco-Latin square is not a feasible design.

4.12 In a chemical plant, five experimental treatments are to be used on a basic raw material in an effort to increase the chemical yield. Since batches of raw material may differ, five batches are chosen at random. Also, the order of the experiments may affect yield as well as may the operator who performs the experiment. Considering order and operators as further restrictions on randomization, set up a suitable design for this experiment that will be least expensive to run. Outline its analysis.

4.13 Three groups of students are to be tested for the percentage of high-level questions asked by each group. As questions can be on various types of material, six lessons are taught to each group and a record is made of the percentage of high-level questions asked by each group on all six lessons. Show a data layout for this situation and outline its ANOVA table.

4.14 Data from the results of Problem 4.13 were as follows:

Lesson	Group A	B	C
1	13	18	7
2	16	25	17
3	28	24	14
4	26	13	15
5	27	16	12
6	23	19	9

Do an analysis of these data and state your conclusions.

4.15 For Problem 4.14 analyze further in an attempt to determine which group asks the highest percentage of high-level questions.

4.16 Thirty students were pretested for science achievement and were subsequently posttested using the same test after 15 weeks of special instruction. A t test on the difference scores of those students was found to be 2.3. Consider this problem as a randomized block design and show a data layout and outline its ANOVA table with the proper degrees of freedom. Also indicate what F value should be exceeded if one were to conclude that the training made a difference at the 5 percent significance level.

4.17 A student decides to investigate the accuracy of five scales in various local commercial places. She decides to account for the variations in her weight at different times of day and also on different days. She will only collect data on Saturdays, which she has free. The design will be a Latin square so she chooses the five times of day as 9:00 a.m., 11:00 a.m., 1:00 p.m., 3:00 p.m., and 5:00 p.m. One possible

Latin square design is as follows:

Day	Time				
	9	11	1	3	5
1	A	B	C	D	E
2	B	C	D	E	A
3	C	D	E	A	B
4	D	E	A	B	C
5	E	A	B	C	D

    a. Give the model and write out the ANOVA table.

    b. Suppose she does a Scheffé test on the times of day and finds that the 3 and 5 are not significantly different. If she is going to do the experiment again on some other scales, should she do a simpler randomized block design with five days as blocks and make observations at 3 p.m. and 5 p.m.? Give reasons for your decision.

4.18 Workers at an agricultural experiment station conducted an experiment to compare the effect of four different fertilizers on the yield of a cane crop. They divided four blocks of soil into four plots of equal size and shape and assigned the fertilizers to the plots at random such that each fertilizer was applied once in each block. Data were collected on yield in hundred weights per plot. The analysis gave a total sum of squares of 540 with fertilizer sum of squares at 210 and block sum of squares at 258. Set up an ANOVA table for these results.

4.19 From the data of Problem 4.18 state the hypothesis to be tested in terms of the mathematical model, run the test, and state your conclusions.

4.20 In Problem 4.18 if the observed mean yield of cane for fertilizer B was 41.25 hundred weight, what is its standard error?

4.21 In Problem 4.18 the original data show that all readings are divisible by 5. If one codes the data by subtracting 40 and dividing by 5, set up an expression for 95 percent confidence limits on the mean of fertilizer B in terms of the coded data.

4.22 Write a contrast to be used in comparing the mean of fertilizers A and B with the mean of fertilizer D and also show its standard error for Problem 4.18.

4.23 Set up three contrasts where one contrast is orthogonal to the other two but these two are *not* orthogonal to each other.

4.24 a. What is the main advantage of a randomized block design over a completely randomized design?

    b. Under what conditions would the completely randomized design have been better?

4.25 Material is analyzed for weight in grams from three vendors—A, B, C—by three different inspectors—I, II, III—and using three different scales—1, 2, 3. The experiment is set up on a Latin square plan with results as follows:

	Scale		
Inspector	1	2	3
I	$A = 16$	$B = 10$	$C = 11$
II	$B = 15$	$C = 9$	$A = 14$
III	$C = 13$	$A = 11$	$B = 13$

Analyze the experiment to test for a difference in weight between vendors, between inspectors, and between scales.

Factorial Experiments

5.1 INTRODUCTION

In the preceding three chapters all of the experiments involved only one factor and its effect on a measured variable. Several different designs were considered, but all of these represented restrictions on the randomization where interest was still centered on the effect of a single factor.

Suppose there are now two factors of interest to the experimenter, for example, the effect of both temperature and altitude on the current flow in a small computer. One traditional method is to hold altitude constant and vary the temperature and then hold temperature constant and change the altitude, or, in general, hold all factors constant except one and take current flow readings for several levels of this one factor, then choose another factor to vary, holding all others constant, and so forth. To examine this type of experimentation, consider a very simple example in which temperature is to be set at 25°C and 55°C only and altitudes of 0 K (K = 10,000 feet) and 3 K (30,000 feet) are to be used. If one factor is to be varied at a time, the altitude may be set for sea level or 0 K and the temperature varied from 25°C to 55°C. Suppose the current flow changed from 210 mA to 240 mA with this temperature increase. Now there is no way to assess whether or not this 30-mA increase is real or due to chance. Unless there is available some previous estimate of the error variability, the experiment must be repeated in order to obtain an estimate of the error or chance variability within the experiment. If the experiment is now repeated and the readings are 205 mA and 230 mA as the temperature is varied from 25°C to 55°C, it seems obvious, without any formal statistical analysis, that there is a real increase in current flow, since for each repetition the increase is large compared with the variation in current flow within a given temperature. Graphically these results appear as in Figure 5.1.

Four experiments have now been run to determine the effect of temperature at 0 K altitude only. To check the effect of altitude, the temperature can be held at 25°C and the altitude varied to 3 K by adjustment of pressure

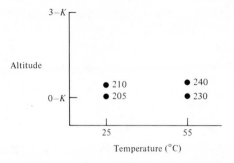

Figure 5.1 Temperature effect on current flow.

in the laboratory. Using the results already obtained at 0 K (assuming they are representative), two observations of current flow are now taken at 3 K with the results shown in Figure 5.2.

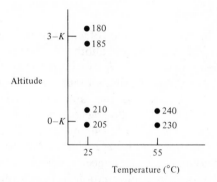

Figure 5.2 Temperature and altitude effect on current flow.

From these experiments the temperature increase is seen to increase the current flow an average of

$$\frac{30 \text{ mA} + 25 \text{ mA}}{2} = 27.5 \text{ mA}$$

and the increase in altitude decreases the current flow on the average of

$$\frac{30 \text{ mA} + 20 \text{ mA}}{2} = 25 \text{ mA}$$

This information is gained after six experiments have been performed and no information is available on what would happen at a temperature of 55°C and an altitude of 3 K.

An alternative experimental arrangement would be a factorial arrangement where each temperature level is combined with each altitude and only four experiments are run. Results of four such experiments might appear as in Figure 5.3.

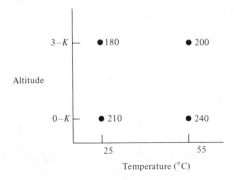

Figure 5.3 Factorial arrangement of temperature and altitude effect on current flow.

With this experiment, one estimate of temperature effect on current flow is $240 - 210$ mA $= 30$ mA at 0 K, and another estimate is $200 - 180$ mA $= 20$ mA at 3 K. Hence two estimates can be made of temperature effect [average $(30 + 20)/2 = 25$ mA] using all four observations without necessarily repeating any observation at the same point. Using the same four observations, two estimates of altitude effect can be determined: $180 - 210 = -30$ mA at 25°C, and $200 - 240 = -40$ mA at 55°C, an average decrease of 35 mA for a 3-K increase in altitude. Here, with just four observations instead of six, valid comparisons have been made on both temperature and altitude, and in addition some information has been obtained as to what happens at 55°C and altitude 3 K.

From this simple example, some of the advantages of a factorial experiment can be seen:

1. More efficiency is possible than with one-factor-at-a-time experiments (here four sixths or two thirds the amount of experimentation).
2. All data are used in computing both effects. (Note that all four observations are used in determining the average effect of temperature and the average effect of altitude.)
3. Some information is gleaned on possible interaction between the two factors. (In the example, the increase in current flow of 20 mA at 3 K was about the same order of magnitude as the 30 mA increase at 0 K. If these increases had differed considerably, interaction might be said to be present.)

These advantages are even more pronounced as the number of levels of the two factors is increased. A *factorial experiment* is one in which all levels of a given factor are combined with all levels of every other factor in the experiment. Thus, if four temperatures are considered at three altitudes, a 4 × 3 factorial experiment would be run requiring 12 different experimental conditions. In the example above, if the current flow were 160 mA for 55°C and 3 K, the results could be shown as in Figure 5.4.

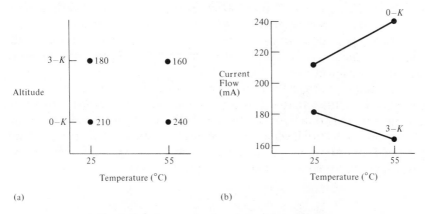

Figure 5.4   Interaction in a factorial experiment.

From Figure 5.4(b) note that as the temperature is increased from 25 to 55°C at 0 K, the current flow increases by 30 mA, but at 3 K for the same temperature increase the current flow *decreases* by 20 mA. When a change in one factor produces a different change in the response variable at one level of another factor than at other levels of this factor, there is an *interaction* between the two factors. This is also observable in Figure 5.4(b) as the two altitude lines are not parallel. If the data of Figure 5.3 are plotted, we get the relations pictured in Figure 5.5.

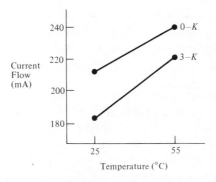

Figure 5.5   No-interaction temperature diagram in a temperature–altitude study.

It is seen that the lines are much more nearly parallel and no interaction is said to be present. An increase in temperature at 0 K produces about the same increase in current flow (30 mA) as at 3 K (20 mA). Now a word of warning is necessary. Lines can be made to look nearly parallel or quite diverse depending on the scale chosen; therefore, it is necessary to run statistical tests in order to determine whether or not the interaction is statistically significant. It is only possible to test the significance of such interaction if more than one observation is taken for each experimental condition. The graphic procedure shown above is given merely to provide some insight into what interaction is and how it might be displayed for explaining the factorial experiment.

In addition to the advantages of a factorial experiment, it may be that when the response variable Y is a yield and one is concerned with which combination of two independent variables X_1 and X_2 will produce a maximum yield, the factorial will give a combination near the maximum whereas the one-factor-at-a-time procedure will not do so. Figure 5.6 shows two cases in which the contours indicate the yields.

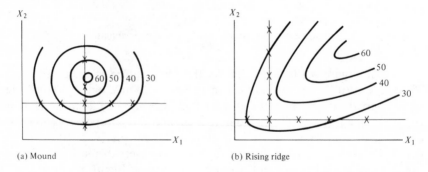

(a) Mound   (b) Rising ridge

Figure 5.6   Yield contours using one-factor-at-a-time method.

In case (a), X_2 is set and X_1 is varied until the maximum yield is found, then X_1 is held there while X_2 is varied and a maximum is found near the top of the mound where the yield is around 60 units. But in case (b), the same procedure misses the maximum by a long way, giving about a 37-unit yield.

If, now, a factorial arrangement is used, one first notes the extremes of each independent variable and may also choose one or more points between these extremes. If this is done for the diagrams in Figure 5.6, Figure 5.7 shows that in both cases the maximum is reached. In some cases the maximum may not have been reached but the factorial will indicate the direction to be followed in the next experiment to get closer to such a maximum. This concept is presented in Chapter 16.

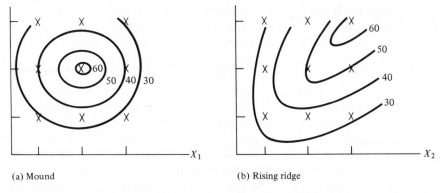

(a) Mound   (b) Rising ridge

Figure 5.7 Yield contours using a factorial method.

5.2 FACTORIAL EXPERIMENTS

To see how a factorial experiment is set up and analyzed, consider Example 5.1.

Example 5.1 To determine the effect of exhaust index (in seconds) and pump heater voltage (in volts) on the pressure inside a vacuum tube (in microns of mercury), three exhaust indexes and two voltages are chosen at fixed levels. It was decided to run two experiments at each of these six treatment conditions (three exhaust indexes × two voltages). The order for running the 12 experiments was completely randomized. This can be accomplished easily by labeling the six treatment conditions numbers 1 through 6 and then tossing a die to decide the order in which the experiments are to be run.

Table 5.1 shows the results of one such complete randomization for the 12 experiments. In such a procedure some treatment combinations may be run twice before others are run once, but this is the nature of complete

Table 5.1 Data Layout for Vacuum Tube Pressure Experiment

Pump Heater Voltage (volts)	Exhaust Index (seconds)					
	60		90		150	
127	1|	4	2|	1	3|	6
		9		11		12
220	4|	2	5|	3	6|	7
		8		5		10

randomization. If one were to randomize six runs for each combination and then randomize six more, the randomization will have been restricted and no longer a completely randomized design. In such a procedure it is also true that some numbers may come up again on the die even though that combination already has been run twice. In these cases the third time is ignored and the flipping continues until all treatment combinations are run twice. This is, of course, a slight restriction to keep the number of observations per treatment constant. The resulting experiment may be labeled a 3×2 factorial experiment with two observations per treatment with all $N = 12$ observations run in a completely randomized order. Of course, this random order can be determined before the experiment is actually run and it is important to supervise the experiment to make sure that it is actually run in the order prescribed.

The mathematical model for this experiment can be written as

$$Y_{ijk} = \mu + \tau_{ij} + \varepsilon_{k(ij)} \tag{5.1}$$

where

$i = 1, 2, 3$ for the three exhaust indexes

$j = 1, 2$ for the two voltages

$k = 1, 2$ for the two observations in each i, j treatment combination

τ_{ij} = the six treatment effects

$\varepsilon_{k(ij)}$ = the error within each of the six treatments

The notation $k(ij)$ indicates that the errors are unique to each i, j combination or are nested within each i, j. Following the analysis procedures of Chapter 3, Table 5.2 shows the resulting data in a between-treatments, within-treatments format after multiplying all results by 1000 to eliminate decimal points.

If the data of Table 5.2 are subjected to a one-way ANOVA, the results are as given in Table 5.3.

Table 5.2   Vacuum Tube Pressure as a One-Way Experiment

	EI_{60}, V_{127}	EI_{90}, V_{127}	EI_{150}, V_{127}	EI_{60}, V_{220}	EI_{90}, V_{220}	EI_{150}, V_{220}
	48	28	7	62	14	6
	58	33	15	54	10	9
$T_{ij\cdot}$	106	61	22	116	24	15
$\sum_k Y_{ijk}^2$	5668	1873	274	676	296	117

$$T_{\cdot\cdot} = 344$$

$$\sum_i \sum_j \sum_k Y_{ijk}^2 = 14{,}988$$

Table 5.3   One-Way ANOVA for Vacuum
Tube Pressure Experiment

Source	df	SS
Between treatments	5	4987.67
Within treatments	6	139.00
Totals	11	5126.67

In Table 5.3 the treatment effects are really combinations of exhaust indexes and voltages, and if rearranged to show these main effects, the data might appear as in Table 5.4.

Table 5.4   Two-Way Layout for Vacuum Tube
Pressure Experiment

Pump Heater Voltage	Exhaust Index			$T_{i..}$
	60	90	150	
127	48	28	7	189
	58	33	15	
220	62	14	9	155
	54	10	6	
$T_{.j.}$	222	85	37	$T_{...} = 344$

If the sums of squares for the two main effects are computed from the marginal totals of Table 5.4, we find

$$SS_{EI} = \frac{(222)^2 + (85)^2 + (37)^2}{4} - \frac{(344)^2}{12} = 4608.17$$

$$SS_V = \frac{(189)^2 + (155)^2}{6} - \frac{(344)^2}{12} = 96.33$$

These two add to 4704.50, which is less than the sum of squares between treatments of 4987.67. It is this difference between the treatment SS and the SS of the two main effects that shows the interaction between the two main effects. Here

$$SS_{EI \times V \text{ interaction}} = SS_{\text{treatments}} - SS_{EI} - SS_V$$
$$= 4987.67 - 4608.17 - 96.33 = 283.17$$

Since the treatments carried 5 df, the interaction has $5 - 2 - 1 = 2$ df. These results may now be displayed in the ANOVA of Table 5.5. In this

Table 5.5 Two-Way ANOVA for Vacuum Tube Pressure Experiment

Source	df	SS	MS	F	Probability
Between EI's	2	4608.17	2304.08	99.5	0.001
Between V's	1	96.33	96.33	4.2	0.088
EI by V interaction	2	283.17	141.58	6.1	0.036
Error (within treatments)	6	139.00	23.17		
Totals	11	5126.67			

final table each main effect (EI and V) can be tested for significance as well as can the interaction by comparing each mean square with the error mean square. Exhaust index is seen to be highly significant. Voltage is not significant at the 5 percent significance level but the interaction is significant at the 5 percent level.

The mathematical model of Equation (5.1) can now be expanded to read

$$Y_{ijk} = \mu + \qquad \tau_{ij} \qquad + \varepsilon_{k(ij)}$$
$$Y_{ijk} = \mu + \overbrace{E_i + V_j + EV_{ij}} + \varepsilon_{k(ij)} \qquad (5.2)$$

where E_i represents the exhaust index effect in this problem, V_j the voltage effect, and EV_{ij} the interaction. It should be noted here that whereas in Chapter 4 randomization restrictions came from the error term, these effects come from the treatment term. In Equation (5.2) one tests the following hypotheses:

$$H_{0_1}: E_i = 0 \qquad \text{for all } i$$
$$H_{0_2}: V_j = 0 \qquad \text{for all } j$$
$$H_{0_3}: EV_{ij} = 0 \quad \text{for all } i \text{ and } j$$

as both main effects are fixed effects.

5.3 INTERPRETATIONS

Since the interaction is significant in this example, one should be very cautious in interpreting the main effects. A significant interaction here means that the effect of exhaust index on vacuum tube pressure at one voltage is different from its effect at the other voltage. This can be seen graphically by plotting the six treatment means as shown in Figure 5.8. Note that the lines in the figure are not parallel. One reasonable procedure for interpreting such an interaction is to run a Newman–Keuls test on the six means. Had the main effects not interacted one could treat each set of main effect means separately. Here testing main effects is not recommended because the results

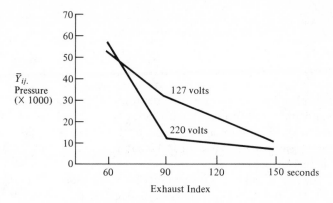

Figure 5.8 Vacuum tube current flow interaction plot.

depend on how these main effects combine. Using the six means, Newman–Keuls gives

1. Means are   7.5   11   12   30.5   53   58
   for treatments: $E_3 V_2$ $E_3 V_1$ $E_2 V_2$ $E_2 V_1$ $E_1 V_1$ $E_1 V_2$

2. $s_\varepsilon^2 = 23.17$ with 6 df from Table 5.5.

3. $s_{\bar{Y}_{ij.}} = \sqrt{\dfrac{23.17}{2}} = 3.40.$

4. For p: 2   3   4   5   6
   5 percent ranges: 3.46   4.34   4.90   5.31   5.63

5. LSR: 11.76, 14.76, 16.66, 18.05, 19.14.

6. Checking means:

$$58 - 7.5 = 50.5 > 19.14^*$$
$$58 - 11 = 47 \ \ > 18.05^*$$
$$58 - 12 = 46 \ \ > 16.66^*$$
$$58 - 30.5 = 27.5 > 14.76^*$$
$$58 - 53 = \ \ 5 \ \ < 11.76$$
$$53 - 7.5 = 45.5 > 18.05^*$$
$$53 - 11 = 42 \ \ > 16.66^*$$
$$53 - 12 = 41 \ \ > 14.76^*$$
$$53 - 30.5 = 22.5 > 11.76^*$$
$$30.5 - 7.5 = 23 \ \ > 16.66^*$$
$$30.5 - 11 = 19.5 > 14.76^*$$
$$30.5 - 12 = 18.5 > 11.76^*$$
$$12 - 7.5 = \ \ 4.5 < 14.76$$

Hence there are three groups of means:

$$7.5, 11, 12 \qquad 30.5 \qquad 53, 58$$

If one is looking for the lowest pressure, any one of the three combinations in the first group will minimize the pressure in the vacuum tube. Thus one can recommend a 150-second exhaust index at either voltage or a 90-second exhaust index at 220 volts. From a practical point of view this really gives two choices: Pump at 127 volts for 150 seconds or at 220 volts for 90 seconds, whichever is cheaper.

It might be noted that in the model for this problem, Equation (5.2), it is assumed that the errors $\varepsilon_{k(ij)}$ are NID $(0, \sigma_e^2)$. This means that the variances within each of the six cells that make up the treatments are assumed to have come from normal populations with equal variances. In the data of Table 5.4 the six ranges are 10, 5, 8, 8, 4, and 3, which average 6.33. From Table 3.16 in Chapter 3, $D_4\bar{R} = 20.7$ and all six ranges are well within this maximum range.

It may be of interest to note that whenever there are but two observations per treatment, the sum of squares for the error may be computed directly. When $n = 2$, the sum of squares of the ranges divided by 2 equals the SS_{error}. Here $\sum R^2 = 278$ and $\sum R^2/2 = 139$, which equals the error sum of squares of Table 5.5.

5.4   COMPUTER PROGRAM

As shown in Chapter 3, a computer program can be used to simplify the arithmetic when analyzing data in a factorial experiment. For the data in Example 5.1, Table 5.6 shows the input for an ANOVA on the data of Table 5.4. The entries should be self-explanatory. On line 11 one requests that the ANOVA be run on Y by $EI(1, 3)$, $PV(1, 2)$, which indicates the effect of exhaust index (EI) at three levels and pump heater voltage (PV) at two levels on Y, which is the pressure in the vacuum tube. The printout of the ANOVA is shown in Table 5.7, which is essentially the same as Table 5.5. (The Y's were first multiplied by 1000 here too.) In addition to the ANOVA table, since all statistics were requested, another printout (Table 5.8) shows the means of the data—the grand mean of all 12 observations, the means for each level of exhaust index, the means for each level of pump heater voltage, and the means for each combination of main effects. Because the interaction was significant in this problem, these last six means are ready for plotting as in Figure 5.8.

Table 5.6 Computer Input for Data of Table 5.4

```
+ + +CREATE PROB3
      1.  000 = 77509, I13
      2.  000 = COMMON (SPSS)
      3.  000 = SPSS
      4.  000 = #EOR
      5.  000 = RUN NAME            STAT 502 EX 3
      6.  000 = VARIABLE LIST       Y, EI, PV
      7.  000 = INPUT FORMAT        (F2.0, 1X, F1.0, 1X, F1.0)
      8.  000 = N OF CASES          12
      9.  000 = VAR LABELS          EI, EXHAUST INDEX/
     10.  000 =                     PV, PUMP HEATER VOLTAGE/
     11.  000 = ANOVA               Y BY EI(1, 3), PV(1, 2)
     12.  000 = STATISTICS          ALL
     13.  000 = READ INPUT DATA
     14.  000 = 48  1  1
     15.  000 = 58  1  1
     16.  000 = 28  2  1
     17.  000 = 33  2  1
     18.  000 =  7  3  1
     19.  000 = 15  3  1
     20.  000 = 62  1  2
     21.  000 = 54  1  2
     22.  000 = 14  2  2
     23.  000 = 10  2  2
     24.  000 =  6  3  2
     25.  000 =  9  3  2
     26.  000 = FINISH
```

Table 5.7 ANOVA Printout for Example 5.1

SOURCE OF VARIATION	SUM OF SQUARES	DF	MEAN SQUARE	F	SIGNIF OF F
MAIN EFFECTS	4704.500	3	1568.167	67.691	.001
EI	4608.167	2	2304.083	99.457	.001
PV	96.333	1	96.333	4.158	.088
2-WAY INTERACTIONS	283.167	2	141.583	6.112	.036
EI PV	283.167	2	141.583	6.112	.036
EXPLAINED	4987.667	5	997.533	43.059	.001
RESIDUAL	139.000	6	23.167		
TOTAL	5126.667	11	466.061		

Table 5.8 Printout of Means for Example 5.1

TOTAL POPULATION

28.67
(12)

EI

1	2	3
55.50	21.25	9.25
(4)	(4)	(4)

PV

1	2
31.50	25.83
(6)	(6)

EI   PV

	1	2
1	53.00	58.00
	(2)	(2)
2	30.50	12.00
	(2)	(2)
3	11.00	7.50
	(2)	(2)

5.5 ANOVA RATIONALE

For a two-factor factorial experiment with n observations per cell, run as a completely randomized design, a general model would be

$$Y_{ijk} = \mu + A_i + B_j + AB_{ij} + \varepsilon_{k(ij)} \tag{5.3}$$

where A and B represent the two factors, $i = 1, 2, \ldots, a$ levels of factor A, $j = 1, 2, \ldots, b$ levels of factor B, and $k = 1, 2, \ldots, n$ observations per cell. In terms of population means this becomes

$$
\begin{aligned}
Y_{ijk} - \mu_{\ldots} \equiv & (\mu_{i..} - \mu_{\ldots}) + (\mu_{.j.} - \mu_{\ldots}) \\
& + (\mu_{ij.} - \mu_{i..} - \mu_{.j.} + \mu_{\ldots}) + (Y_{ijk} - \mu_{ij.})
\end{aligned} \tag{5.4}
$$

where $\mu_{ij.}$ represents the true mean of the i, j cell or treatment combination. Justification for the interaction term in the model comes from subtracting A and B main effects from the cell effect as follows:

$$(\mu_{ij.} - \mu_{\ldots}) - (\mu_{i..} - \mu_{\ldots}) - (\mu_{.j.} - \mu_{\ldots}) = \mu_{ij.} - \mu_{i..} - \mu_{.j.} + \mu_{\ldots}$$

If each mean is now replaced by its sample estimate, the resulting sample model is

$$
\begin{aligned}
Y_{ijk} - \bar{Y}_{\ldots} = & (\bar{Y}_{i..} - \bar{Y}_{\ldots}) + (\bar{Y}_{.j.} - \bar{Y}_{\ldots}) \\
& + (\bar{Y}_{ij.} - \bar{Y}_{i..} - \bar{Y}_{.j.} + \bar{Y}_{\ldots}) + (Y_{ijk} - \bar{Y}_{ij.})
\end{aligned}
$$

If this expression is now squared and summed over i, j, and k, all cross products vanish, and the results give

$$\sum_i^a \sum_j^b \sum_k^n (Y_{ijk} - \bar{Y}_{...})^2 = \sum_i^a \sum_j^b \sum_k^n (\bar{Y}_{i..} - \bar{Y}_{...})^2$$

$$+ \sum_i^a \sum_j^b \sum_k^n (\bar{Y}_{.j.} - \bar{Y}_{...})^2$$

$$+ \sum_i^a \sum_j^b \sum_k^n (\bar{Y}_{ij.} - \bar{Y}_{i..} - \bar{Y}_{.j.} + \bar{Y}_{...})^2$$

$$+ \sum_i^a \sum_j^b \sum_k^n (Y_{ijk} - \bar{Y}_{ij.})^2$$

which again expresses the idea that the total sum of squares can be broken down into the sum of squares between means of factor A, plus the sum of squares between means of factor B, plus the sum of squares of $A \times B$ interaction, plus the error sum of squares (or within cell sum of squares). Each sum of squares is seen to be independent of the others. Hence, if any such sum of squares is divided by its associated degrees of freedom, the results are independently chi-square distributed, and F tests may be run.

The degree-of-freedom breakdown would be

$$(abn - 1) \equiv (a - 1) + (b - 1) + (a - 1)(b - 1) + ab(n - 1)$$

the interaction being cell df $= (ab - 1)$ minus the main effect df, $(a - 1)$ and $(b - 1)$ or $(ab - 1) - (a - 1) - (b - 1) = ab - a - b + 1 = (a - 1) \times (b - 1)$, and within each cell the degrees of freedom are $n - 1$ and there are ab such cells giving $ab(n - 1)$ df for error. An ANOVA table can now be set up expanding and simplifying the sum of squares expressions using totals (Table 5.9).

The formulas for sum of squares in Table 5.9 provide good computational formulas for a two-way ANOVA with replication. The error sum of squares might be rewritten as

$$SS_{error} = \sum_i^a \sum_j^b \left[\sum_k^n Y_{ijk}^2 - \frac{T_{ij.}^2}{n} \right]$$

which points up the fact that the sum of squares within each of the $a \times b$ cells is being pooled or added for all such cells. This depends on the assumption that the variance within all cells came from populations with equal variance. The interaction sum of squares can also be rewritten as

$$\left(\sum_i^a \sum_j^b \frac{T_{ij.}^2}{n} - \frac{T_{...}^2}{nab} \right) - \left(\sum_i^a \frac{T_{i..}^2}{nb} - \frac{T_{...}^2}{nab} \right) - \left(\sum_j^b \frac{T_{.j.}^2}{na} - \frac{T_{...}^2}{nab} \right)$$

Table 5.9   General ANOVA for Two-Factor Factorial with n Replications per Cell

Source	df	SS	MS
Factor A_i	$a - 1$	$\displaystyle\sum_i^a \frac{T_{i..}^2}{nb} - \frac{T_{...}^2}{nab}$	Each SS divided by its df
Factor B_j	$b - 1$	$\displaystyle\sum_j^b \frac{T_{.j.}^2}{na} - \frac{T_{...}^2}{nab}$	
$A \times B$ interaction	$(a - 1)(b - 1)$	$\displaystyle\sum_i^a \sum_j^b \frac{T_{ij.}^2}{n} - \sum_i^a \frac{T_{i..}^2}{nb}$ $\displaystyle\qquad\qquad - \sum_j^b \frac{T_{.j.}^2}{na} + \frac{T_{...}^2}{nab}$	
Error $\varepsilon_{k(ij)}$	$ab(n - 1)$	$\displaystyle\sum_i^a \sum_j^b \sum_k^n Y_{ijk}^2 - \sum_i^a \sum_j^b \frac{T_{ij.}^2}{n}$	
Totals	$abn - 1$	$\displaystyle\sum_i^a \sum_j^b \sum_k^n Y_{ijk}^2 - \frac{T_{...}^2}{nab}$	

which shows again that interaction is calculated by subtracting the main effect sum of squares from the cell sum of squares.

Example 5.2   To extend the factorial idea a bit further, consider a problem with three factors. Such a problem was presented in Chapter 1 on the effect of tool type, angle of bevel, and type of cut on power consumption for ceramic tool cutting. Reference to this problem will point out the phases of experiment, design, and analysis as followed in Example 5.1.

It is a $2 \times 2 \times 2$ factorial experiment with four observations per cell run in a completely randomized manner. The mathematical model is

$$Y_{ijkm} = \mu + T_i + B_j + TB_{ij} + C_k + TC_{ik} + BC_{jk} + TBC_{ijk} + \varepsilon_{m(ijk)}$$

where TBC_{ijk} represents a three-way interaction.

The data for this example are given in Table 1.2 and the ANOVA table in Table 1.3. That this analysis is a simple extension of the methods used on Example 5.1 will be shown with the coded data from Table 1.2 (see Table 5.10).

Table 5.10 shows the total for each small cell and the sum of the squares of the cell observations. These results will be useful in doing the ANOVA. By this time we should be able to set up the steps in the analysis without recourse to formulas in dot notation.

First the total sum of squares: Add the squares of all readings (the circled numbers) and subtract a correction term. The grand total is -13

Table 5.10 Coded Ceramic Tool Data of Table 1.2, Code: 2(X–28.0)

Type of Cut	Tool Type				
	1		2		
	Bevel Angle		Bevel Angle		
	15°	30°	15°	30°	
Continuous	2	1	0	3	
	–3	1	1	8	
	5 (42)	4 (99)	0 (37)	2 (77)	25
	–2	9	–6	0	
	2	15	–5	13	
Interrupted	0	–2	–7	–1	
	–6	2	–6	0	
	–3 (54)	–1 (10)	0 (101)	–2 (21)	–38
	–3	–1	–4	–4	
	–12	–2	–17	–7	
Totals	–10	13	–22	+6	–13

squared, divided by the number of observations (32):

$$SS_{total} = 441 - \frac{(-13)^2}{32} = 435.72$$

Tool type sum of squares: Add for each tool type. The totals are $+3$ and -16; square these and divide by the number of observations per type (16), add these results for both types, and subtract the correction term. Thus,

$$SS_{tool\ type} = \frac{3^2 + (-16)^2}{16} - \frac{(-13)^2}{32} = 11.28$$

Bevel angle sum of squares: Same procedure on the totals for each bevel angle, -32 and 19:

$$SS_{bevel\ angle} = \frac{(-32)^2 + (19)^2}{16} - \frac{(-13)^2}{32} = 81.28$$

Type of cut sum of squares: Same procedure, with cut totals of 25 and -38:

$$SS_{type\ of\ cut} = \frac{(25)^2 + (-38)^2}{16} - \frac{(-13)^2}{32} = 124.03$$

For the $T \times B$ interaction, ignore type of cut and use cell totals for the $T \times B$ cells. These are $-10, 13, -22, +6$:

$$SS_{T \times B \text{ interaction}} = \frac{(-10)^2 + (13)^2 + (-22)^2 + (6)^2}{8} - \frac{(-13)^2}{32} - 11.28 - 81.28$$

$$= 0.78$$

For $T \times C$ interaction, ignore bevel angle and the cell totals become 17, $-14, 8, -24$:

$$SS_{T \times C \text{ interaction}} = \frac{(17)^2 + (-14)^2 + (8)^2 + (-24)^2}{8} - \frac{(-13)^2}{32} - 11.28 - 124.03$$

$$= 0.03$$

For $B \times C$ interaction, ignore tool type and the cell totals become $-3, 28, -29, -9$:

$$SS_{B \times C \text{ interaction}} = \frac{(-3)^2 + (28)^2 + (-29)^2 + (-9)^2}{8} - \frac{(-13)^2}{32} - 81.28 - 124.03$$

$$= 3.78$$

For the three-way interaction $T \times B \times C$, consider the totals of the smallest cells, 2, 15, -5, 13, -12, -2, -17, and -7. From this cell sum of squares subtract *not only* the main effect sum of squares *but also* the three two-way interaction sums of squares. Thus,

$$SS_{T \times B \times C \text{ interaction}}$$

$$= \frac{(2)^2 + (15)^2 + (-5)^2 + (13)^2 + (-12)^2 + (-2)^2 + (-17)^2 + (-7)^2}{4}$$

$$- \frac{(-13)^2}{32} - 11.28 - 81.28 - 124.03 - 0.78 - 0.03 - 3.78 = 0.79$$

By subtraction

$$SS_{\text{error}} = 213.75$$

These results are displayed in Table 5.11. If they are compared with those in Table 1.3, they appear to differ considerably. Actually they give the same F test results, but in Table 5.11 the data were coded involving multiplication by 2; the data of Table 1.3 are uncoded. Multiplication by 2 will multiply the variance or mean square by 4; thus if all mean squares in Table 5.11 are divided by 4, the results are the same as in Table 1.3. For example, on tool types $11.28/4 = 2.82$, and error $213.75/4 = 53.44$. It is worth noting that any decoding is unnecessary for determining the F ratios. However, if one wishes confidence limits on the original data or components of variance on the original data, it may be necessary to decode the results.

Table 5.11 ANOVA for Ceramic Tool Problem

Source	df	SS	MS
Tool type T_i	1	11.28	11.28
Bevel angle B_j	1	81.28	81.28
$T \times B$ interaction TB_{ij}	1	0.78	0.78
Type of cut C_k	1	124.03	124.03
$T \times C$ interaction TC_{ik}	1	0.03	0.03
$B \times C$ interaction BC_{jk}	1	3.78	3.78
$T \times B \times C$ interaction TBC_{ijk}	1	0.79	0.79
Error $\varepsilon_{m(ijk)}$	24	213.75	8.91
Totals	31	435.72	

The interpretation of the results of this example is given in Chapter 1. The purpose of presenting it again in this chapter is to show that factorial experiments with three or more factors can easily be analyzed by simple extension of the methods of this chapter.

This problem, of course, could be fed into the computer to determine Table 5.11.

5.6 REMARKS

Since the examples in this chapter have contained several replications within a cell, it would be well to examine a situation involving only one observation per cell. In this case $k = 1$, and the model is written as

$$Y_{ij} = \mu + A_i + B_j + AB_{ij} + \varepsilon_{ij}$$

A glance at the last two terms indicates that we cannot distinguish between the interaction and the error—they are hopelessly confounded. Then the only reasonable situation for running one observation per cell is one in which past experience generally assures us that there is no interaction. In such a case, the model is written as

$$Y_{ij} = \mu + A_i + B_j + \varepsilon_{ij}$$

It may also be noted that this model looks very much like the model for a randomized block design for a single-factor experiment (Chapter 4). In Chapter 4 that model was written as

$$Y_{ij} = \mu + \tau_j + \beta_i + \varepsilon_{ij}$$

Even though the models do look alike and an analysis would be run in the same way, this latter is a single-factor experiment—treatments are the factor—and β_i represents a restriction on the randomization. In the factorial model, there are two factors of interest, A_i and B_j, and the design is completely randomized. It is, however, assumed in the randomized-block situation that there is no interaction between treatments and blocks. This is often a more reasonable assumption for blocks and treatments since blocks are often chosen at random. For a two-factor experiment an interaction between A and B may very well be present, and some external information must be available in order to assume that no such interaction exists. An experimenter who is not sure about interaction must take more than one observation per cell and test the hypotheses of no interaction.

As the number of factors increases, however, the presence of higher order interactions is much more unlikely, so it is fairly safe to assume no four-way, five-way, . . . , interactions. Even if these were present, they would be difficult to explain in practical terms.

5.7   SUMMARY

Experiment	Design	Analysis
I. Single factor		
	1. Completely randomized	1. One-way ANOVA
	$Y_{ij} = \mu + \tau_j + \varepsilon_{ij}$	
	2. Randomized block	2.
	$Y_{ij} = \mu + \tau_j + \beta_i + \varepsilon_{ij}$	
	a. Complete	a. Two-way ANOVA
	b. Incomplete, balanced	b. Special ANOVA
	c. Incomplete, general	c. Regression method
	3. Latin square	3.
	$Y_{ijk} = \mu + \beta_i + \tau_j$	
	$\quad + \gamma_k + \varepsilon_{ijk}$	
	a. Complete	a. Three-way ANOVA
	b. Incomplete,	b. Special ANOVA
	Youden square	(like 2b)
	4. Graeco-Latin square	4. Four-way ANOVA
	$Y_{ijkm} = \mu + \beta_i + \tau_j$	
	$\quad + \gamma_k + \omega_m + \varepsilon_{ijkm}$	
II. Two or more factors		
A. Factorial (crossed)		
	1. Completely randomized	1.
	$Y_{ijk} = \mu + A_i + B_j$	
	$\quad + AB_{ij} + \varepsilon_{k(ij)} \cdots$	
	for more factors	
	a. General case	a. ANOVA with interactions

PROBLEMS

5.1 To determine the effect of two glass types and three phosphor types on the light output of a television tube, light output is measured by the current required in series with the tube to produce 30 foot-lamberts of light output. Thus the higher the current is in microamperes, the poorer the tube is in light output. Three observations were taken under each of the six treatment conditions and the experiment was completely randomized. The following data were recorded.

	Phosphor Type		
Glass Type	A	B	C
1	280	300	270
	290	310	285
	285	295	290
2	230	260	220
	235	240	225
	240	235	230

Do an analysis of variance on these data and test the effect of glass type, phosphor types, and interaction on the current flow.

5.2 Plot the results of Problem 5.1 to show that your conclusions are reasonable.

5.3 For any significant effects in Problem 5.1 test further between the levels of the significant factors.

5.4 Based on the results in Problems 5.1–5.3, what glass type and phosphor type (or types) would you recommend if a low current is most desirable?

5.5 Adhesive force on gummed material was determined under three fixed humidity and three fixed temperature conditions. Four readings were made under each set of conditions. The experiment was completely randomized and the results set out in an ANOVA table as follows:

Source	df	SS	MS
Humidity		9.07	
Temperature		8.66	
$H \times T$ interaction		6.07	
Error			
Total		52.30	

Complete this table.

5.6 For the data in Problem 5.5 test all indicated hypotheses and state your conclusions.

5.7 Set up a mathematical model for the experiment in Problem 5.5 and indicate the hypotheses to be tested in terms of your model.

5.8 The object of an experiment is to determine thrust forces in drilling at different speeds and feeds, and in different materials. Five speeds and three feeds are used, and two materials with two samples tested under each set of conditions. The order of the experiment is completely randomized and the levels of all factors are fixed. The following data are recorded on thrust forces after subtracting 200 from all readings.

Material	Feed	Speed				
		100	220	475	715	870
B_{10}	0.004	122	108	108	66	80
		110	85	60	50	60
	0.008	332	276	248	248	276
		330	310	295	275	310
	0.014	640	612	543	612	696
		500	500	450	610	610
V_{10}	0.004	192	136	122	108	136
		170	130	85	75	75
	0.008	386	333	318	472	499
		365	330	330	350	390
	0.014	810	779	810	893	1820
		725	670	750	890	890

Do a complete analysis of this experiment and state your conclusions.

5.9 Plot any results in Problem 5.8 that are significant.

5.10 Set up tests on means where suitable and draw conclusions from Problem 5.8.

5.11 In an experiment for testing rubber materials interest centered on the effect of the mix (A, B, or C), the laboratory involved (1, 2, 3, or 4), and the temperature (145, 155, 165°C) on the time in minutes to a 2-in.-lb rise above the minimum time. Assuming a completely randomized design, do an ANOVA on the following data.

	Temperature (°C)								
	145			155			165		
	Mix			Mix			Mix		
Laboratory	A	B	C	A	B	C	A	B	C
1	11.2	11.2	11.5	6.7	6.8	7.0	4.8	4.8	5.0
	11.1	11.5	11.4	6.8	6.7	7.0	4.8	4.9	4.9
2	11.8	12.3	12.3	7.3	7.5	7.5	5.3	5.4	5.3
	11.8	12.3	11.9	7.2	7.7	7.3	5.3	5.2	5.3
3	11.5	12.3	12.7	6.6	7.1	7.8	5.0	5.3	5.2
	11.6	12.0	12.5	6.9	7.2	7.3	5.0	5.0	5.0
4	11.5	11.8	12.7	7.2	6.7	7.1	4.5	4.7	4.5
	11.3	11.7	12.7	6.9	7.0	7.0	4.6	4.5	4.5

5.12 What is there in the results of Problem 5.11 that would lead you to question some assumption about the experiment?

5.13 Tomato plants were grown in a greenhouse under treatments consisting of combinations of soil type (factor A) and fertilizer type (factor B). A completely randomized two-factor design was used with two replications per cell. The following data on the yield Y (in kilograms) of tomatoes were obtained for the 30 plants under study.

Soil Type (A)	Fertilizer Type (B)			
	1	2	3	$T_{i.}$
I	5, 7	5, 5	3, 5	30
II	5, 9	1, 3	2, 2	22
III	6, 8	4, 8	2, 4	32
IV	7, 11	7, 9	3, 7	44
V	6, 9	4, 6	3, 5	33
$T_{.j}$	73	52	36	161

Note: $\sum_k \sum_j \sum_i Y_{ijk}^2 = 1043$.

Complete the following ANOVA table.

Source	df	Sum of Squares	Mean Square
Soil type (A)			
Fertilizer type (B)			
Interaction (AB)			
Error			

5.14 Carry out tests on main effects and interactions *in an appropriate order* using levels of significance for each test that seem appropriate to you. For each test state H_0 and H_1, show the rejection region, and give your conclusion. After doing all tests, summarize your results in words.

5.15 On the basis of your results in Problem 5.14, state how you would carry out Newman–Keuls comparisons to find out the best combination (or combinations) of soil type and fertilizer type to achieve maximum mean yield of tomatoes. Then carry out these comparisons (at an α of 0.05) and state your conclusions.

5.16 Each of the following tables represents the cell means ($\bar{Y}_{ij.}$) for a two-factor completely randomized balanced experiment. For each table tell which (if any) of the following SS's would be zero for that table: SS_A, SS_B, SS_{AB}. (There may be none; there may be more than one.)

a.

B	A	
	1	3
	5	3

$= 0$

b.

	A	
B	1	3
	5	7

———————————————— = 0

c.

	A	
B	1	3
	5	5

———————————————— = 0

d.

	A	
B	5	3
	5	3

———————————————— = 0

5.17 In each two-factor design table below the numbers in the cells of the table are the population means of the observations for those cells. For each table, indicate whether or not there is interaction between the factors and justify your assertion.

a.

		Factor B			
		1	2	3	4
Factor A	1	7	6	5	2
	2	9	8	7	4

Is there interaction? *Yes/No* (circle one) Explain:

b.

		Factor B		
		1	2	3
Factor A	1	7	6	5
	2	5	6	7

Is there interaction? *Yes/No* (circle one) Explain:

c.

		Factor B	
		1	2
Factor A	1	8	10
	2	7	9
	3	4	7

Is there interaction? *Yes/No* (circle one) Explain:

5.18 An industrial engineer presented two types of stimuli (two-dimensional and three-dimensional films) of two different jobs (1 and 2) to each of five analysts. Each analyst was presented each job-stimulus film twice and the order of the whole experiment was considered completely randomized. The engineer was interested in the consistency of analyst ratings of four sequences within each job stimulus presentation. The variable recorded is the log variance of the four sequences since log variance is more likely to be normally distributed than the variance. Data showed:

| | Stimulus | | | |
| | Two-Dimensional | | Three-Dimensional | |
Analyst	Job 1	Job 2	Job 1	Job 2
1	1.42	1.40	1.00	0.92
	1.25	1.44	0.90	0.93
2	1.59	1.83	1.46	1.43
	2.09	2.02	1.68	1.02
3	1.26	1.80	0.85	1.33
	1.48	1.75	1.43	1.48
4	1.76	0.98	1.21	0.73
	1.47	1.23	1.25	1.22
5	1.43	1.17	1.60	1.31
	1.72	1.18	2.03	1.12

Do a complete analysis and summarize your results.

5.19 Make further tests on Problem 5.18 as suggested by the ANOVA results.

5.20 The following results were reported on a study of "factors that affect the salary of high-school teachers of commercial subjects":

Source	df	SS
Sex	1	239,763
Size of school	2	423,056
Years in position	5	564,689
Sex by size interaction	2	18,459
Sex × years interaction	5	85,901
Size × years interaction	10	240,115
Sex × size × years interaction	10	151,394
Within classes	153	1,501,642

Indicate the mathematical model for the above study. Show a possible data layout. Complete the ANOVA table and comment on any significant results.

5.21 Four factors are studied for their effect on the luster of plastic film. These factors are (1) film thickness (1 or 2 mils); (2) drying conditions (regular or special); (3)

length of wash (20, 30, 40, or 60 minutes); and (4) temperature of wash (92°C or 100°C). Two observations of film luster are taken under each set of conditions. Assuming complete randomization, analyze the following data.

	Regular Dry		Special Dry	
Minutes	92°C	100°C	92°C	100°C
1-mil Thickness				
20	3.4 3.4	19.6 14.5	2.1 3.8	17.2 13.4
30	4.1 4.1	17.5 17.0	4.0 4.6	13.5 14.3
40	4.9 4.2	17.6 15.2	5.1 3.3	16.0 17.8
60	5.0 4.9	20.9 17.1	8.1 4.3	17.5 13.9
2-mil Thickness				
20	5.5 3.7	26.6 29.5	4.5 4.5	25.6 22.5
30	5.7 6.1	31.6 30.2	5.9 5.9	29.2 29.8
40	5.6 5.6	30.5 30.2	5.5 5.8	32.6 27.4
60	7.2 6.0	31.4 29.6	8.0 9.9	33.5 29.5

5.22 Plot and discuss any significant interactions found in Problem 5.21. What conditions would you recommend for maximum luster and why?

5.23 To study the effect of four types of wax applied to floors for three different lengths of polishing times it was decided to try each combination twice and completely to randomize the order in which floors received which treatments. The criterion was a gloss index.
a. Outline an ANOVA table for this problem with proper degrees of freedom.
b. Explain, in words, what a significant interaction would mean in this situation.

5.24 A manufacturer of combustion engines is concerned about the percentage of smoke emitted by the engine in order to meet EPA standards. An experiment was conducted involving engines with three timing levels (T), three throat diameters (D), two volume ratios in the combustion chamber (V), and two injection systems (I). Thirty-six engines were designed to represent all possible combinations of these four factors and the engines were then operated in a completely random order and percent smoke recorded to the nearest 0.1 percent. Results showed:

			T								
			1 D			2 D			3 D		
			1	2	3	1	2	3	1	2	3
	1 I	1	0.4	3.8	7.0	1.8	4.8	8.1	1.7	4.9	8.2
		2	0.5	1.6	4.0	0.7	3.0	5.6	1.2	3.2	10.0
V											
	2 I	1	0.7	1.8	6.0	1.4	1.0	1.5	7.0	3.1	3.4
		2	0.1	0.1	0.2	0.6	0.4	1.4	4.0	1.8	2.0

Do a complete analysis of these data and make recommendations as to which combination or combinations of factor levels should give a minimum percent of smoke emitted.

5.25 Students' scores on a psychomotor test were recorded for various sized targets at which the student must aim, different machines used, and the level of background illumination. Results are shown in the following ANOVA table.

Source	df	SS	MS
Target size (A)	3	235.200	78.400
Machine (B)	2	86.467	43.233
Level of illumination (C)	1	76.800	76.800
$A \times B$	6	104.200	17.367
$A \times C$	3	93.867	31.289
$B \times C$	2	12.600	6.300
$A \times B \times C$	6	174.333	29.056
Within sets	96	1198.000	12.478
Totals	119	1981.467	

a. Using a 5 percent significance level indicate which effects are statistically significant.

b. Assuming that you had the original data at your disposal, explain briefly the next steps you would take in analyzing the data from this experiment.

c. Devise a data layout sheet that could be used to collect each item of data for this experiment.

d. Assuming the mean score for target size A_3, machine B_2, and level of illumination C_1 is 9.6, set 90 percent confidence limits on the true mean for this treatment combination.

6

2^f Factorial Experiments

6.1 INTRODUCTION

In the last chapter factorial experiments were considered and a general method for their analysis was given. There are a few special cases that are of considerable interest in future designs. One of these is the case of f factors where each factor is at just two levels. These levels might be two extremes of temperature, two extremes of pressure, two time values, two machines, and so on. Although this may seem like a rather trivial case since only two levels are involved, it is, nevertheless, very useful for at least two reasons: to introduce notation and concepts useful when more involved designs are discussed and to illustrate what main effects and interactions there really are in this simple case. It is also true that in practice many experiments are run at just two levels of each factor. The ceramic tool-cutting example of Chapters 1 and 5 is a $2 \times 2 \times 2$, or 2^3 factorial, with four observations per cell. Throughout this chapter the two levels will be considered as fixed levels. This is quite reasonable because the two levels are chosen at points near the extremes, rather than at random.

6.2 2^2 FACTORIAL

The simplest case to consider is one in which two factors are of interest and each factor is set at just two levels. This is a $2 \times 2 = 2^2$ factorial, and the design will be considered as completely randomized. The example in Section 5.1 is of this type. The factors are temperature and altitude, and each is set at two levels: temperature at 25°C and 55°C and altitude at 0 K and 3 K. This gives four treatment combinations, displayed in Figure 5.3 where the response variable is the current flow in milliamperes. To generalize a bit, consider temperature as factor A and altitude as factor B. The model for this completely randomized design would be

$$Y_{ij} = \mu + A_i + B_j + AB_{ij} + \varepsilon_{ij} \tag{6.1}$$

where $i = 1, 2$ and $j = 1, 2$ in this case. Unless there is some replication, of course, no assessment of interaction can be made independent of error. From the data of Figure 5.3 when both temperature A and altitude B are at their low levels, the response is 210 mA. This may be designated by subscripts on AB as follows: $A_0B_0 = 210$ mA. Following this notation, A_1B_0 means A at its high level and B at its low level, or temperature 55°C and altitude 0 K. The response is $A_1B_0 = 240$ mA. Likewise, $A_0B_1 = 180$ mA and $A_1B_1 = 200$ mA. Since a 2^f experiment is encountered so often in the literature, most authors have adopted another notation for these treatment combinations. For this new notation, just the subscripts on AB are used as exponents on the small letters ab. If both factors are at their low levels, $a^0b^0 = (1)$, and (1) represents the response of both factors at their low level. $a^0b^1 = b$ represents B at its high level and A at its low level, $a^1b^0 = a$ represents the high level of A and low level of B, and $a^1b^1 = ab$ represents the response when both factors are at their high levels. This notation can easily be extended to more factors, provided only two levels of each factor are involved.

For the example of Figure 5.3 the treatment combinations can be represented by the vertices of a square as in Figure 6.1.

In Figure 6.1 the low and high levels of factors A and B are represented by 0 and 1, respectively, on the A and B axes. The intersection of these levels in the plane of the figure shows the four treatment combinations. For example: $00 = (1)$ represents both factors at their low levels, $10 = a$ represents A high, B low, and so on, $01 = b$, $11 = ab$. These expressions are only symbolic and are to be considered as merely a mnemonic device to simplify the design and its analysis. The normal order for writing these treatment combinations is (1), a, b, ab. Note that (1) is written first, then the

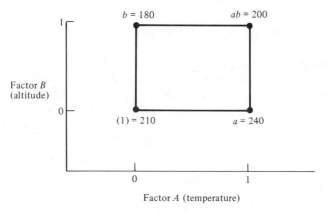

Figure 6.1 2^2 factorial experiment.

high level of each factor with the low level of the other (a, b), and the fourth term is the algebraic product of the second and third (ab). When a third factor is introduced, it is placed at the end of this sequence and then multiplied by all of its predecessors. For example, if factor C is also present at two levels, the treatment combinations are

$$(1), a, b, ab, c, ac, bc, abc$$

which can be represented as the vertices of a cube.

Returning to Figure 6.1 we find that the *effect of a factor* is defined as the change of response produced by a change in the level of that factor. At the low level of B, the effect of A is then $240 - 210$ or $a - (1)$, whereas the effect of A at the high level of B is $200 - 180$ or $ab - b$. The average effect of A is then

$$A = \tfrac{1}{2}[a - (1) + ab - b]$$

or

$$A = \tfrac{1}{2}[-(1) + a - b + ab]$$

Note that the coefficients on this A effect are all $+1$ when A is at its high level in the treatment combinations $(+a, +ab)$ and the coefficients are all -1 when A is at its low level as in (1) and b. Note also that

$$2A = -(1) + a - b + ab$$

is a contrast as defined in Section 3.4 (the sum of its coefficients are $-1 + 1 - 1 + 1 = 0$). This concept will be useful, as the sum of squares due to this contrast or effect can easily be determined.

The average effect of B, based on low level of A $[180 - 210 = b - (1)]$ and high level of A $(200 - 240 = ab - a)$, is

$$B = \tfrac{1}{2}[b - (1) + ab - a]$$

or

$$B = \tfrac{1}{2}[-(1) - a + b + ab]$$

and again it is seen that the same four responses are used, but $+1$ coefficients are on the treatment combinations for the high level of B and -1 coefficients for the low level of B. Also, $2B$ is a contrast.

To determine the effect of the interaction between A and B, note that at the high level of B the A effect is $ab - b$, and at the low level of B the A effect is $a - (1)$. If these two effects differ, there is an interaction between A and B. Thus the interaction is the average *difference* between these two differences

$$AB = \tfrac{1}{2}\{(ab - b) - [a - (1)]\}$$
$$= \tfrac{1}{2}[ab - b - a + (1)]$$

or

$$= \tfrac{1}{2}[(1) - a - b + ab]$$

Here again the same four treatment combinations are used with a different combination of coefficients. Note that $2AB$ is also a contrast, and all contrasts $2A$, $2B$, $2AB$ are orthogonal to each other. Summarizing in normal order gives

$$2A = -(1) + a - b + ab$$
$$2B = -(1) - a + b + ab$$
$$2AB = +(1) - a - b + ab$$

Note that the interaction effect takes the responses on one diagonal of the square with $+1$ coefficients and the responses on the other diagonal with -1 coefficients. Note also that the coefficients for the interaction effect can be found by multiplying the corresponding coefficients of the two main effects. As the only coefficients used are $+1$'s and -1's, the proper coefficients on the treatment combinations for each main effect and interaction can be determined from Table 6.1.

Table 6.1 Coefficients for Effects in a 2^2 Factorial Experiment

Treatment Combination	A	B	AB
(1)	−	−	+
a	+	−	−
b	−	+	−
ab	+	+	+

From Table 6.1 the orthogonality of the effects is easily seen, as well as the generation of the interaction coefficients from the main effect coefficients.

Another approach to the interaction between A and B would be to consider that at the high level of A the effect of B is $ab - a$, and at the low level of A the effect of B is $b - (1)$, so that the average difference is

$$AB = \tfrac{1}{2}\{(ab - a) - [b - (1)]\}$$
$$= \tfrac{1}{2}[ab - a - b + (1)]$$
$$= \tfrac{1}{2}[+(1) - a - b + ab]$$

which is the same expression as given before.

For the response data of Figure 6.1 the effects are

$$A = \tfrac{1}{2}(-210 + 240 - 180 + 200) = 25 \text{ mA}$$
$$B = \tfrac{1}{2}(-210 - 240 + 180 + 200) = -35 \text{ mA}$$
$$AB = \tfrac{1}{2}(+210 - 240 - 180 + 200) = -5 \text{ mA}$$

Since $2A$, $2B$, and $2AB$ are contrasts, the sum of squares due to a contrast is

$$SS_{C_m} = \frac{(\text{contrast})^2}{n \sum c_{jm}^2}$$

where n is the number of observations in each total (here $n = 1$). Also $\sum c_{jm}^2 = 1 + 1 + 1 + 1 = 4$ (or 2^2). From this definition

$$SS_A = \frac{[2(25)]^2}{4} = 625$$

$$SS_B = \frac{[2(-35)]^2}{4} = 1225$$

$$SS_{AB} = \frac{[2(-5)]^2}{4} = 25$$

$$SS_{\text{total}} = 1875$$

Since each effect and the interaction has but 1 df and there is no measure of error as only one observation was taken in each cell, this ANOVA is quite trivial, but it does show another approach to analysis based on effects. The simple ANOVA table would be that shown by Table 6.2.

Table 6.2   ANOVA for a 2^2 Factorial with No Replication

Source	df	SS	MS
A_i	1	625	625
B_j	1	1225	1225
Error or AB_{ij}	1	25	25
Totals	3	1875	

If the general methods of Chapter 5 are used on the data in Figure 6.1 (coded by subtracting 200), the results are those in Table 6.3.

$$SS_{\text{total}} = (10)^2 + (-20)^2 + (40)^2 + (0)^2 - \frac{(30)^2}{4} = 1875$$

$$SS_A = \frac{(-10)^2 + (40)^2}{2} - \frac{(30)^2}{4} = 850 - 225 = 625$$

$$SS_B = \frac{(50)^2 + (-20)^2}{2} - \frac{(30)^2}{4} = 1450 - 225 = 1225$$

$$SS_{AB} \text{ (by subtraction)} = 1875 - 625 - 1225 = 25$$

Table 6.3 Coded Data of Figure 6.1

	Factor A		
Factor B	0	1	Totals
0	$+10$	$+40$	$+50$
1	-20	0	-20
Totals	-10	$+40$	$+30$

which are the same sums of squares as given in Table 6.2. Even though no separate measure of error is available, a glance at the mean squares of Table 6.2 shows that both main effects are large compared with the interaction effect.

The results of this section can easily be extended to cases where there are n replications in the cells, by using cell totals for the responses at (1), a, b, and ab and adjusting the sum of squares for the effects accordingly (see Example 6.1).

Example 6.1 An example of a 2^2-factorial experiment with two replications per cell is considered here, using hypothetical responses to illustrate the principles of the previous section. Consider the data of Table 6.4.

Using the general methods of Chapter 5, the sums of squares are

$$\text{SS}_{\text{total}} = 4^2 + 6^2 + 3^2 + 7^2 + 2^2 + (-2)^2 + (-4)^2 + (-6)^2$$
$$- \frac{(10)^2}{8} = 170 - 12.5 = 157.5$$

$$\text{SS}_A = \frac{(20)^2 + (-10)^2}{4} - 12.5 = 112.5$$

$$\text{SS}_B = \frac{(10)^2 + (0)^2}{4} - 12.5 = 12.5$$

$$\text{SS}_{A \times B \text{ interaction}} = \frac{(10)^2 + (10)^2 + (0)^2 + (-10)^2}{2} - 12.5 - 112.5 - 12.5$$
$$= 12.5$$

$$\text{SS}_{\text{error}} = 157.5 - 112.5 - 12.5 - 12.5 = 20.0$$

These results could be displayed in an ANOVA table. Using the methods of Section 6.2, we get, with the treatment combinations, the total response

$$(1) = 10 \qquad a = 0 \qquad b = 10 \qquad ab = -10$$

Table 6.4 2^2 Factorial with Two Replications

Factor B	Factor A 0	Factor A 1	Totals
0	4 6 10	2 -2 0	+10
1	3 7 10	-4 -6 -10	0
Totals	+20	-10	+10

and the contrasts

$$4A = -10 + 0 - 10 + (-10) = -30$$
$$4B = -10 - 0 + 10 + (-10) = -10$$
$$4AB = +10 - 0 - 10 + (-10) = -10$$

The coefficient 4 used with each response represents the two individual responses at each level. In general the coefficient of these effects is $n \cdot 2^{f-1}$ where n is the number of replications and f the number of factors. Here $f = 2$ and $n = 2$. The sum of squares for these three contrasts are

$$SS_A = \frac{(4A)^2}{n \cdot 2^2} = \frac{(-30)^2}{2 \cdot 4} = \frac{900}{8} = 112.5$$

$$SS_B = \frac{(4B)^2}{n \cdot 2^2} = \frac{(-10)^2}{2 \cdot 4} = \frac{100}{8} = 12.5$$

$$SS_{AB} = \frac{(4AB)^2}{n \cdot 2^2} = \frac{(-10)^2}{2 \cdot 4} = \frac{100}{8} = 12.5$$

since

$$2^2 = \sum_{i=1}^{4} c_{jm}^2 = 1 + 1 + 1 + 1 = 4$$

These results are the same as given by the general method. The total sum of squares must be calculated as usual in order to get the error sum of squares by subtraction.

For this special case of a 2^f factorial experiment, Yates [26] developed a rather simple scheme for computing these contrasts. The method can best be illustrated on Example 6.1, using Table 6.5.

Table 6.5 Yates' Method on a 2^2 Factorial

Treatment Combination	Response	(1)	(2)	SS
(1)	10	10	10 = total	12.5
a	0	0	$-30 = 4A$	112.5
b	10	-10	$-10 = 4B$	12.5
ab	-10	-20	$-10 = 4AB$	12.5

In Table 6.5 list all treatment combinations in the first column. Place the total response to each of these treatment combinations in the second column. For the third column, labeled (1), add the responses in pairs, for example, $10 + 0 = 10$ for the first two and $10 - 10 = 0$ for the next two; this completes half of column (1). For the second half, subtract the responses in pairs, always subtracting the first from the second, for example, $0 - 10 = -10$; $-10 - (10) = -20$. This completes column (1). Column (2) is determined in the same manner as column (1) using the column (1) results: $10 + 0 = 10$; $-10 - 20 = -30$; $0 - 10 = -10$; $-20 - (-10) = -10$. Proceed in this same manner until the fth column is reached: (1), (2), (3), ..., (f). In this case $f = 2$, so there are just two columns. The values in column f are the contrasts, where the first entry is the grand total of all readings; the one corresponding to a is $n \cdot 2^{f-1} \cdot A$, b is $n \cdot 2^{f-1} \cdot B$, and ab is $n \cdot 2^{f-1} \cdot AB$.

When the results of the last column [column (f)] are squared and divided by $n \cdot 2^f$, the results are the sums of squares as shown above. The first sum of squares $(\text{total})^2/8$ is the correction term for the grand mean given in Chapter 5. The Yates method reduces the analysis to simply adding and subtracting numbers. It is very useful provided the experiment is a 2^f factorial with n observations per treatment.

As proof for the Yates method on a 2^2 factorial go through the steps above using the treatment combination symbols for the responses as in Table 6.6. It is obvious that the last column does give the proper treatment contrasts.

Table 6.6 Yates' Method on a 2^2 in General; One Observation per Cell

Treatment Combination	(1)	(2)
(1)	$(1) + a$	$(1) + a + b + ab$ = total
a	$b + ab$	$-(1) + a - b + ab = 2A$
b	$a - (1)$	$-(1) - a + b + ab = 2B$
ab	$ab - b$	$(1) - a - b + ab = 2AB$

6.3 2^3 FACTORIAL

Considering a third factor C, also at two levels, the experiment will be a $2 \times 2 \times 2$ or 2^3 factorial, again run in a completely randomized manner. The treatment combinations are now (1), a, b, ab, c, ac, bc, abc, and they may be represented as vertices of a cube as in Figure 6.2.

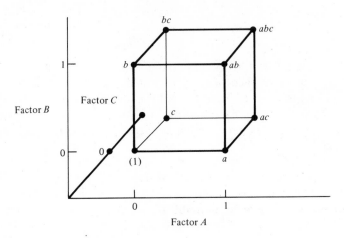

Figure 6.2 2^3 factorial arrangements.

With these $8 = 2^3$ observations, or $8n$ if there are replications, the main effects and each interaction may be expressed by using the proper coefficients $(-1$ or $+1)$ on these eight responses. For the effect of factor A, consider all responses in the right-hand plane (a, ab, ac, abc) with plus signs and all in the left-hand plane $[(1), b, bc, c]$ with minus signs as these will show the effect of increasing the level of A. Or,

$$4A = -(1) + a - b + ab - c + ac - bc + abc$$

For factor B consider responses in the lower and higher planes of the cube

$$4B = -(1) - a + b + ab - c - ac + bc + abc$$

The AB interaction is determined by the difference in the A effect from level 0 of B to level 1 of B, regardless of C

$$\text{at } B_0: 2A \text{ effect } a + ac - c - (1)$$
$$\text{at } B_1: 2A \text{ effect } abc + ab - b - bc$$

The difference in these is the interaction

$$4AB = [abc + ab - b - bc] - [a + ac - c - (1)]$$
$$4AB = abc + ab - b - bc - a - ac + c + (1)$$

or

$$4AB = +(1) - a - b + ab + c - ac - bc + abc$$

which gives the same signs as the products of corresponding signs on $4A$ and $4B$.

For factor C compare the responses in the back plane of the cube with those in the front plane of the cube

$$4C = c + bc + ac + abc - (1) - b - a - ab$$

or

$$4C = -(1) - a - b - ab + c + ac + bc + abc$$

The AC and BC interactions can be determined as AB was, and it can be seen that the resulting interaction effects are

$$4AC = +(1) - a + b - ab - c + ac - bc + abc$$
$$4BC = +(1) + a - b - ab - c - ac + bc + abc$$

To determine the ABC interaction, consider the BC interaction at level 0 of A versus the BC interaction at level 1 of A. Any difference in these is an ABC interaction

$$2BC \text{ interaction at } A_0: +(1) - b - c + bc$$
$$2BC \text{ interaction at } A_1: +a - ab - ac + abc$$

Their difference is

$$4ABC = (a - ab - ac + abc) - [(1) - b - c + bc]$$

or

$$4ABC = -(1) + a + b - ab + c - ac - bc + abc$$

which can also be obtained from multiplying the coefficients of A and BC, or B and AC, or C and AB. These others can also be used to get the ABC interaction, but the results are the same. Summarizing gives us Table 6.7.

Table 6.7 illustrates once again the orthogonality of the effects and can easily be extended to four, five, and more factors if each factor is at two levels only. In a 2^3-factorial experiment, the sum of squares is given by

$$SS_{contrast} = \frac{(contrast)^2}{n \cdot 2^3} = \frac{(contrast)^2}{8n}$$

and again the Yates method leads to an easy computation of the contrasts.

Table 6.7   Coefficients for Effects in a 2^3 Factorial Experiment

Treatment Combination	Total	A	B	AB	C	AC	BC	ABC
					Effect			
(1)	+	−	−	+	−	+	+	−
a	+	+	−	−	−	−	+	+
b	+	−	+	−	−	+	−	+
ab	+	+	+	+	−	−	−	−
c	+	−	−	+	+	−	−	+
ac	+	+	−	−	+	+	−	−
bc	+	−	+	−	+	−	+	−
abc	+	+	+	+	+	+	+	+

Example 6.2   In Chapters 1 and 5 the problem on power requirements for cutting with ceramic tools was analyzed in detail. It is readily seen that this is a $2 \times 2 \times 2 = 2^3$ factorial with four replications per cell. This problem could be analyzed by the special methods of this chapter. From Table 5.10 on coded data the treatment combinations might be summarized as in Table 6.8.

Table 6.8   Ceramic Tool Data in 2^3 Form

	Tool Type			
	1		2	
	Bevel Angle		Bevel Angle	
Type of Cut	15°	30°	15°	30°
Continuous	(1)	b	a	ab
	2	15	−5	13
Interrupted	c	bc	ac	abc
	−12	−2	−17	−7

Tool type = factor A, bevel = factor B, cut = factor C.

In Table 6.8 the totals of the four observations for each treatment combination have been entered in their corresponding cells. By the Yates method we get the entries in Table 6.9.

The sum of squares is seen to be in substantial agreement with those of Table 5.10. We must resort to the individual readings, however, to determine the total sum of squares and then the error sum of squares. This was done in Table 5.10 and the interpretation of this problem is given in Chapter 5. The purpose of repeating the problem here was merely to show the use of the Yates method on a 2^3-factorial experiment.

Table 6.9 Yates' Method on Ceramic Tool Data

Treatment Combination	Response	(1)	(2)	(3)	SS
(1)	2	− 3	25	− 13 = total	5.28
a	− 5	28	− 38	− 19 = $16A$	11.28
b	15	− 29	− 9	51 = $16B$	81.28
ab	13	− 9	− 10	5 = $16AB$	0.78
c	− 12	− 7	31	− 63 = $16C$	124.03
ac	− 17	− 2	20	− 1 = $16AC$	0.03
bc	− 2	− 5	5	− 11 = $16BC$	3.78
abc	− 7	− 5	0	− 5 = $16ABC$	0.78

6.4 2^f—REMARKS

The methods shown for 2^2 and 2^3 factorials may easily be extended to 2^f factorials where f factors are each considered at two levels. The contrasts are determined from the responses to the treatment combinations by associating plus signs with high levels of a factor and minus signs with low levels; the contrasts for interactions are found by multiplication of corresponding coefficients.

Table 6.10 ANOVA for a 2^f Factorial with n Replications

Source		df	
Main effects	A	1	
	B	1	f
	C	1	
	$\vdots$	$\vdots$	
Two-factor interactions	AB	1	
	AC	1	$C(f, 2) = \dfrac{f(f-1)}{2}$
	BC	1	
	$\vdots$	$\vdots$	
Three-factor interactions	ABC	1	
	ABD	1	$C(f, 3) = \dfrac{f(f-1)(f-2)}{6}$
	BCD	1	
	$\vdots$	$\vdots$	
Four-factor interactions, and so on			
Sum of all treatment combinations		$2^f - 1$	
Residual or error		$2^f \cdot (n - 1)$	
Total		$n \cdot 2^f - 1$	

The general relationships for 2^f factorials with n observations per treatment are

$$\text{contrast} = n \cdot 2^{f-1} \text{ (effect)}$$

or

$$\text{effect} = \frac{1}{n2^{f-1}} \text{ (contrast)}$$

and

$$SS_{\text{contrast}} = \frac{(\text{contrast})^2}{n2^f}$$

A general ANOVA would be as in Table 6.10, but not in "normal" order.

6.5 SUMMARY

Experiment	Design	Analysis
I. Single factor		
	1. Completely randomized	1. One-way ANOVA
	$Y_{ij} = \mu + \tau_j + \varepsilon_{ij}$	
	2. Randomized block	2.
	$Y_{ij} = \mu + \beta_i + \tau_j + \varepsilon_{ij}$	
	a. Complete	a. Two-way ANOVA
	b. Incomplete, balanced	b. Special ANOVA
	c. Incomplete, general	c. Regression method
	3. Latin square	
	$Y_{ijk} = \mu + \beta_i + \tau_j$	
	$\quad + \gamma_k + \varepsilon_{ijk}$	
	a. Complete	a. Three-way ANOVA
	b. Incomplete,	b. Special ANOVA
	Youden square	(like 2b)
	4. Graeco-Latin square	4. Four-way ANOVA
	$Y_{ijkm} = \mu + \beta_i + \tau_j$	
	$\quad + \gamma_k + \omega_m + \varepsilon_{ijkm}$	
II. Two or more factors		
A. Factorial (crossed)		
	1. Completely randomized	1.
	$Y_{ijk} = \mu + A_i + B_j$	
	$\quad + AB_{ij} + \varepsilon_{k(ij)} \cdots$	
	for more factors	
	a. General case	a. ANOVA with interactions
	b. 2^f case	b. Yates' method or general ANOVA; use (1), a, b, ab, $\ldots$

PROBLEMS

6.1 For the 2^2 factorial with three observations per cell given below (hypothetical data), do an analysis by the method of Chapter 5.

	Factor A	
Factor B	A_1	A_2
B_1	0	4
	2	6
	1	2
B_2	-1	-1
	-3	-3
	1	-7

6.2 Redo Problem 6.1 by the Yates method and compare results.

6.3 In an experiment on chemical yield three factors were studied, each at two levels. The experiment was completely randomized and the factors were known only as A, B, and C. The results were

A_1				A_2			
B_1		B_2		B_1		B_2	
C_1	C_2	C_1	C_2	C_1	C_2	C_1	C_2
1595	1745	1835	1838	1573	2184	1700	1717
1578	1689	1823	1614	1592	1538	1815	1806

Analyze by the methods of Chapter 5.

6.4 Analyze Problem 6.3 by the Yates method and compare your results.

6.5 Plot any results from Problem 6.3 that might be meaningful from a management point of view.

6.6 The results of Problem 6.3 lead to another experiment with four factors each at two levels, with the following data.

A_1							
B_1				B_2			
C_1		C_2		C_1		C_2	
D_1	D_2	D_1	D_2	D_1	D_2	D_1	D_2
1985	2156	1694	2184	1765	1923	1806	1957
1592	2032	1712	1921	1700	2007	1758	1717

A_2							
B_1				B_2			
C_1		C_2		C_1		C_2	
D_1	D_2	D_1	D_2	D_1	D_2	D_1	D_2
1595	1578	2243	1745	1835	1863	1614	1917
2067	1733	1745	1818	1823	1910	1838	1922

Analyze these data by the general methods of Chapter 5.

6.7   Analyze Problem 6.6 using the Yates method.

6.8   Plot from Problem 6.6 any results you think are meaningful.

6.9   Consider a four-factor experiment with each factor at two levels.
   a. Write out the treatment combinations in this experiment in normal order.
   b. Assuming there is only one observation per treatment, outline an ANOVA for this example and argue what you would use for an error term to test for significance of various factors and interactions.
   c. Explain how you would determine the AD interaction from the responses to the treatment combinations given in part (a).

6.10   An experimenter is interested in the effects of five factors:—A, B, C, D, and E—each at two levels on some response variable Y. Assuming that data are available for each of the possible treatment combinations in this experiment, answer the following.
   a. How would you determine the effect of factor C and its sum of squares based on data taken at each of the treatment combinations?
   b. Set up an ANOVA table for this problem assuming n observations for each treatment combination.
   c. If $n = 2$ for part (b), at what numerical value of F would you reject the hypothesis of no ACE interaction.
   d. How would you propose testing for ACE interaction and what F would be necessary for rejection in this problem if $n = 1$?

6.11   Consider an experiment with five factors A, B, C, D, and E, each at two levels.
   a. How many treatment combinations are there?
   b. Outline an ANOVA table to test all main effects, all two-way and all three-way interactions, if there is only one observation per treatment combination.
   c. Explain how you would find the proper signs on the treatment combinations for the ACE interaction.
   d. Some experimenter comments to you that "since all factors are at only two levels, it is unnecessary to run a Newman–Keuls test as one only has to look and see which level gives the greater mean response." Explain situations in which this comment would be true for your results and in which situations it would not be true.

6.12   Three factors are studied for their effect on the horsepower necessary to remove a cubic inch of metal per minute. The factors are feed rate at two levels, tool condition at two levels, and tool type at two levels. For each treatment combination three observations are taken and all 24 observations of horsepower are made in a completely randomized order. The data appear below.

	Feed Rate			
	0.011		0.015	
Tool	Tool Condition		Tool Condition	
Type	Same	New	Same	New
Utility	0.576	0.548	0.514	0.498
	0.576	0.555	0.515	0.519
	0.565	0.540	0.518	0.504
Precision	0.526	0.547	0.494	0.521
	0.542	0.524	0.504	0.480
	0.548	0.525	0.530	0.494

Do a complete ANOVA on these data as a 2^f experiment and state your conclusions.

6.13 If minimum horsepower is desired for the experiment given in Problem 6.12, what are your recommendations as to the best set of factor conditions on the basis of your results in Problem 6.12.

6.14 Any factor whose levels are powers of 2 such as 4, 8, 16, . . . , may be considered as pseudo factors in the form 2^2, 2^3, 2^4,. . . . Yates' method can then be used if all factor levels are 2 or powers of 2 and the resulting sums of squares for the pseudo factors can be added to give the sums of squares for the actual factors and appropriate interactions. Try out this scheme on the data of Problem 5.21.

6.15 A systematic test was made to determine the effects on the coil breakdown voltage of the following six variables, each at two levels as indicated.

1. Firing furnace: Number 1 or 3
2. Firing temperature: 1650°C or 1700°C
3. Gas humidification: Yes or no
4. Coil outside diameter: Large or small
5. Artificial chipping: Yes or no
6. Sleeve: Number 1 or 2

Assuming complete randomization, devise a data sheet for this experiment, outline its ANOVA, and explain your error term if only one coil is to be used for each of the treatment combinations.

6.16 If seven factors are to be studied at two levels each, outline the ANOVA for this experiment in a table similar to Table 6.10, and, assuming $n = 1$, suggest a reasonable error term.

6.17 In a steel mill an experiment was designed to study the effect of drop-out temperature, back-zone temperature, and type of atmosphere on heat slivers in samples of steel. The response variable is a mean surface factor obtained for each type of sliver. Results are

	Drop-out Temperature			
	2145°		2165°	
	Back-Zone Temperature			
Atmosphere	2040–2140°	2060–2160°	2040–2140°	2060–2160°
Oxidizing	7.8	4.3	4.5	8.2
Reducing	6.8	5.9	11.0	6.0

Assuming complete randomization, analyze these results. To get an error term for testing, three of the data points were repeated to give a SS_{error} of 1.11 based on 3 degrees of freedom.

6.18 Further analyze the results of Problem 6.17 and make any recommendations you can for minimizing heat slivers.

6.19 At the same time that data were collected on heat slivers in Problem 6.17, data were also collected on grinder slivers. The results were found to be as follows:

	Drop-out Temperature			
	2145°		2165°	
	Back-Zone Temperature			
Atmosphere	2040–2140°	2060–2160°	2040–2140°	2060–2160°
Oxidizing	0.0	4.3	9.8	4.0
Reducing	0.0	4.4	10.1	4.4

For these data the $SS_{error} = 13.22$ based on three repeated measurements. Completely analyze these data.

6.20 Sketch rough graphs for the following situations where each factor is at two levels only:

1. Main effects A and B both significant, AB interaction not significant.

2. Main effect A significant, B not significant, and AB significant.

3. Both main effects not significant but interaction significant.

6.21 For a 2^3 factorial can you sketch a graph (or graphs) to show no significant two-factor interactions, but a significant three-factor interaction?

7

Qualitative and Quantitative Factors

7.1 INTRODUCTION

In Chapter 3, Figure 3.1, we noted several possible procedures to follow after an analysis of variance. In that chapter fixed levels of a single factor were considered when such levels were qualitative levels. If the levels are quantitative, such as temperature, dosage, humidity, or time, a different route is suggested after an ANOVA yields significant differences between response variable averages. It is also noted that the path to follow depends upon whether or not these quantitative levels are equispaced. It will be seen in this chapter that there are definite advantages in designing the experiment so that the levels of any quantitative variables are equispaced.

It is assumed that the reader has had some introduction to linear regression in which one tries to write a mathematical model of a straight line that will predict some response variable Y from one independent variable X. After reviewing this linear "fit," the concept will be extended to fitting a curve when a straight line is not an adequate model. These notions will then be extended using the method of orthogonal polynomials to test for the highest order polynomial necessary to do a decent prediction of Y from X. The analysis will be made only if the analysis of variance shows a significant difference in the Y averages due to levels of X and we wish to go further and see how Y may be related to X.

After building some methodology to pull out various trend lines, the notions will be extended to more than one factor where the factors in the experiment may be all at quantitative levels or where some factors may be at qualitative levels and some at quantitative levels. The next chapter will extend these ideas to a multiple regression model where prediction equations are sought and the experiment may not necessarily have all quantitative levels equispaced.

7.2 LINEAR REGRESSION

To review linear regression and to become familiar with the notation used, consider an example where an experiment was designed to determine the effect of the amount of drug dosage on a person's reaction time to a given stimulus.

Example 7.1 To study the effect of drug dosage in milligrams on a person's reaction time in milliseconds, 15 subjects were randomly assigned one of five dosages of a drug—0.5, 1.0, 1.5, 2.0, or 2.5 mg—with three subjects assigned to each level of drug dosage. The reaction times were recorded as shown in Table 7.1. An analysis of variance of these data yields Table 7.2.

Table 7.1 Reaction Time Data

	Dosage (mg)					
	0.5	1.0	1.5	2.0	2.5	
Reaction	26	28	28	32	38	
Time (ms)	28	26	30	33	39	
Y_{ij}	29	30	31	31	38	
n	3	3	3	3	3	$N = 15$
$T_{.j}$	83	84	89	96	115	$T_{..} = 467$
$\sum_i Y_{ij}^2$	2301	2360	2645	3074	4409	$\sum_i \sum_j Y_{ij}^2 = 14{,}789$

Table 7.2 Reaction Time ANOVA

Source	df	SS	MS	F	Significance
Between dosages	4	229.73	57.42	28.72	0.000
Error	10	20.00	2.00		
Totals	14	249.73			

Obviously dosage has a highly significant effect on reaction time. But since reaction time is a quantitative factor, the next question might well be: How does reaction time vary with drug dosage? Can one find a functional relationship between these two variables that might enable one to predict reaction time from drug dosage?

As a first step one might plot a scattergram of all 15 X, Y pairs of points. This is shown in Figure 7.1.

In Figure 7.1 the small x's denote the average Y for each X_j or the $\bar{Y}_{.j}$'s. The plot shows that Y increases with increases in X and one may try a straight-line "fit" as a first approximation for predicting Y from X. Such

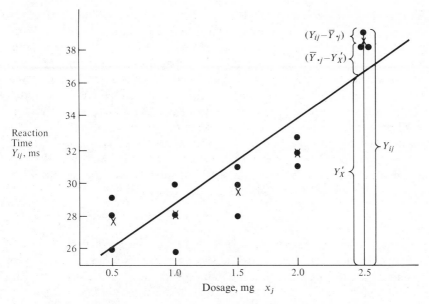

Figure 7.1   Reaction time versus drug dosage.

a line is shown in Figure 7.1. One can see how close the line comes to the average Y values for each X. Note that for a given X_j (say 2.5 mg) the observed Y_{ij} can be partitioned into three parts: Y'_X is the predicted value of Y_{ij} for a given X_j as found on the straight line, $\bar{Y}_{.j} - Y'_X$ is the amount by which the mean Y for a given X_j departs from its predicted value Y'_X (referred to as the departure from linear regression in this model), and $Y_{ij} - \bar{Y}_{.j}$ is the amount by which an observed Y_{ij} varies from its treatment mean.

In the analysis of variance model

$$Y_{ij} = \mu + (\mu_j - \mu) + (Y_{ij} - \mu_j) \tag{7.1}$$

and now one tries to predict μ_j from some function based on X_j. If this predicted mean is labeled $\mu_{Y/X}$, the model can be expanded to read

$$Y_{ij} = \mu + (\mu_{Y/X} - \mu) + (\mu_j - \mu_{Y/X}) + (Y_{ij} - \mu_j) \tag{7.2}$$

where $\mu_{Y/X}$ is the true predicted mean based on X and $\mu_j - \mu_{Y/X}$ is the amount by which the true mean μ_j departs from its predicted value.

For the sample model Equation (7.2) becomes

$$Y_{ij} = \bar{Y}_{..} + (Y'_X - \bar{Y}_{..}) + (\bar{Y}_{.j} - Y'_X) + (Y_{ij} - \bar{Y}_{.j}) \tag{7.3}$$

If the second and third terms on the right of this equation are combined, we have the ANOVA sample model. Thus, we attempt here to partition the mean of a given treatment into two parts: that which can be predicted by regression on X and the amount by which the mean shows a departure from such a regression model. The purpose, of course, is to find a model for predicting the response means such that the departures from these means are very small. By choosing a polynomial of high enough degree one can find a model that actually goes through all the Y means. However, it is hoped that an adequate lower degree model can be found whose departures are small and insignificant when compared with the error of individual Y's around their means.

In Equation (7.3) Y'_X represents the predicted Y for a given X for any assumed model. The straight-line model is only one possible attempt to "fit" a curve to the data. The straight-line model is given by

$$Y'_X = b_0 + b_1 X_j \tag{7.4}$$

where b_0 is the Y intercept and b_1 is its slope. The usual method for determining the values of b_0 and b_1 is the *method of least squares*. The values of b_0 and b_1 are determined from the data in such a way as to minimize the sum of squares of the deviations of each Y_{ij} from its predicted value Y'_X. [One could also find b_0 and b_1 such that the sum of squares of the departures from regression $(\bar{Y}_{.j} - Y'_X)$ is minimized, but this leads to the same results.] Here

$$Y_{ij} - Y'_X = Y_{ij} - b_0 - b_1 X_j$$

and

$$\text{SS}_{\text{deviations}} = \sum_i \sum_j (Y_{ij} - Y'_X)^2 = \sum_i \sum_j (Y_{ij} - b_0 - b_1 X_j)^2$$

Since the summations of X's and Y's can be found from a given set of data, $\text{SS}_{\text{deviations}}$ is a function of b_0 and b_1. To minimize $\text{SS}_{\text{deviations}}$ then, one differentiates $\text{SS}_{\text{deviations}}$ with respect to b_0 and b_1

$$\frac{\delta(\text{SS}_{\text{deviations}})}{\delta b_0} = 2 \sum_i \sum_j (Y_{ij} - b_0 - b_1 X_j)(-1) = 0$$

$$\frac{\delta(\text{SS}_{\text{deviations}})}{\delta b_1} = 2 \sum_i \sum_j (Y_{ij} - b_0 - b_1 X_j)(-X_j) = 0$$

Dividing through by (-2) and simplifying

$$\left.\begin{array}{l} \sum_i \sum_j Y_{ij} = b_0 N + b_1 n \sum_j X_j \\[2mm] \sum_i \sum_j X_j Y_{ij} = b_0 n \sum_j X_j + b_1 n \sum_j X_j^2 \end{array}\right\} \tag{7.5}$$

These equations are often called the least squares normal equations, which can now be solved for b_0 and b_1.

From Table 7.1 $\sum_i \sum_j Y_{ij} = T_{..} = 467$, $N = 15$, $n = 3$.

$$\sum_j X_j = 0.5 + 1.0 + 1.5 + 2.0 + 2.5 = 7.5$$

$$\sum_j X_j^2 = (0.5)^2 + (1.0)^2 + (1.5)^2 + (2.0)^2 + (2.5)^2 = 13.75$$

To find the cross product $\sum_i \sum_j X_j Y_{ij}$ note that $\sum_i Y_{ij} = T_{.j}$, so $\sum_j X_j T_{.j}$ gives

$$(0.5)(83) + (1.0)(84) + (1.5)(89) + (2.0)(96) + (2.5)(115) = 738.5$$

Substituting in Equations (7.5)

$$467 = 15b_0 + 22.5b_1$$
$$738.5 = 22.5b_0 + 41.25b_1$$

To solve one might multiply the first equation by 1.5, giving

$$700.5 = 22.5b_0 + 33.75b_1$$
$$738.5 = 22.5b_0 + 41.25b_1$$

Subtracting and solving for b_1,

$$b_1 = \frac{700.5 - 738.5}{33.75 - 41.25} = \frac{-38.0}{-7.5} = 5.07$$

and from the original first equation

$$b_0 = \frac{467 - (22.5)(5.07)}{15} = 23.53$$

and the linear model for this set of sample data is

$$Y'_X = 23.53 + 5.07X_j$$

Some previous work on a straight-line fit may have used the formula for slope as

$$b_1 = \frac{SP_{XY}}{SS_X}$$

or the ratio of the sum of cross products of X and Y to the sum of the squares of X. And b_0 as

$$b_0 = \bar{Y} - b_1 \bar{X}$$

Here

$$b_1 = \frac{SP_{XY}}{SS_X} = \frac{\sum_i \sum_j X_j Y_{ij} - \dfrac{\sum_i \sum_j X_j \sum_i \sum_j Y_{ij}}{N}}{\sum_i \sum_j X_j^2 - \dfrac{(\sum_i \sum_j X_j)^2}{N}}$$

In terms of Table 7.1 notation

$$b_1 = \frac{\sum_j X_j T_{.j} - \dfrac{n \sum_j X_j T_{..}}{N}}{n \sum_j X_j^2 - \dfrac{(n \sum_j X_j)^2}{N}} = \frac{738.5 - \dfrac{3(7.5)(467)}{15}}{3(13.75) - \dfrac{3(7.5)^2}{15}} = 5.07$$

as before and

$$b_0 = \bar{Y}_{..} - b_1 \bar{X}_{..} = \frac{467}{15} - (5.07)\frac{3(7.5)}{15} = 23.53$$

as before.

Table 7.3   Departures from Linear Regression

X_j	$\bar{Y}_{.j}$	Y'_X	$\bar{Y}_{.j} - Y'_X$	$(\bar{Y}_{.j} - Y'_X)^2$
0.5	27.667	26.067	1.600	2.560
1.0	28.000	28.600	−0.600	0.360
1.5	29.667	31.134	−1.467	2.152
2.0	32.000	33.667	−1.667	2.779
2.5	38.333	36.200	2.133	4.550
Totals			0	12.401

To see how well this linear model predicts the Y's from the X's, we construct Table 7.3. With the predicted values taken from the model

$$Y'_X = 23.53 + 5.07 X_j$$

as, for example, when $X_j = 1.0$,

$$Y'_X = 23.53 + 5.07(1.0) = 28.600$$

The table shows the departures from linear regression, for example,

$$\bar{Y}_{.j} - Y'_X = 28.000 - 28.600 = -0.600$$

Note that the sum of the departures adds to zero and the sum of squares of departures from linear regression is

$$SS_{departures} = n \sum_j (\bar{Y}_{.j} - Y'_X)^2 = 3(12.40) = 37.20$$

as each departure of a mean must be weighted with three observed values for each X_j. Knowing the sum of squares of departures from linear regression, the ANOVA Table 7.2 can be expanded to give Table 7.4.

Table 7.4 shows a highly significant linear effect but also a significant departure from the linear model at the 5 percent significance level. With

Table 7.4 ANOVA on Reaction Time with Linear Regression

Source	df	SS	MS	F	Significance
Between dosages	4	229.73			
Linear	1		192.53	192.53	96.3***	0.000
Departure from linear	3		37.20	12.40	6.2*	0.012
Error	10	20.00		2.00	
Totals	14	249.73			

such a departure it would be appropriate to try a second-degree equation or a quadratic of the form

$$Y'_X = b_0 + b_1 X_j + b_2 X_j^2$$

Since there is a strong linear effect in this example and, in some problems, the linear model is adequate to explain most of the variation in the Y variable, some useful statistics can be determined from Table 7.4.

The proportion of the total sum of squares that can be accounted for by linear regression is sometimes called the *coefficient of determination r^2*, and its positive square root r the *Pearson product–moment correlation coefficient*. Here

$$r^2 = 192.53/249.73 = 0.7712$$

and

$$r = 0.88$$

Thus linear regression will account for about 77 percent of the variation seen in the reaction time Y.

By taking the ratio of the sum of squares between means to the total sum of squares, one finds the maximum amount of the total variation that could be accounted for by a curve or model that passes through all the mean Y's for each X. This is called *eta squared* (η^2). Here

$$\eta^2 = 229.73/249.73 = 0.9199$$

or approximately 92 percent of the variation in reaction time could be accounted for by a model through all the means for each dosage X_j.

Another statistic used with a linear model is the standard error of estimate ($s_{Y.X}$). It is the standard deviation of the deviations of the Y_{ij}'s from their predicted values

$$Y_{ij} - Y'_X = (Y_{ij} - \bar{Y}_{.j}) + (\bar{Y}_{.j} - Y'_X)$$

This expression adds the error about the mean and the departure of the mean from the predicted curve, which is appropriate only if the departure is nonsignificant. Then

$$s_{Y.X} = \sqrt{\frac{SS_{\text{departure}} + SS_{\text{error}}}{N - 2}}$$

$$s_{Y.X} = \sqrt{\frac{37.20 + 20.00}{13}} = 2.10$$

This statistic is often used to set confidence limits around a line of best fit when linear regression is appropriate.

In this discussion of Example 7.1 no use has been made of the fact that the dosages (X_j's) are equispaced. When this is true, one can code the X_j's by considering their mean $\bar{X}_{..}$ and the width of the interval between them c. Let

$$u_j = \frac{X_j - \bar{X}_{..}}{c} = \frac{X_j - 1.5}{0.5}$$

and our model becomes

$$Y'_u = b'_0 + b'_1 u \tag{7.6}$$

whose least squares normal equations are given by

$$\left. \begin{array}{l} \sum_i \sum_j Y_{ij} = b'_0 N + b'_1 n \sum_j u_j \\[2mm] \sum_i \sum_j u_j Y_{ij} = b'_0 n \sum_j u_j + b'_1 n \sum_j u_j^2 \end{array} \right\} \tag{7.7}$$

By equispacing the X_j's, the u_j's are $-2, -1, 0, 1, 2$, and $\sum_j u_j = 0$. Since this is always the case, Equations (7.7) become

$$\left. \begin{array}{l} b'_0 = \dfrac{\sum_i \sum_j Y_{ij}}{N} = \bar{Y}_{..} \\[4mm] b'_1 = \dfrac{\sum_i \sum_j u_j Y_{ij}}{n \sum_j u_j^2} = \dfrac{\sum_j u_j T_{.j}}{n \sum_j u_j^2} \end{array} \right\} \tag{7.8}$$

and they are relatively easy to solve. Note, too, that since $\sum_j u_j = 0$, the expression $\sum_j u_j T_{.j}$ is a contrast and its sum of squares is $(\sum_j u_j T_{.j})^2 / n \sum_j u_j^2$. For our data

$$b'_0 = \bar{Y}_{..} = 467/15 = 31.133$$

$$b'_0 = \frac{-2(83) - 1(84) + 0(89) + 1(96) + 2(115)}{3(10)} = \frac{76}{30} = 2.533$$

and

$$Y'_u = 31.13 + 2.53 u_j$$

That this is the same model, substitute for u_j and

$$Y'_X = 31.133 + 2.533(X_j - 1.5)/0.5$$
$$= 23.53 + 5.07X_j$$

as before. Note also that the sum of squares due to linear regression can be found from the linear contrast

$$SS_{\substack{\text{due to linear} \\ \text{regression}}} = \frac{(76)^2}{3(10)} = 192.53$$

as shown in Table 7.4.

7.3 CURVILINEAR REGRESSION

When a linear regression model is not sufficient to explain all the significant variation in the means, the next logical step is to consider a second-degree (or quadratic) model:

$$Y'_X = b_0 + b_1 X_j + b_2 X_j^2 \tag{7.9}$$

By the methods of least squares, the normal equations are

$$\sum_i \sum_j Y_{ij} = b_0 N + b_1 n \sum_j X_j + b_2 n \sum_j X_j^2$$

$$\sum_i \sum_j X_j Y_{ij} = b_0 n \sum_j X_j + b_1 n \sum_j X_j^2 + b_2 n \sum_j X_j^3 \tag{7.10}$$

$$\sum_i \sum_j X_j^2 Y_{ij} = b_0 n \sum_j X_j^2 + b_1 n \sum_j X_j^3 + b_2 n \sum_j X_j^4$$

These equations can be solved for b_0, b_1, and b_2, but it is difficult. Again coded by u_j's,

$$Y'_u = b'_0 + b'_1 u_j + b'_2 u_j^2 \tag{7.11}$$

$$\sum_i \sum_j Y_{ij} = b'_0 N + b'_1 n \sum_j u_j + b'_2 n \sum_j u_j^2$$

$$\sum_i \sum_j u_j Y_{ij} = b'_0 n \sum_j u_j + b'_1 n \sum_j u_j^2 + b'_2 n \sum_j u_j^3 \tag{7.12}$$

$$\sum_i \sum_j u_j^2 Y_{ij} = b'_0 n \sum_j u_j^2 + b'_1 n \sum_j u_j^3 + b'_2 n \sum_j u_j^4$$

Because of the choice of the u_j's, the sums of all odd powers of the u_j's are zero ($\sum u_j = \sum u_j^3 = 0$). The equations become

$$\sum_i \sum_j Y_{ij} = b'_0 N + b'_2 n \sum_j u_j^2$$

$$\sum_i \sum_j u_j Y_{ij} = b'_1 n \sum_j u_j^2 \tag{7.13}$$

$$\sum_i \sum_j u_j^2 Y_{ij} = b'_0 n \sum_j u_j^2 + b'_2 n \sum_j u_j^4$$

For the data of Example 7.1, the middle equation of (7.13) gives

$$b_1' = \frac{\sum_i \sum_j u_j Y_{ij}}{n \sum_j u_j^2} = \frac{76}{30} = 2.53$$

The other two equations of (7.13) give:

$$467 = 15b_0' + 30b_1'$$
$$972 = 30b_0' + 102b_1'$$

since

$$\sum_i \sum_j u_j^2 Y_{ij} = \sum_j u_j^2 T_{.j} = 4(83) + 1(84) + 0(89) + 1(96) + 4(115) = 972$$

and

$$\sum_j u_j^4 = (-2)^4 + (-1)^4 + (0)^4 + (1)^4 + (2)^4 = 34$$

and

$$n \sum_j u_j^4 = 102$$

Multiplying the first equation by 2 and subtracting it from the second:

$$972 = 30b_0' + 102b_2'$$
$$934 = 30b_0' + 60b_2'$$
$$38 = 42b_2'$$

and

$$b_2' = \frac{38}{42} = 0.90$$

and

$$b_0' = \frac{467 - (30)(0.90)}{15} = 29.33$$

and the quadratic model in terms of u_j is

$$Y_u' = 29.33 + 2.53u_j + 0.90u_j^2$$

Table 7.5　Departures from Quadratic

X_j	u_j	$\bar{Y}_{.j}$	Y_X'	$\bar{Y}_{.j} - Y_X'$	$(\bar{Y}_{.j} - Y_X')^2$
0.5	−2	27.67	27.87	−0.20	0.040
1.0	−1	28.00	27.70	0.30	0.090
1.5	0	29.67	29.33	0.34	0.116
2.0	1	32.00	32.76	−0.76	0.578
2.5	2	38.33	37.99	0.34	0.116
Totals				$0.02 \approx 0$	0.940

One need not write this in terms of X_j to see how well it predicts the Y's from the X's. Table 7.5 shows the predicted Y's and the departure of the means from the quadratic model.

From Table 7.5 the sum of the squares of the departure from the quadratic is

$$SS_{departure} = 3(0.94) = 2.82$$

and this step can be added to refine Table 7.4 further and give Table 7.6. Now the departure from the quadratic is not significant at the 5 percent level so it is appropriate to stop with the quadratic for predicting reaction time from dosage. One may also note from Table 7.6 that linear and quadratic terms together account for $192.53 + 34.38 = 226.91$ of the total sum of squares in Y of 249.73. This is $226.91/249.73 = 0.9086$ or approximately 91 percent of the variability accounted for. The square root of this statistic is sometimes referred to as a *correlation ratio* R and $R^2 = 0.91$. The error and nonsignificant departure term could be 'pooled' to give a standard error of estimate for the quadratic

$$s_{Y.X, X^2} = \sqrt{\frac{20.00 + 2.82}{10 + 2}} = 1.38$$

which might be used to set confidence limits on the Y's around the quadratic curve.

Table 7.6   ANOVA with Quadratic Regression

Source	df	SS	MS	F	Significance
Between dosages	4	229.73			
Linear	1	192.53	192.53	96.3***	0.000
Quadratic	1	34.38	34.38	17.2**	0.002
Departure	2	2.82	1.41	. <1	0.517
Error	10	20.00	2.00		
Totals	14	249.73			

7.4   ORTHOGONAL POLYNOMIALS

In the above discussions on linear and quadratic regression it was noted that the problem could be solved much more easily if the X_j values were equi-spaced and coded such that $\sum_j u_j = \sum_j u_j^3 = 0$. The advantage of this type of curve fitting is even greater as the degree of the polynomial increases. If

$$Y'_x = f(X, X^2, X^3, \ldots, X^k)$$

is a polynomial in X, it has been shown that it may be rewritten in the form

$$Y'_u = A_0\xi'_0 + A_1\xi'_1 + A_2\xi'_2 + \cdots + A_k\xi'_k \qquad (7.14)$$

where each ξ'_j is a polynomial of degree j and all polynomials, such as ξ'_m and ξ'_q, are orthogonal to each other. The advantage of writing the model in this form is that additional polynomials of higher degree may be added which are independent of the ones already considered. The values of these polynomials are given for the first five powers of u as

$$\xi'_0 = 1$$

$$\xi'_1 = \lambda_1 u$$

$$\xi'_2 = \lambda_2\left[u^2 - \frac{k^2 - 1}{12}\right]$$

$$\xi'_3 = \lambda_3\left[u^3 - u\left(\frac{3k^2 - 7}{20}\right)\right] \qquad (7.15)$$

$$\xi'_4 = \lambda_4\left[u^4 - \frac{u^2}{14}(3k^2 - 13) + \frac{3}{560}(k^2 - 1)(k^2 - 9)\right]$$

$$\xi'_5 = \lambda_5\left[u^5 - \frac{5u^3}{18}(k^2 - 7) + \frac{u}{1008}(15k^4 - 230k^2 + 407)\right]$$

where k is the number of levels of the factor X, and the λ's are chosen so that the ξ''_j's are integers for all u. Because of the complications of these formulas and the practical need to test for significant high-order regression models, the values of the ξ''_j's and λ's have been tabled by Fisher and Yates [11], and a section is reproduced as Appendix Table F.

The A_j's are given by

$$A_j = \frac{\sum_{i,j} Y_{ij}\xi'_j}{\sum_{i,j} (\xi'_j)^2} \qquad (7.16)$$

To see how these formulas and tables apply, consider a quadratic model with just three points on the X scale. The u_j's would be $-1, 0, +1$. As these are already integers, $\lambda_1 = 1$ and ξ'_1 takes on values $-1, 0, +1$ for each j. For ξ'_2, $(k^2 - 1)/12 = (9 - 1)/12 = 2/3$ and the u^2 values are $+1, 0, +1$, which makes $[u^2 - (k^2 - 1)/12] = 1/3, -2/3, 1/3$ so λ_2 is taken as equal to 3 in order to make the ξ'_2 integers. Then the ξ'_2 are $+1, -2, +1$. Appendix Table F gives these values for $k = 3$. Also given in the table are $\sum (\xi'_j)^2$ for each set of ξ'_j and the λ_j. In the expression for A_j, each Y_{ij} is multiplied by the corresponding ξ'_j, and since $\sum_j \xi'_j = 0$, this expression is a contrast. Since the coefficients $-1, 0, +1$ are orthogonal to $1, -2, 1$, these provide two orthogonal contrasts on the Y_{ij}'s (or their totals, $T_{.j}$) and the sum of squares for such contrasts is given by

$$SS_C = \frac{(\sum_i \sum_j Y_{ij} \, \xi'_j)^2}{n \sum_j (\xi'_j)^2}$$

One needs only to apply the coefficients given in Appendix Table F to the treatment totals of a single-factor problem to determine the sum of squares for linear, quadratic, cubic, quartic, and so on, effects up to the $(k-1)$th-degree equation. Equations (7.14), (7.15), and (7.16) are only needed if the prediction model $Y'_u = f(u, u^2, u^3, \ldots, u^{k-1})$ is desired.

To illustrate this, consider the example of Sections 7.2 and 7.3. Since $k = 5$ dosage levels, Table F gives the following coefficients for the linear, quadratic, cubic, and quartic effects.

$T_{.j}$	83	84	89	96	115	$\sum_j (\xi'_j)^2$	λ_j	Contrast	SS
Linear	-2	-1	0	1	2	10	1	76	192.53
Quadratic	2	-1	-2	-1	2	14	1	38	34.38
Cubic	-1	2	0	-2	1	10	5/6	8	2.13
Quartic	1	-4	6	-4	1	70	35/12	12	0.69
								Total	229.73

Applying each set of coefficients to the treatment totals gives the contrasts as listed and their corresponding sum of squares. For example, the cubic contrast is

$$-1(83) + 2(84) + 0(89) - 2(96) + 1(115) = 8$$

and its sum of squares

$$SS_{cubic} = (8)^2/3(10) = 2.13$$

as shown. The complete ANOVA can now be written as in Table 7.7. The F column again shows that there are no significant effects after the quadratic and the quadratic should be adequate for prediction purposes.

If the equation is desired, one needs to find the values of ξ'_j and A_j for each degree of the polynomial desired. For our data note that

$$A_j = \frac{\sum_i \sum_j Y_{ij} \xi'_j}{\sum_i \sum_j (\xi'_j)^2} = \frac{\sum_j T_{.j} \xi'_j}{n \sum_j (\xi'_j)^2} = \frac{\text{Contrast}}{n \sum_{j'} (\xi'_j)^2}$$

This expression is very similar to the expression for the sum of squares of a contrast, except that the contrast is not squared and it will retain its sign.

Table 7.7 ANOVA for Polynomial Model

Source	df	SS	MS	F	Significance
Between dosages	4	229.73			
Linear	1	192.53	192.53	96.3	0.000
Quadratic	1	34.38	34.38	17.2	0.002
Cubic	1	2.13	2.13	1.1	0.325
Quartic	1	0.69	0.69	<1	0.570
Error	10	20.00	2.00		
Totals	14	249.73			

The A_j's and ξ_j's in terms of u are then

j	A_j	ξ_j'
0	$A_0 = \bar{Y}_{..} = 31.13$	$\xi_0' = 1$

$$1 \quad A_1 = \frac{76}{3(10)} = 2.53 \quad \xi_1' = 1u$$

$$2 \quad A_2 = \frac{38}{3(14)} = 0.90 \quad \xi_2' = 1(u^2 - 2)$$

$$3 \quad A_3 = \frac{8}{3(10)} = 0.27 \quad \xi_3' = 5/6(u^3 - 3.4u)$$

$$4 \quad A_4 = \frac{12}{3(70)} = 0.06 \quad \xi_4' = 35/12\left[u^4 - \frac{u^2}{14}(62) + \frac{3}{560}(24)(16)\right]$$

If the A_j's and ξ_j's are multiplied together, they give a polynomial in u to as high a degree as desired up to the $(k-1)$th degree. Hence the best fitting straight line is

$$Y_u' = A_0\xi_0' + A_1\xi_1'$$
$$= 31.13 + 2.53u$$

as given in Section 7.2.
    For a quadratic

$$Y_u' = A_0\xi_0' + A_1\xi_1' + A_2\xi_2'$$
$$= 31.13 + 2.53u + 0.90(u^2 - 2)$$

which simplifies to

$$Y_u' = 29.33 + 2.53u + 0.90u^2$$

as given in Section 7.3.

One has only to add more terms as each polynomial in u, ξ'_j is orthogonal to those terms already in the model. Continuing to the quartic, even though unnecessary, we find

$$Y'_u = 31.13 + 2.53u + 0.90(u^2 - 2) + (0.27)5/6(u^3 - 3.4u)$$
$$+ 0.06(35/12)[u^4 - 62/14u^2 + (3/560)(24)(16)]$$

which simplifies to

$$Y'_u = 29.69 + 1.76u + 0.12u^2 + 0.22u^3 + 0.18u^4$$

"Fitting" this quartic polynomial gives Table 7.8. This table shows the quartic to go through the five means except for slight rounding errors.

Table 7.8 Departures from Quartic Model

X_j	u_j	$\bar{Y}_{.j}$	Y'_u
0.5	-2	27.67	27.77
1.0	-1	28.00	28.01
1.5	0	29.67	29.69
2.0	1	32.00	31.97
2.5	2	38.33	38.33

This method of orthogonal polynomials is then very useful in finding the degree of the prediction model needed to "fit" a set of data, provided the X'_j's are equispaced. One usually stops after the highest degree polynomial that shows significance and then writes its equation. In Example 7.1 it would be the quadratic or second-degree equation.

Some readers may wonder how the coefficients in Table F were arrived at. Without attempting any derivation, let us consider three equispaced X's and the corresponding three totals: $T_{.j}$'s as shown in Figure 7.2. From Figure 7.2, as one increases X from X_1 to X_3 the increase in $\bar{Y}_{.j}$ is reflected

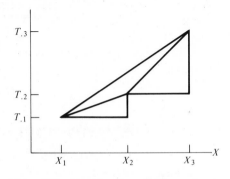

Figure 7.2 Linear and quadratic contrasts.

in the increase in $T_{.j}$ or $T_{.3} - T_{.1}$, which can be considered a linear contrast. The linear effect from X_1 to X_2 is $T_{.2} - T_{.1}$ and from X_2 to X_3 is $T_{.3} - T_{.2}$; so the cumulative or average linear effect is

$$T_{.3} - T_{.2} + T_{.2} - T_{.1} = T_{.3} - T_{.1}$$

If, however, the response curve does not have the same slope from 1 to 2 as from 2 to 3, the slope changes and there is a quadratic effect present. This is measured by a difference in the two linear effects, if they do differ. Hence the quadratic effect is

$$(T_{.3} - T_{.2}) - (T_{.2} - T_{.1})$$

or

$$T_{.3} - 2T_{.2} + T_{.1}$$

which is a contrast and which is also orthogonal to the linear contrast $T_{.3} - T_{.1}$. We note that these are the same coefficients as given in Table F for $k = 3$.

If there are four points X_1, X_2, X_3, and X_4, a cubic effect is present if the quadratic from 1 to 3 is different from the quadratic from 2 to 4. Thus

$$(T_{.4} - 2T_{.3} + T_{.2}) - (T_{.3} - 2T_{.2} + T_{.1})$$

or

$$T_{.4} - 3T_{.3} + 3T_{.2} - T_{.1}$$

whose coefficients of -1, 3, -3, and 1 agree with Table F for $k = 4$. This reasoning may give some notion as to the reasonableness of the orthogonal polynomial coefficients.

7.5   COMPUTER PROGRAM

As in previous chapters a computer program could be used to analyze the results of a problem involving quantitative levels of a factor. For Example 7.1, in addition to asking for a ONEWAY ANOVA of RT (reaction time) by DOS (dosage) in Table 7.9, on the next line one asks for POLYNOMIAL = 4 for the four possible curves that might fit the data. After entering the 15 data points (steps 13 through 27) the printout is given as in Table 7.10.

These results are seen to be the same as in Table 7.7 if one notes that the deviation from cubic is the quartic effect.

7.6   QUANTITATIVE AND QUALITATIVE FACTORS

In Sections 7.1 to 7.5 a single factor was considered at quantitative levels only. Qualitative levels of a single factor were discussed in Chapter 3 and for two or more factors in Chapter 5. When two or more factors are involved in an experiment, some may be at qualitative levels and some at quantitative levels. Two factors are considered in the sections that follow.

7.7 TWO FACTORS—ONE QUALITATIVE, ONE QUANTITATIVE

A problem will now be considered where one factor is set at qualitative levels and the other at quantitative (and equispaced) levels.

Example 7.2 We wish to determine the effect of both depth and position in a tank (Figure 7.3) on the concentration of a cleaning solution in ounces per gallon. Concentrations are measured at three depths from the surface of the tank, 0 inch, 15 inches, and 30 inches. At each depth measurements

Table 7.9 Computer Input for Data of Table 7.1

```
+ + +CREATE PROB8
         1. 000 = 77509, I13
         2. 000 = COMMON (SPSS)
         3. 000 = SPSS
         4. 000 = #EOR
         5. 000 = RUN NAME            STAT 502 PROB8
         6. 000 = VARIABLE LIST       RT, DOS
         7. 000 = INPUT FORMAT        (F2.0, 1X, F1.0)
         8. 000 = N OF CASES          15
         9. 000 = ONE WAY             RT BY DOS (1, 5)/
        10. 000 =                     POLYNOMIAL = 4/
        11. 000 = STATISTICS          ALL
        12. 000 = READ INPUT DATA
        13. 000 = 26  1
        14. 000 = 28  1
        15. 000 = 29  1
        16. 000 = 28  2
        17. 000 = 26  2
        18. 000 = 30  2
        19. 000 = 28  3
        20. 000 = 30  3
        21. 000 = 31  3
        22. 000 = 32  4
        23. 000 = 33  4
        24. 000 = 31  4
        25. 000 = 38  5
        26. 000 = 39  5
        27. 000 = 38  5
        28. 000 = FINISH
```

Table 7.10 ANOVA Printout for Example 7.1

| SOURCE | D.F. | ANALYSIS OF VARIANCE | | F RATIO | F PROB. |
		SUM OF SQUARES	MEAN SQUARES		
BETWEEN GROUPS	4	229.7333	57.4333	28.717	.0000
LINEAR TERM	1	192.5333	192.5333	96.267	.0000
DEV. FROM LINEAR	3	37.2000	12.4000	6.200	.0119
QUAD. TERM	1	34.3810	34.3810	17.190	.0020
DEV. FROM QUAD.	2	2.8190	1.4095	.705	.5172
CUBIC TERM	1	2.1333	2.1333	1.067	.3260
DEV. FROM CUBIC	1	.6857	.6857	.343	.5712
WITHIN GROUPS	10	20.0000	2.0000		
TOTAL	14	249.7333			

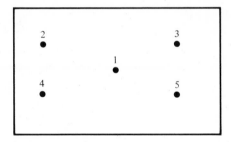

Figure 7.3   Positions in tank at each depth.

are taken at five different lateral positions in the tank. These are considered as five qualitative positions, although probably some orientation measure might be made on them. At each depth and position, two observations are taken. This is then a 5×3 factorial with two replications per cell (total of 30 observations). The design is a completely randomized design and the model is

$$Y_{ijk} = \mu + D_i + P_j + DP_{ij} + \varepsilon_{k(ij)}$$

with

$$i = 1, 2, 3 \qquad j = 1, 2, \ldots, 5 \qquad k = 1, 2$$

where D_i represents the depth and P_j the position. The data collected are shown in Table 7.11.

In this case where one factor is set at quantitative levels (D_i) and the other at qualitative levels (P_j), the first step is to run a two-factor analysis of variance just as if both were qualitative factors. Coding the data of Table 7.11 by subtracting 5.90 and multiplying by 100 gives Table 7.12.

Table 7.11   Cleaning-Solution Concentration Data

	Depth from Top of Tank D_i		
Position P_j	0 in.	15 in.	30 in.
1	5.90	5.90	5.94
	5.91	5.89	5.80
2	5.90	5.89	5.75
	5.91	5.89	5.83
3	5.94	5.91	5.86
	5.90	5.91	5.83
4	5.93	5.94	5.83
	5.91	5.90	5.89
5	5.90	5.94	5.83
	5.87	5.90	5.86

From these coded data,

$$SS_{total} = 644 - \frac{(-44)^2}{30} = 579.47$$

$$SS_D = \frac{(7)^2 + (7)^2 + (-58)^2}{10} - \frac{(-44)^2}{30} = 281.67$$

$$SS_P = \frac{(-6)^2 + (-23)^2 + (-5)^2 + (0)^2 + (-10)^2}{6}$$

$$- \frac{(-44)^2}{30} = 50.47$$

$$SS_{D \times P \text{ interaction}} = \frac{(1)^2 + (1)^2 + (4)^2 + \cdots + (-11)^2}{2} - \frac{(-44)^2}{30}$$

$$- 281.67 - 50.47 = 58.33$$

$$SS_{error} = 579.47 - 281.67 - 50.47 - 58.33 = 189.00$$

giving the ANOVA table of Table 7.13.

Table 7.12 Coded Cleaning-Solution Concentration Data with Totals

Position P_j	Depth from Top of Tank D_i			$T_{.j.}$
	0 in.	15 in.	30 in.	
1	0	0	4	
	1	−1	−10	
	1	−1	−6	−6
2	0	−1	−15	
	1	−1	−7	
	1	−2	−22	−23
3	4	1	−4	
	0	1	−7	
	4	2	−11	−5
4	3	4	−7	
	1	0	−1	
	4	4	−8	0
5	0	4	−7	
	−3	0	−4	
	−3	4	−11	−10
$T_{i..}$	7	7	−58	$T_{...} = -44$
$\sum\limits_{k=1}^{2} \sum\limits_{j=1}^{5} Y_{ijk}^2$	37	37	570	$\sum\limits_{i}^{3} \sum\limits_{j}^{5} \sum\limits_{k}^{2} Y_{ijk}^2 = 644$

Table 7.13 ANOVA for Cleaning-Solution Concentration Problem

Source	df	SS	MS
Depths D_i	2	281.67	140.83
Positions P_j	4	50.47	12.62
$D \times P$ interaction DP_{ij}	8	58.33	7.29
Error $\varepsilon_{k(ij)}$	15	189.00	12.60
Totals	29	579.47	

From Table 7.13 it is seen that only depth produced a significant effect on concentration of solution as

$$F_{2,15} = \frac{140.83}{12.60} = 11.24$$

which is significant at the 1 percent level of significance. Since the depth effect is significant and since the three depths are equispaced, it may be worthwhile to extract a linear and quadratic depth effect to learn how concentration varies with depth. The coefficients for $k = 3$ levels are shown in Table 7.14.

Applying these coefficients to the depth totals $T_{i..}$ gives

$$\text{linear effect of depth} = -1(7) + 0(7) + 1(-58) = -65$$
$$\text{quadratic effect of depth} = +1(7) - 2(7) + 1(-58) = -65$$

Since these are orthogonal contrasts, their sums of squares are

$$\text{SS}_{\text{linear}} = \frac{(-65)^2}{10(2)} = 211.25$$

$$\text{SS}_{\text{quadratic}} = \frac{(-65)^2}{10(6)} = \frac{70.42}{281.67}$$

and their total is the depth sum of squares.

Even though the $D \times P$ interaction is not significant, there may be an interaction between the linear effect of depth and positions or between the

Table 7.14 Orthogonal Coefficients

				$\sum (\xi'_j)^2$	λ
Linear	-1	0	$+1$	2	1
Quadratic	$+1$	-2	$+1$	6	3
$T_{i..}$	7	7	-58		

quadratic effect of depth and positions. To compute these interactions, the linear effect of depth is determined at each position and these effects are then compared to see whether or not they differ. The same procedure is followed for the quadratic effect of depth at each position.

Applying the linear coefficients of $-1, 0, +1$ at each position gives

$$P_1: -1(1) + 0(-1) + 1(-6) \; = \; -7$$
$$P_2: -1(1) + 0(-2) + 1(-22) = -23$$
$$P_3: -1(4) + 0(2) + 1(-11) \;\; = -15$$
$$P_4: -1(4) + 0(4) + 1(-8) \;\;\; = -12$$
$$P_5: -1(-3) + 0(4) + 1(-11) = \;\; -8$$
$$\overline{-65}$$

To compare these five effects, determine the sum of squares between them

$$\frac{(-7)^2 + (-23)^2 + (-15)^2 + (-12)^2 + (-8)^2}{2(2)} - \frac{(-65)^2}{10(2)} = 41.50$$

Similarly, for the quadratic effect of depth at the five positions, we have

$$P_1: +1(1) - 2(-1) + 1(-6) \; = \; -3$$
$$P_2: +1(1) - 2(-2) + 1(-22) = -17$$
$$P_3: +1(4) - 2(2) + 1(-11) \;\; = -11$$
$$P_4: +1(4) - 2(4) + 1(-8) \;\;\; = -12$$
$$P_5: +1(-3) - 2(4) + 1(-11) = -22$$
$$\overline{-65}$$

Comparing these five quadratic effects gives

$$\frac{(-3)^2 + (-17)^2 + (-11)^2 + (-12)^2 + (-22)^2}{2(6)} - \frac{(-65)^2}{10(6)} = 16.83$$

Note that the sum of these two sums of squares $(41.50 + 16.83)$ equals the $D \times P$ interaction sum of squares (58.33) as it should. Summarizing for this problem with one quantitative and one qualitative factor, the complete ANOVA breakdown is as shown in Table 7.15.

The results in Table 7.15 indicate a strong depth effect, with the linear depth effect significant at the 1 percent level (**) and the quadratic depth effect significant at the 5 percent level (*). There is no position nor interaction between depth and position. These results seem reasonable from a graph of cell totals versus depth and positions (Figure 7.4).

This graph shows little interaction, as the curves are quite parallel (statistically speaking) and there is little difference between the five position curves. The depth effect is quite obvious, and concentration is seen to drop

Table 7.15   Complete ANOVA for Cleaning-Solution
Concentration Problem

Source	df	SS	MS
Depths	2	281.67	
Linear	1	211.25	211.25**
Quadratic	1	70.42	70.42*
Positions	4	50.47	12.62
$D \times P$ interaction	8	58.33	
$D_{linear} \times P$	4	41.50	10.38
$D_{quadratic} \times P$	4	16.83	4.21
Error	14	189.00	12.60
Totals	29	579.47	

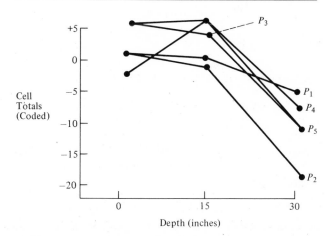

Figure 7.4   Graph of cleaning-solution concentration problem.

off with increasing depth but on more of a curve than a straight line. Further
investigation is suggested at depths between those already studied. The lack
of any position effect or interaction should mean that this new experiment
could be run at only one position; the results should then be the same at all
five positions.

If one wishes the equation for predicting concentration (coded) from
depth,

$$u = \frac{X - 15}{15}$$

and

$$A_0 = \bar{Y}_{..} = -44/30 = -1.47$$
$$A_1 = -65/10(2) = -3.25$$
$$A_2 = -65/10(6) = -1.08$$

and

$$\xi_0' = 1$$
$$\xi_1' = 1u$$
$$\xi_2' = 3(u^2 - 2/3) = 3u^2 - 2$$

and

$$Y_u' = -1.47 - 3.25u - 1.08(3u^2 - 2)$$
$$= 0.69 - 3.25u - 3.24u^2$$

giving predictions as follows:

X_j	u_j	Y_u'	$\bar{Y}.$
0	−1	0.70	0.70
15	0	0.69	0.70
30	1	−5.80	−5.80

which match exactly as they should.

7.8   TWO FACTORS—BOTH QUANTITATIVE

If both factors in a two-factor factorial are at quantitative levels, each factor can be broken down into its linear, quadratic, or cubic effects, and all combinations of interaction can be determined, such as linear by linear, linear by quadratic, and quadratic by quadratic. To illustrate the procedure for analyzing a factorial experiment where both factors are at quantitative (and equispaced) levels, consider Example 7.3.

Example 7.3   Data were collected in a completely randomized design on the effect of lacquer concentrations at four equispaced levels and standing times at three equispaced levels on screen quality as indicated on a 20-point scale. Two observations were made for each of the $4 \times 3 = 12$ treatment combinations. The resulting data are shown in Table 7.16.

Table 7.16   Screen Quality Data

Standing Time	Lacquer Concentration			
	$\frac{1}{2}$	1	$1\frac{1}{2}$	2
30	16	12	17	13
	14	11	19	11
20	15	14	15	12
	15	17	18	14
10	10	7	10	9
	9	6	14	13

The ANOVA model is

$$Y_{ijk} = \mu + L_i + T_j + LT_{ij} + \varepsilon_{k(ij)}$$
$$i = 1, 2, 3, 4 \qquad j = 1, 2, 3 \qquad k = 1, 2$$

and the analysis can be shown to be as given in Table 7.17. These results indicate strong main effects—lacquer concentration and standing time—but no significant interaction. One could then extract linear, quadratic, and cubic effects for the lacquer concentrations and linear and quadratic effects for the standing times as two independent factors. However, since some examples may have significant quantitative by quantitative interactions, we will go through the steps to partition the interaction as well as the main effects into component parts. For lacquer concentrations there are four levels, or $k = 4$ in Table F. To apply these coefficients one needs the four lacquer totals. These are seen, along with other useful totals, in Table 7.18 based on Table 7.16.

Table 7.17   Screen Quality ANOVA

Source	df	SS	MS	F	Probability
Lacquer (L_i)	3	63.79	21.26	7.38	0.005
Time (T_j)	2	126.58	63.29	21.98	0.000
$L \times T$ interaction	6	38.07	6.35	2.20	0.115
Error	12	34.51	2.88		
Totals	23	262.95			

Table 7.18   Screen Quality Data with Totals

Standing Times	Lacquer Concentration								$T_{.j}$
	$\frac{1}{2}$		1		$1\frac{1}{2}$		2		
30	16		12		17		13		
	14		11		19		11		
		30		23		36		24	113
20	15		14		15		12		
	15		17		18		14		
		30		31		33		26	120
10	10		7		10		9		
	9		6		14		13		
		19		13		24		22	78
$T_{i.}$		79		67		93		72	$311 = T_{..}$

Applying the orthogonal coefficients gives the following.

$T_{i.}$	79	67	93	72	$(\xi_j')^2$	λ_j	Contrast	SS
Linear	-3	-1	1	3	20	2	5	0.21
Quadratic	1	-1	-1	1	4	1	-9	3.37
Cubic	-1	3	-3	1	20	10/3	-85	60.21
							Total $SS_L =$	63.79

Applying appropriate orthogonal coefficients to the standing time totals (here $k = 3$) gives the following.

$T_{.j}$	78	120	113	$(\xi_j')^2$	λ_j	Contrast	SS
Linear	-1	0	1	2	1	35	76.56
Quadratic	1	-2	1	6	3	-49	50.02
						Total $SS_T =$	126.58

These results partition the main effects into components with 1 df each and they can be tested for significance. To break down the interaction term, note that there are six single-degree-of-freedom terms

$$L_L \times T_L \qquad L_Q \times T_L \qquad L_C \times T_L$$
$$L_L \times T_Q \qquad L_Q \times T_Q \qquad L_C \times T_Q$$

where subscripts L, Q, and C represent linear, quadratic, and cubic respectively. A simple computing scheme is to apply the product of the main effect coefficients to the treatment totals as shown in Table 7.19 for linear by linear, $L_L \times T_L$, interaction.

Table 7.19   Linear by Linear Interaction

		L_L			
		-3	-1	1	3
	1	-3	-1	1	3
		30	23	36	24
T_L	0	0	0	0	0
		30	31	33	26
	-1	3	1	-1	-3
		19	13	24	22

In Table 7.19 the number in the upper-left-hand corner of each cell is the product of the coefficients of the main effects for its row and column. The number in the right-hand corner is the cell total of $T_{ij.}$. Applying the coefficients to these $T_{ij.}$'s gives the $L_L \times T_L$ contrast as follows:

$$-3(30) - 1(23) + 1(36) + 3(24) + 0(30) + 0(31) + 0(33)$$
$$+ 0(26) + 3(19) + 1(13) - 1(24) - 3(22) = -25$$

and

$$SS_{L_L \times T_L} = \frac{(-25)^2}{2(20)(2)} = 7.81$$

The sum of squares of coefficients in the denominator is the product of the two $(\xi_j')^2$ from the two factors.

Table 7.20   Linear by Quadratic Interaction

		L_L						
		-3		-1		1		3
	1	-3		-1		1		3
			30		23		36	24
T_Q -2		6		2		-2		-6
			30		31		33	26
	1	-3		-1		1		3
			19		13		24	22

To find the $L_L \times T_Q$ use the same procedure with quadratic coefficients on the standing times as shown in Table 7.20. The $L_L \times T_Q$ contrast is now

$$-3(30) - 1(23) + 1(36) + 3(24) + 6(30) + 2(31) - 2(33)$$
$$- 6(26) - 3(19) - 1(13) + 1(24) + 3(22) = 35$$

and

$$SS_{L_L \times T_Q} = \frac{(35)^2}{2(20)(6)} = 5.10$$

If this same procedure is now applied to $L_Q \times T_L$, $L_Q \times T_Q$, $L_C \times T_L$, and $L_C \times T_Q$, the resulting sums of squares are

$$SS_{L_Q \times T_L} = 5.06$$
$$SS_{L_Q \times T_Q} = 4.69$$
$$SS_{L_C \times T_L} = 2.81$$
$$SS_{L_C \times T_Q} = 12.60$$

Table 7.21 Complete ANOVA for Screen Quality Example

Source	df	SS	MS	F	Probability
Lacquer	3	63.79			
Linear	1	0.21	0.21	<1	0.791
Quadratic	1	3.37	3.37	1.17	0.301
Cubic	1	60.21	60.21	20.91	0.001
Time	2	126.58			
Linear	1	76.56	76.56	26.58	0.000
Quadratic	1	50.02	50.02	17.37	0.001
$L \times T$ interaction	6	38.07			
$L_L \times T_L$	1	7.81	7.81	2.71	0.125
$L_L \times T_Q$	1	5.10	5.10	1.77	0.208
$L_Q \times T_L$	1	5.06	5.06	1.77	0.208
$L_Q \times T_Q$	1	4.69	4.69	1.59	0.231
$L_C \times T_L$	1	2.81	2.81	<1	0.343
$L_C \times T_Q$	1	12.60	12.60	4.38	0.058
Error	12	34.51	2.88		
Totals	23	262.95			

Adding the six components gives 38.07, the interaction sum of squares reported in Table 7.17. The complete breakdown and tests of significance can be summarized as in Table 7.21. This table shows three significant effects: a cubic effect of lacquer concentration on screen quality, and a linear and quadratic effect of standing time on screen quality. Since there is no significant interaction one might plot the mean responses to each main effect as in Figure 7.5.

Because the two main effects do not interact one might find an equation to predict the lacquer means (a cubic) and another equation to predict standing time means (a quadratic). These two equations could be combined to predict the cell or treatment means. Use Equation 7.15 and the contrasts found in the data to give

For lacquer ($k = 4$):

$$
\begin{array}{ll}
A_j\text{'s} & \xi'_j \\
A_0 = \bar{Y}_{..} = 311/24 = \quad 12.9583 & \xi'_0 = 1 \\
A_1 = 5/6(20) \quad\quad = \quad\; 0.0417 & \xi'_1 = 2u \\
A_2 = -9/6(4) \quad\;\; = -0.3750 & \xi'_2 = 1(u^2 - 5/4) \\
A_3 = -85/6(20) \;\; = -0.7083 & \xi'_3 = 10/3(u^3 - 41/20u)
\end{array}
$$

and

$$Y'_u = 12.9583 + 0.0834u - 0.375(u^2 - 1.25) - 0.7083(10/3)(u^3 - 2.05u)$$

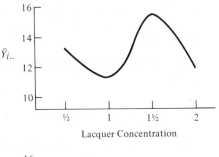

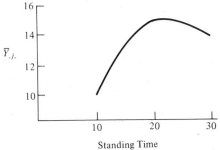

Figure 7.5   Screen quality averages versus lacquer concentration and standing times.

which simplifies to

$$Y'_u = 13.4270 + 4.9234u - 0.375u^2 - 2.361u^3$$

For standing times $(k = 3)$:

$$
\begin{array}{lll}
\qquad\qquad A_j\text{'s} & & \xi'_j \\[4pt]
A_0 = \bar{Y}_{..} & = 12.9583 & \xi'_0 = 1 \\
A_1 = 35/8(2) & = 2.1875 & \xi'_1 = 1u \\
A_2 = -49/8(6) & = -1.0208 & \xi'_2 = 3(u^2 - 2/3)
\end{array}
$$

and

$$Y'_u = 12.9583 + 2.1875u - 1.0208(3u^2 - 2)$$

or

$$Y'_u = 14.9999 + 2.1875u - 3.0624u^2$$

Now the lacquer concentration equation will have u values as coded for the lacquer data and the standing time equation for coded standing times. Calling u_L the lacquer variable

$$u_L = \frac{X_L - \bar{X}_L}{c_L} = \frac{X_L - 1.25}{0.5}$$

and u_T for standing time

$$u_T = \frac{X_T - \bar{X}_T}{c_T} = \frac{X_T - 20}{10}$$

As no interaction is assumed the two equations could be combined by addition, remembering, however, that the grand mean is in both equations. The equation for predicting the mean screen quality for any combination of L and T is then

$$Y' = 13.4270 + 4.9234u_L - 0.375u_L^2 - 2.361u_L^3 + 14.9999 + 2.1875u_T$$
$$- 3.0624u_T^2 - 12.9583$$

or

$$Y' = 15.4686 + 4.9234u_L - 0.375u_L^2 - 2.361u_L^3 + 2.1875u_T - 3.0624u_T^2$$

Table 7.22 Screen Quality Based on Prediction Equation

Standing Times		Lacquer Concentration			
		$X_L = \frac{1}{2}$	1	$1\frac{1}{2}$	2
X_T	u_T	$u_L = -\frac{3}{2}$	$-\frac{1}{2}$	$\frac{1}{2}$	$\frac{3}{2}$
30	1	14.33	12.33	16.67	13.17
		15.0	11.5	18.0	12.0
20	0	15.21	13.21	17.54	14.04
		15.0	15.5	16.5	13.0
10	−1	9.96	7.96	12.29	8.79
		9.5	6.5	12.0	11.0

To see how good a predictor this is, Table 7.22 gives the predicted value at the upper left of each cell using this equation and the observed mean of the cell at the lower right of that cell. If one now examines the departures of the cell mean in Table 7.22 from their predicted values, the sum of squares of the departures is

$$(15.0 - 14.33)^2 + (11.5 - 12.33)^2 + \cdots + (11.0 - 8.79)^2 = 19.04$$

Since each mean represents two observations

$$SS_{\text{departure}} = 2(19.04) = 38.08$$

which compares well with the interaction sum of squares of 38.07 reported in Table 7.17 or Table 7.21.

7.9   SUMMARY

The examples of this chapter may easily be extended to higher order factorials whenever one or more factors are considered at quantitative levels. The use of orthogonal polynomials makes the analysis rather simple, provided the experiment is designed with equispaced quantitative levels. The summary of designs at the end of Chapter 6 has not been changed by this chapter, as these methods can be used in all experiments where quantitative levels are involved.

Chapter 8 presents a more general method for handling many independent variables in determining their effect on some dependent variable Y.

PROBLEMS

7.1   For the following data on X and Y plot a scattergram and determine the least squares straight line of best fit.

		X		
3	4	5	6	7
7	8	10	11	10
Y 8	8	9	9	10
9	9	9	10	9

7.2   For Problem 7.1 present your results in an ANOVA table and comment on these results.

7.3   From your table in Problem 7.2 find r^2, η^2, and the standard error of estimate.

7.4   Use the method of orthogonal polynomials on the data of Problem 7.1 and find the best fitting polynomial.

7.5   An experiment to determine the effect of planting rate in thousands of plants per acre on yield in bushels per acre of corn gave the following results.

Planting Rate X	Yield Y
12	130.5, 129.6, 129.9
16	142.5, 140.3, 143.4
20	145.1, 144.8, 144.1
24	147.8, 146.6, 148.4
28	134.8, 135.1, 136.7

Plot a scattergram of these data and comment on what degree polynomial might prove appropriate for predicting yield from planting rate.

7.6　From the data of Problem 7.5 use orthogonal polynomials and determine what degree polynomial is appropriate here.

7.7　Find the equation of the polynomial in Problem 7.6.

7.8　From an ANOVA table of the results of Problem 7.6 find r^2, R^2, η^2 and comment on what these statistics tell you.

7.9　Seasonal indexes of hog prices taken in two different years for five months gave the following results.

	Month				
Year	1	2	3	4	5
1	93.9	94.3	101.2	96.7	107.5
2	95.4	96.5	94.7	97.1	107.2

a.　Sketch a scatterplot of these data and comment on what type of curve might be used to predict hog prices when the month is known.

b.　Do a complete analysis of these data and find the equation of the best fitting curve for predicting hog price from month.

7.10　A bakery is interested in the effect of six baking temperatures on the quality of its cakes. The bakers agree on a measure of cake quality, Y. A large batch of cake mix is prepared and 36 cakes poured. The baking temperature is assigned at random to the cakes such that six cakes are baked at each temperature. Results are as follows:

	Temperature				
175°	185°	195°	205°	215°	225°
22	4	10	22	32	20
10	21	22	14	27	36
18	18	18	21	22	33
26	28	32	25	37	33
21	21	28	26	27	20
21	28	25	25	31	25

Do a complete analysis of these data and write the 'best' fitting model for predicting cake quality from temperature.

7.11　Compute statistics that will indicate how good your fit is in Problem 7.10 and comment.

7.12 Data on the effect of age on retention of information gave

Ages:	6–14	15–23	24–32	33–41
Total score of ten people, $T_{.j}$:	80	92	103	90

Source	df	SS		MS
Between ages	3	26.675		8.892
Linear	1		8.405	8.405
Quadratic	1		15.625	15.625
Cubic	1		2.645	2.645
Error	36	72.000		2.000
Total	39	98.675		

a. Determine the significance of each prediction equation and state which degree equation (linear, quadratic, or cubic) will adequately predict retention score from age.
b. Write the equation for part (a).
c. Find the departure from regression of the mean retention score of a 37-year-old based on your answer to part (b).

7.13 A study of the effect of chronological age on history achievement scores gave the following results.

History Achievement Score Coded by $\dfrac{Y-18}{15}$	Chronological Age				
	8	9	10	11	12
5			1	6	5
4		1	7	4	5
3		6	2		
2	3	2			
1	6	1			
0	1				

Do a complete analysis justifying your curve of best fit.

7.14 In studying the effect of the distance a road sign stands from the edge of the road on the amount by which a driver swerves from the edge of the road, five distances were chosen. Then the five distances were set at random, observing four cars at each set distance. The results gave

$$\eta^2 = 0.70$$
$$r^2 = 0.61$$

a. Test whether or not a linear function will "fit" these data. (You may assume a total SS of 10,000 if you wish, but this is not necessary.)

b. In fitting the proper curve, what are you assuming about the four readings at each specified distance?

7.15 Thickness of a film is studied to determine the effect of two types of resin and three gate settings on the thickness. Results showed:

	Gate Setting (mm)		
Resin Type	2	4	6
1	1.5	2.5	3.6
	1.3	2.5	3.8
2	1.4	2.6	3.7
	1.3	2.4	3.6

Do an analysis of variance of these data following the methods of Chapter 5.

7.16 For Problem 7.15 extract a linear and quadratic effect of gate setting and test for significance. Also partition the interaction and test its components.

7.17 Write any appropriate prediction equation or equations from the data of Problem 7.15.

7.18 If the experiment of Problem 7.15 is extended to include a third factor, weight fraction at three levels, the data are as follows:

	Resin Type					
	1			2		
Gate Setting (mm)	Weight Fraction			Weight Fraction		
	0.20	0.25	0.30	0.20	0.25	0.30
2	1.6	1.5	1.5	1.5	1.4	1.6
	1.5	1.3	1.3	1.4	1.3	1.4
4	2.7	2.5	2.4	2.4	2.6	2.2
	2.7	2.5	2.3	2.3	2.4	2.1
6	3.9	3.6	3.5	4.0	3.7	3.4
	4.0	3.8	3.4	4.0	3.6	3.3

Compile an ANOVA table for this three-factor experiment as in Chapter 5.

7.19 Outline a further breakdown of the ANOVA table in Problem 7.18 based on how the independent variables are set. Show the proper degrees of freedom.

7.20 For the significant ($\alpha = 0.05$) effects in Problem 7.18 complete the ANOVA breakdown suggested in Problem 7.19.

7.21 Graph any significant effects found in Problem 7.20 and discuss.

7.22 Data on the effect of knife edge radius R in inches and feedroll force F in pounds per inch on the energy necessary to cut 1-inch lengths of alfalfa are given as follows:

Feedroll Force F_j (lb/in)	Knife-Edge Radius R_i (inches)				$T_{.j.}$
	0.000	0.005	0.010	0.015	
5	29	98	44	84	
	30	128	81	100	
	20	67	77	63	
	79	293	202	247	821
10	22	35	53	103	
	26	80	93	90	
	16	29	59	98	
	64	144	205	291	704
15	18	49	58	80	
	17	68	103	91	
	11	61	128	77	
	46	178	289	248	761
20	38	68	87	86	
	31	74	116	113	
	21	47	90	81	
	90	189	293	280	852
$T_{i..}$	279	804	989	1066	$T_{...} = 3138$

Write the mathematical model for this experiment and do an ANOVA on the two factors and their interaction.

7.23 Outline a further ANOVA based on Problem 7.22 with single-degree-of-freedom terms.

7.24 Do an analysis for each of the terms in Problem 7.23.

7.25 Plot some results of Problem 7.22 and try to argue that they confirm some results found in Problem 7.24.

8
Multiple Regression

8.1 INTRODUCTION

In Chapter 7 we considered a prediction model to predict some dependent variable Y from one independent variable X where the model may be some polynomial in X. We also considered examples of two or more independent, quantitative variables for cases where the levels of each such variable were set at equispaced values. We now consider a more general situation in which Y may be a function of several independent variables $X_1, X_2, \ldots, X_k$ with no restrictions on the settings of these k independent variables. In fact, in most such multiple regression situations the X variables have already acted and we simply record their values along with the dependent variable Y. This is, of course, *ex-post-facto* research as opposed to experimental research where one manipulates the X's and observes the effect on Y.

In practice there are many studies of this type. For example, one may wish to predict the surface finish of steel from drop-out temperature and back-zone temperature. Here Y is a function of two recorded temperatures X_1 and X_2. It may be of interest to predict college grade-point average (GPA) in the freshman year for students whose input data include rank in high school, high school regents' average, SAT (Scholastic Aptitude Test) verbal score, and SAT mathematics score. Here Y, the freshman year GPA, is to be predicted from four independent variables: X_1—high school rank, X_2—high school regents' average, X_3—SAT verbal, and X_4—SAT mathematical.

As in the previous chapter, a mathematical model is written and the coefficients in the model are determined from the observed sample data by the method of least squares, making the sum of squares of deviations from this model a minimum.

8.2 THE MULTIPLE REGRESSION MODEL

To predict the value of a dependent variable (Y) from its regression on several independent variables $(X_1, X_2, \ldots, X_k)$ the linear population model is given as

$$Y' = \beta_0 + \beta_1 X_1 + \beta_2 X_2 + \cdots + \beta_k X_k$$

where the β's are the true coefficients to be used to weight the observed X's. In practice a sample of N is chosen and all variables $(Y, X_1, X_2, \ldots, X_k)$ are recorded for each of the N items in the sample and the corresponding sample model is

$$Y' = B_0 + B_1 X_1 + B_2 X_2 + \cdots + B_k X_k \tag{8.1}$$

where the B's are estimators of the β's. (Capital B's are used here because computer programs do not usually provide for lowercase letters.) Note that if $k = 1$, Equation (8.1) gives

$$Y' = B_0 + B_1 X_1$$

or the straight-line model of Chapter 7.

Since each observed Y can be expressed as

$$Y_i = Y_i' + e_i \qquad i = 1, 2, \ldots, N$$

by making the sum of the squares of the errors of estimate $(\sum e_i^2)$ a minimum, one can derive the following least squares equations.

$$\sum Y = B_0 N + B_1 \sum X_1 + B_2 \sum X_2 + \cdots + B_k \sum X_k$$
$$\sum X_1 Y = B_0 \sum X_1 + B_1 \sum X_1^2 + B_2 \sum X_1 X_2 + \cdots + B_k \sum X_1 X_k$$
$$\sum X_2 Y = B_0 \sum X_2 + B_1 \sum X_2 X_1 + B_2 \sum X_2^2 + \cdots + B_k \sum X_2 X_k \tag{8.2}$$
$$\vdots \qquad \vdots \qquad \vdots \qquad \vdots \qquad \vdots$$
$$\sum X_k Y = B_0 \sum X_k + B_1 \sum X_k X_1 + B_2 \sum X_k X_2 + \cdots + B_k \sum X_k^2$$

This set of $k + 1$ equations in $k + 1$ unknowns $(B_0, B_1, \ldots, B_k)$ can be solved for the B's. The solution is tedious if more than two or three independent variables are involved and so these equations are usually solved by a computer program that can handle a large number of independent variables.

To see how such problems are handled consider a problem from Burr [8].

Example 8.1 Twenty-four samples of a steel alloy are checked as to their percent elongation under stress and for each sample a chemical analysis is made and the percentages of five specific chemical elements are recorded.

These are identified only as X_1, X_2, X_3, X_4, and X_5. The data are given in Table 8.1.

Table 8.1   Percent Elongation Data

Item No.	Y	X_1	X_2	X_3	X_4	X_5
1	11.3	0.50	1.3	0.4	3.4	0.010
2	10.0	0.47	1.2	0.3	3.6	0.012
3	9.8	0.48	3.1	0.7	4.3	0.000
4	8.8	0.54	2.6	0.7	4.0	0.022
5	7.8	0.45	2.8	0.7	4.2	0.000
6	7.4	0.41	3.2	0.7	4.7	0.000
7	6.7	0.62	3.0	0.6	4.7	0.026
8	6.3	0.53	4.1	0.9	4.6	0.035
9	6.3	0.57	3.7	0.8	4.6	0.000
10	6.3	0.67	2.7	0.6	4.8	0.013
11	6.0	0.54	3.1	0.7	4.2	0.000
12	6.0	0.42	3.1	0.7	4.4	0.000
13	5.8	0.33	2.6	0.6	4.7	0.008
14	5.5	0.51	3.9	0.9	4.4	0.000
15	5.5	0.54	3.1	0.7	4.2	0.000
16	4.7	0.48	4.0	1.1	3.7	0.024
17	4.1	0.38	3.3	0.8	4.1	0.000
18	4.1	0.39	3.2	0.7	4.6	0.016
19	3.9	0.60	2.9	0.7	4.3	0.025
20	3.5	0.54	3.2	0.7	4.9	0.022
21	3.1	0.33	2.9	2.9	1.0	0.063
22	1.6	0.40	3.2	3.2	1.0	0.059
23	1.1	0.64	2.5	0.7	3.8	0.018
24	0.6	0.34	5.0	1.3	3.9	0.044

To enter Equations (8.2) the following sums are computed from Table 8.1.

$$N = 24$$

$\sum X_1 = 11.68$ $\sum X_1^2 = 5.9038$ $\sum X_1 X_2 = 35.553$

$\sum X_2 = 73.7$ $\sum X_2^2 = 240.85$ $\sum X_1 X_3 = 10.101$

$\sum X_3 = 22.1$ $\sum X_3^2 = 31.17$ $\sum X_1 X_4 = 47.644$

$\sum X_4 = 96.1$ $\sum X_4^2 = 408.09$ $\sum X_1 X_5 = 0.18405$

$\sum X_5 = 0.397$ $\sum X_5^2 = 0.0145$ $\sum X_1 Y = 67.318$

$\sum Y = 136.2$ $\sum Y^2 = 943.78$ $\sum X_2 X_3 = 71.17$

$\sum X_2 X_4 = 297.57$ $\sum X_3 X_4 = 74.58$ $\sum X_4 X_5 = 1.2881$

$\sum X_2 X_5 = 1.2886$ $\sum X_3 X_5 = 0.5945$ $\sum X_4 Y = 563.35$

$\sum X_2 Y = 390.32$ $\sum X_3 Y = 102.53$ $\sum X_5 Y = 1.6384$

and the equations to be solved would be

$$136.2 = 24B_0 + 11.68B_1 + 73.7B_2 + 22.1B_3 + 96.1B_4 + 0.397B_5$$
$$67.318 = 11.68B_0 + 5.9038B_1 + 35.553B_2 + 10.101B_3 + 47.644B_4 + 0.18405B_5$$
$$390.32 = 73.7B_0 + 35.553B_1 + 240.85B_2 + 71.17B_3 + 297.57B_4 + 1.2886B_5$$
$$102.53 = 22.1B_0 + 10.101B_1 + 71.17B_2 + 31.17B_3 + 74.58B_4 + 0.5945B_5$$
$$1.2881 = 96.1B_0 + 47.644B_1 + 297.57B_2 + 74.58B_3 + 408.09B_4 + 1.2881B_5$$
$$1.6384 = 0.397B_0 + 0.18405B_1 + 1.2886B_2 + 0.5945B_3 + 1.2881B_4 + 0.0145B_5$$

Many computer programs are available to solve such a set of equations.

An SPSS BACKWARD solution will give the B values for all variables in the equation as follows (to two decimal places):

$$Y' = 14.92 - 1.58X_1 - 1.28X_2 - 1.69X_3 - 0.58X_4 - 41.14X_5$$

This prediction equation could be used to predict elongation Y from the five variables X_1 through X_5.

The SPSS program also gives the following printout.

MULTIPLE R   0.71634
R SQUARE   0.51314
STD DEVIATION   2.14964

ANOVA	df	SS	MS	F	Significance
Regression	5	87.66795	17.53359	3.7943	0.016
Residual	18	83.17705	4.62045		

The value of R SQUARE indicates that 51.314 percent of the variation (SS) in Y is accounted for by the above equation. Note: Total SS = 170.845 and $87.66795/170.845 = 0.51314$. The F statistic is significant at the 0.016 level, indicating the equation accounts for a significant part of the total variation. The STD DEVIATION reported is really the standard error of estimate—the standard deviation of the errors remaining after the prediction equation has been "fitted" to the data. The standard deviation of Y prior to prediction is 2.73:

$$S_Y = \sqrt{\frac{SS_Y}{N-1}} = \sqrt{\frac{\sum Y^2 - \frac{(\sum Y)^2}{N}}{N-1}} = \sqrt{\frac{943.78 - \frac{(136.2)^2}{24}}{23}} = 2.73$$

If $R^2 = 0.51314$ of the variation in Y is accounted for by the regression of Y on $X_1, X_2, \ldots, X_5$, then $1 - R^2 = 0.48686$ is unaccounted for or residual. So the SS of the error or residual should be

$$SS_{error} = SS_Y(1 - R^2_{Y.12345})$$
$$= 170.845(0.48686)$$
$$= 83.177$$

and

$$S_{Y.12345} = \sqrt{\frac{SS_{error}}{N - k - 1}} = \sqrt{\frac{83.177}{18}} = 2.15$$

as given.

Sometimes the multiple R is designated as $R_{Y.12345}$ to show the joint effect of all five X's on Y. In like manner, $s_{Y.12345}$ denotes the standard error of estimate in Y from the regression model based on five X's.

The question might now be raised as to whether or not one needs all five independent variables to predict Y. If a Stepwise Regression program is used, the computer will enter the X's in the order of their contribution to the prediction of Y. At each step in the process the input can be examined to see whether or not adding another variable will contribute significantly to the prediction of Y. If not, one may end up with a reduced regression equation based on a subset of the X's that will be adequate in predicting Y.

Our data were entered into an SPSS Stepwise Regression as shown in Table 8.2.

The printout will be considered in several stages to see what one can determine from the computer. The first part of the printout gives the mean and standard deviation of each variable.

Variable	Mean	Standard Deviation	Cases
Y	5.6750	2.7254	24
X_1	.4867	.0977	24
X_2	3.0708	.7948	24
X_3	.9208	.6859	24
X_4	4.0042	1.0063	24
X_5	.0165	.0186	24

Note that the standard deviation of Y is 2.73 as we computed it above. Next, the correlations (Pearson's r's) between Y and each of the X variables are given along with the intercorrelations such as r_{13} between the independent variables. This form is referred to as the intercorrelation matrix.

X_1	.16884				
X_2	−.56054	−.17600			
X_3	−.53234	−.42456	.26356		
X_4	.28508	.38712	.13389	−.87641	
X_5	−.52747	−.21924	.20448	.78077	−.70099
	Y	X_1	X_2	X_3	X_4

In examining the correlations between Y and each X, one notes that the highest numerical r is Y with X_2 ($r_{y2} = -0.56054$) and thus a Stepwise Regression will enter X_2 first into the regression equation and give the resulting simple linear equation and other useful information as seen in Table 8.3.

Table 8.2   Computer Input for Elongation Data

```
+ + +CREATE PROB4
      1.000 = 77509,I13
      2.000 = COMMON (SPSS)
      3.000 = SPSS
      4.000 = #EOR
      5.000 = RUN NAME              STAT 502 PROB4
      6.000 = VARIABLE LIST         Y, X1, X2, X3, X4, X5
      7.000 = INPUT FORMAT          (5F4.0, F5.0)
      8.000 = N OF CASES            24
      9.000 = REGRESSION            METHOD = STEPWISE/
     10.000 =                       VARIABLES = Y, X1, X2, X3, X4, X5/
     11.000 =                       REGRESSION = Y WITH X1, X2,
                                               X3, X4, X5/
     12.000 = STATISTICS            ALL
     13.000 = READ INPUT DATA
     14.000 = 11.3 .50 1.3   .4 3.4 .010
     15.000 = 10.0 .47 1.2   .3 3.6 .012
     16.000 =  9.8 .48 3.1   .7 4.3 .000
     17.000 =  8.8 .54 2.6   .7 4.0 .022
     18.000 =  7.8 .45 2.8   .7 4.2 .000
     19.000 =  7.4 .41 3.2   .7 4.7 .000
     20.000 =  6.7 .62 3.0   .6 4.7 .026
     21.000 =  6.3 .53 4.1   .9 4.6 .035
     22.000 =  6.3 .57 3.7   .8 4.6 .000
     23.000 =  6.3 .67 2.7   .6 4.8 .013
     24.000 =  6.0 .54 3.1   .7 4.2 .000
     25.000 =  6.0 .42 3.1   .7 4.4 .000
     26.000 =  5.8 .33 2.6   .6 4.7 .008
     27.000 =  5.5 .51 3.9   .9 4.4 .000
     28.000 =  5.5 .54 3.1   .7 4.2 .000
     29.000 =  4.7 .48 4.0 1.1 3.7 .024
     30.000 =  4.1 .38 3.3   .8 4.1 .000
     31.000 =  4.1 .39 3.2   .7 4.6 .016
     32.000 =  3.9 .60 2.9   .7 4.3 .025
     33.000 =  3.5 .54 3.2   .7 4.9 .022
     34.000 =  3.1 .33 2.9 2.9 1.0 .063
     35.000 =  1.6 .40 3.2 3.2 1.0 .059
     36.000 =  1.1 .64 2.5   .7 3.8 .018
     37.000 =   .6 .34 5.0 1.3 3.9 .044
     38.000 = FINISH
     39.000 = #S
```

Table 8.3 Printout of STEP 1 Regression on Elongation Data

VARIABLE(S) ENTERED ON STEP NUMBER   1..   X2

MULTIPLE R	.56054
R SQUARE	.31420
ADJUSTED R SQUARE	.28303
STD DEVIATION	2.30775

VARIABLES IN THE EQUATION

VARIABLE	B	STD ERROR B	F	SIGNIFICANCE	BETA	ELASTICITY
X2	-1.9221129	.60542644	10.079408	.004	-.5605371	-1.04009
(CONSTANT)	11.577488	1.9179138	36.439343	.000		

ANALYSIS OF VARIANCE	DF	SUM OF SQUARES	MEAN SQUARE	F	SIGNIFICANCE
REGRESSION	1.	53.67981	53.67981	10.07941	.004
RESIDUAL	22.	117.16519	5.32569		
COEFF OF VARIABILITY	40.7 PCT				

VARIABLES NOT IN THE EQUATION

VARIABLE	PARTIAL	TOLERANCE	F	SIGNIFICANCE
X1	.08609	.96902	.15681280	.696
X3	-.48145	.93053	6.3364618	.020
X4	.43882	.98207	5.0083015	.036
X5	-.50930	.95819	7.3548356	.013

The "Variables in the Equation" portion of this printout gives the B values (here B_2), the coefficient of X_2, and B_0, the constant term. The equation at this step is then

$$Y' = 11.577488 - 1.9221129X_2$$

or

$$Y' = 11.58 - 1.92X_2$$

as the best fitting straight line for predicting Y from X_2 alone. With each B is recorded its standard error and a test as to whether it is significantly greater than zero. Here both B_0 and B_2 are highly significant (probabilities of 0.000 and 0.004 respectively).

This printout also shows an ANOVA for the single X's effect of Y and the multiple R and R^2, which, in this first step, is the simple Pearson r of Y versus X_2. Thus this equation will account for 0.3142 of the variation in Y. Note from the ANOVA that the linear regression on X_2 with 1 df has an SS of 53.68 out of a total (53.68 + 117.17) of 170.85, or 53.68/170.85 = 0.3142. The STD DEVIATION is actually the standard error of estimate or standard deviation of the errors of estimate from this first equation. Here $s_{Y.2} = 2.31$, which is some reduction from the original s_Y of 2.73.

The "Variables Not in the Equation" portion of the printout records partial correlation coefficients. These are correlations between Y and each X after X_2 has been removed. They are sometimes designated as $r_{Y1.2}$, and so forth, indicating the correlation between Y and X_1 with X_2 removed. These four partials are now examined to see whether any are significantly greater than zero. Here three of the four are (at the 0.05 level) and the largest partial is $r_{Y5.2} = -0.50903$, so X_5 is next entered into the regression equation as STEP 2 as printed out in Table 8.4.

The resulting equation is now

$$Y' = 11.69 - 1.62X_2 - 63.18X_5$$

At this stage $R^2_{y.25} = 0.49209$ and the standard error of estimate is 2.03. An examination of the next set of partial r's

$$r_{y1.25} = -0.01274$$
$$r_{y3.25} = -0.16275$$
$$r_{y4.25} = 0.09925$$

shows none significant at the 0.05 significance level. Hence one may stop in this problem after X_2 and X_5 have been entered and there is no added significance in going to an equation with the other three variables included.

One might also note that the SS for the error after X_2 and X_5 have been accounted for is 86.77 whereas it is only 83.18 when all three X's are included. One might also note that the standard error of estimate appears to

Table 8.4 Printout of STEP 2 Regression on Elongation Data

VARIABLE(S) ENTERED ON STEP NUMBER 2... X5

MULTIPLE R	.70149
R SQUARE	.49209
ADJUSTED R SQUARE	.44372
STD DEVIATION	2.03276

VARIABLES IN THE EQUATION

VARIABLE	B	STD ERROR B	F SIGNIFICANCE	BETA ELASTICITY
X2	−1.6199958	.54479609	8.8421893 .007	−.4724320 −.87661
X5	−63.179169	23.296325	7.3548356 .013	−.4308697 −.18416
(CONSTANT)	11.694826	1.6899321	47.890463 0	

ANALYSIS OF VARIANCE	DF	SUM OF SQUARES	MEAN SQUARE	F	SIGNIFICANCE
REGRESSION	2.	84.07077	42.03539	10.17287	.001
RESIDUAL	21.	86.77423	4.13211		
COEFF OF VARIABILITY	35.8 PCT				

VARIABLES IN THE EQUATION

VARIABLE	PARTIAL	TOLERANCE	F SIGNIFICANCE
X1	−.01274	.93398	.3245936E-02 .955
X3	−.16275	.37913	.54415907 .469
X4	.09925	.42841	.19898894 .660

172

increase from the two-variable equation to the five-variable equation. This is simply due to the fact that we are dividing by a smaller number of degrees of freedom and using three nonsignificant variables.

Even though only X_2 and X_5 are needed to predict Y quite adequately, the stepwise program will continue and add X_3, X_4, and finally X_1, unless programmed to stop and give the equation shown above as the BACK-WARD solution.

To see that this is a reasonable prediction equation we might insert X_2 and X_5 values into it and examine the errors of estimate, for example, if $X_2 = 1.3$ and $X_5 = 0.010$,

$$Y' = 11.69 - 1.62(1.3) - 63.18(.010)$$

$$= 8.957 \text{ versus an observed } Y = 11.3$$

giving an error of estimate $Y - Y' = 11.3 - 8.95 = 2.343$. For all 24 points we find as in Table 8.5. Here $\sum e = -0.003$ (close to zero) and $\sum e^2 = 86.77$ as given in the ANOVA table after X_2 and X_5 are accounted for.

One might also note in this table that the residuals or errors of estimate seem to be nonrandom as in general they are positive for large Y's and negative for small Y's. This may indicate a need to consider some higher order terms such as X_2^2, X_5^2, or $X_2 X_5$. These could be entered into the equation by proper instructions at the start of a program. Some computer programs will give a plot of these residuals when requested.

If one wishes only to check the significance of a given R^2, the proper test is

$$F_{k, N-k-1} = \frac{R_{y.12...k}^2 / K}{1 - R_{y.12...k}^2 / N - k - 1} \tag{8.3}$$

Table 8.5 Deviations from Regression on Elongation Data

Y	Y'	e	Y	Y'	e
11.3	8.957	2.343	5.8	6.977	−1.177
10.0	8.993	1.007	5.5	5.377	+0.123
9.8	6.673	3.127	5.5	6.673	−1.173
8.8	6.093	2.707	4.7	3.699	1.001
7.8	7.159	0.641	4.1	6.349	−2.149
7.4	6.511	0.889	4.1	5.500	−1.400
6.7	5.192	1.508	3.9	5.417	−1.517
6.3	2.842	3.458	3.5	5.121	−1.621
6.3	5.701	0.599	3.1	3.017	0.0831
6.3	6.500	−0.200	1.6	2.783	−1.183
6.0	6.673	−0.673	1.1	6.508	−5.408
6.0	6.673	−0.673	0.6	0.815	−0.215

Table 8.6 Summary Table for Elongation Data

				SUMMARY TABLE		
STEP	VARIABLE ENTERED REMOVED	F TO ENTER OR REMOVE	SIGNIFICANCE	MULTIPLE R	R SQUARE	R SQUARE CHANGE
1	X2	10.07941	.004	.56054	.31420	.31420
2	X5	7.35484	.013	.70149	.49209	.17789
3	X3	.54416	.469	.71101	.50554	.01345
4	X4	.20090	.659	.71464	.51071	.00517
5	X1	.08978	.768	.71634	.51314	.00243

To test whether any more variables are needed after a subset r has been found significant the test is

$$F_{k-r,\, N-k-1} = \frac{R_k^2 - R_r^2/k - r}{1 - R_k^2/N - k - 1} \tag{8.4}$$

where R_k represents the correlation of all k values with Y and R_r the correlation of a subset r of the k variables with Y.

In our example $R_k^2 = 0.51314$ and $R_r^2 = 0.49209$. Testing for significance of all five variables:

$$F_{5,18} = \frac{0.51314/5}{0.48686/18} = 3.79$$

as in ANOVA. Testing whether one needs to go beyond X_2 and X_5,

$$F_{3,18} = \frac{0.51314 - 0.49203/3}{0.48686/18} < 1$$

hence nonsignificant.

Many computer programs give a SUMMARY TABLE at the end of their printout, which shows the variables entered in order of their entry and the multiple R and R^2 increase as more variables added will account for more and more of the variability in the Y variable. In this example Table 8.6 gives this information. One notes that after adding X_5, 49 percent of the variation in Y is accounted for and after that additional variables increase this only slightly. The increase can be tested for significance at any stage by Equation (8.4).

8.3   COMPUTER PROGRAMS

Many computer programs are available to solve multiple regression problems. However, there is no way to be sure that a given reduced equation is the best predictor of a given Y unless all possible regressions are considered. This would mean, in our example, Y as a function of X_1, Y as a function of X_2, and so forth. Then Y as a function of the X's taken two at a time, three at a time, four at a time, and finally with all five X's. Some programs will handle all possible regressions. In our example the number of these is $2^k - 1$ or $2^5 - 1 = 31$ possible equations.

We have used SPSS programs in our example. There are three such programs available. The BACKWARD method enters all k independent variables at first as seen above. Then it removes any variables that are not making a significant contribution to the prediction equation. One can designate the level of significance desired to remove such variables by giving the desired or smaller F to be removed. The procedure continues until no more variables can be removed and all those left are contributing significantly to the prediction equation.

A FORWARD method will enter the variables one at a time until the equation is found such that entering any more variables will not improve the prediction significantly. Again, one may designate the size of F necessary for addition to the equation.

A STEPWISE method is similar to the FORWARD method except that at each stage all variables already entered are examined to see whether they are all needed after entering the last variable. In some cases an earlier entry is dropped. For example, if X_4, X_6, and X_7 were already selected in that order, when X_2, the next most significant variable, is entered X_6 might be removed. This may be unnecessary, now that X_2 has been added, and the equation will include X_4, X_7, and X_2 only.

In all these methods default values are in the programs that will operate in adding or removing variables if one does not instruct the computer as to the level desired. It is also possible to include certain specified X's in the equation regardless of their significance.

In the problems at the end of this chapter it is assumed that some multiple regression program is available to the reader.

8.4   REMARKS

In examining a multiple regression equation one must be careful not to assume that because a B value is numerically large the corresponding X has a greater influence in predicting Y than an X with a smaller regression coefficient. These regression coefficients (B's) are all based on their respective means and standard deviations as are the X's and Y. If one desires an equation whose coefficients do reflect the relative effect of the associated variable, the basic measurements ($Y, X_1, X_2, \ldots, X_k$) can be converted to standardized values or Z values where

$$Z'_Y = \frac{Y' - \bar{Y}}{s_Y} \quad \text{and} \quad Z_j = \frac{X_j - \bar{X}_j}{s_j}$$

If this is done, the resulting equation for predicting the standardized Y can be shown to be given by

$$Z'_Y = B'_1 Z_1 + B'_2 Z_2 + \cdots + B'_k Z_k$$

and the least squares equations become

$$\left.\begin{aligned}
r_{Y1} &= B'_1 + B'_2 r_{12} + \cdots + B'_k r_{1k} \\
r_{Y2} &= B'_1 r_{21} + B'_2 + \cdots + B'_k r_{2k} \\
&\vdots \\
r_{Yk} &= B'_1 r_{k1} + B'_2 r_{k2} + \cdots + B'_k
\end{aligned}\right\} \tag{8.5}$$

Here

$$B'_j = \frac{s_j}{s_Y} B_j \quad \text{or} \quad B_j = \frac{s_Y}{s_j} B'_j$$

and the r's are simple Pearson r's. Solving these equations for the B''s is still difficult for very many independent variables. In the special case where $k = 1$,

$$B'_1 = r_{Y1} \quad \text{and} \quad B_1 = \frac{s_Y}{s_1} \cdot r_{Y1}$$

(8.6)

and

$$B_0 = \bar{Y} - B_1 \bar{X}_1$$

In Example 8.1 the first variable entered was X_2 and the preliminary data gave the following.

$$\bar{Y} = 5.6750 \qquad s_Y = 2.7254$$
$$\bar{X}_2 = 3.0708 \qquad s_2 = 0.7948$$

and

$$r_{Y2} = -0.56054$$

Hence

$$B_2 = \frac{s_Y}{s_2} r_{Y2} = \frac{2.7254}{0.7948}(-0.56054) = -1.92$$

and

$$B_0 = \bar{Y} - B_2\bar{X}_2 = 5.6750 - (-1.92)(3.0708) = 11.58$$

giving $Y' = 11.58 - 1.92X_2$ as before.

A word of caution seems necessary also when one plans a regression study. Often experimenters seem inclined to include every X value that they can think of as computers can deal with so many X's. However, it is difficult to believe that all X's would be deemed important in their effect on Y by those knowledgeable in the field. Thus the use of multiple regression should not substitute for careful "soaking" in a problem and searching out of past studies and theoretical considerations before including a given X in a multiple regression study. Sometimes simple scattergrams of Y versus a given X will give some idea as to whether they may be related and the scattergram may also indicate an association that is nonlinear. This latter may indicate a need for a term such as X^2. The literature in the area may also suggest that some X is logarithmically related to Y and one should use log X as the independent variable instead of X. Careful consideration of all these notions before the experiment is essential to good experimentation and increases the prospect of meaningful prediction.

PROBLEMS

8.1 The sales volume Y of a product in thousands of dollars, price X_1 per unit in dollars, and advertising expense X_2 in hundreds of dollars were recorded for $n =$

8 cases. The results are as follows:

				Case				
	1	2	3	4	5	6	7	8
Y	10.1	6.5	5.1	11.0	9.9	14.7	4.8	12.2
X_1	1.3	1.9	1.7	1.5	1.6	1.2	1.6	1.4
X_2	8.8	7.1	5.5	13.8	18.5	9.8	6.4	10.2

a. Find the means and standard deviations of each variable and the intercorrelation matrix.

b. On the basis of the discussion in Section 8.4 find the regression equation Y from X_1 alone.

8.2 Submit the data of Problem 8.1 to a Stepwise Regression program and find the regression equation and the value of $R^2_{Y.12}$ and decide whether or not both X's are necessary.

8.3 In studying the percent conversion (Y) in a chemical process as a function of time in hours (X_1) and average fusion temperature in degrees Celsius (X_2) the following data were collected.

Y	X_1	X_2
62.7	3	297.5
76.2	3	322.5
80.8	3	347.5
80.8	6	297.5
89.2	6	322.5
78.6	6	347.5
90.1	9	297.5
88.0	9	322.5
76.1	9	347.5

Analyze these data using multiple linear regression. Reexamine these data and comment on any pecularities noted.

8.4 A management consulting firm attempted to predict annual salary of executives from the executives' years of experience (X_1), years of education (X_2), sex (X_3), and number of employees supervised (X_4). A sample of 25 executives gave an average salary of $29,700 with a standard deviation of $1300. From a computer program the following statistics were recorded.

$$r^2_{Y4} = 0.42$$
$$R^2_{Y.24} = 0.78$$
$$R^2_{Y.124} = 0.90$$
$$R^2_{Y.1234} = 0.95$$

a. Explain how you would interpret these statistics.
b. Test whether one could stop after variables 2 and 4 have been entered into the equation.
c. If variables 1, 2, and 4 were used in the prediction equation, what would the limits on the salaries expected for a predicted salary have to be in order to be correct about 95 percent of the time?

8.5 Consider a prediction situation where some dependent variable Y is to be predicted from four independent variables X_1, X_2, X_3, and X_4 and assume you have a stepwise regression printout on this problem involving 25 observations. Explain briefly each of the following with respect to this problem and its printout.
a. How does the computer decide which independent variable to enter first into a regression equation?
b. How does it decide which variable to enter next in the equation?
c. How do you decide when to stop adding variables based on the printout?
d. Do you need to add any more variables if $R^2_{Y.1234} = 0.7$ and $R^2_{Y.24} = 0.6$?

8.6 A study was made to determine the effect of four variables on the grade-point average of 55 high school students who attended a junior college after graduation. These variables were high school rank (X_1), SAT score (X_2), IQ (X_3), and age in months (X_4). From the computer printout the following statistics were recorded.

$$R^2_{Y.1234} = 0.45 \qquad R^2_{Y.12} = 0.33$$
$$R^2_{Y.123} = 0.42 \qquad r^2_{Y1} = 0.19$$

Use this information to test for the significance of each added variable and determine whether or not an ordered subset of these four will make a satisfactory prediction. If you wish, you may assume a total of sum of squares in this study of 10,000 units. Also explain how you would predict with approximate 2 to 1 odds the maximum GPA expected for an individual student based on his or her scores of the variables in the appropriate equation.

8.7 In a study of several variables that might affect freshman GPA (Y) a sample of 55 students reported their college entrance arithmetic test (X_1), their analogies test score on an Ohio battery (X_2), their high school average (X_3), and their interest score on an interest inventory (X_4). Results of this study are given in the following (oversimplified) ANOVA table.

Source*	df	SS	MS
Due to X_2	1	1360	1360
Due to X_3/X_2	1	480	480
Due to $X_1/X_2, X_3$	1	80	80
Due to $X_4/X_1, X_2, X_3$	1	80	80
Error	50	2000	40
Total	54	4000	

* / means "given."

Assuming a stepwise procedure was used so that variables were entered in their order of importance, answer the following.

a. Determine $R_{Y.1234}$ and explain its meaning in this problem.

b. Determine which of the four variables are sufficient to predict GPA (Y) and show that your choice is sufficient.

c. If a regression equation based on your answer to (b) is used to predict student A's GPA based on test scores, how close do you think this prediction will be to the actual GPA? Justify your answer and note any assumptions you are making.

8.8   In an attempt to predict the average value per acre of farm land in Iowa in 1920 the following data were collected.

Y = average value in dollars per acre
X_1 = average corn yield in bushels per acre for ten preceding years
X_2 = percentage of farm land in small grain
X_3 = percentage of farm land in corn

Y	X_1	X_2	X_3
87	40	11	14
133	36	13	30
174	34	19	30
385	41	33	39
363	39	25	33
274	42	23	34
235	40	22	37
104	31	9	20
141	36	13	27
208	34	17	40
115	30	18	19
271	40	23	31
163	37	14	25
193	41	13	28
203	38	24	31
279	38	31	35
179	24	16	26
244	45	19	34
165	34	20	30
257	40	30	38
252	41	22	35
280	42	21	41
167	35	16	23
168	33	18	24
115	36	18	21

Analyze these data using multiple linear regression and comment on the results.

8.9   In a study involving handicapped students an attempt was made to predict GPA from two demographic variables, sex and ethnicity, and two independent measured variables, interview score and contact hours used in counseling and/or tutoring. A dummy variable for sex was taken as 1 = male, 2 = female. For ethnicity: 1 = black, 2 = Hispanic, and 3 = white, non-Hispanic. Do a complete analysis of the data below and justify the regression equation that will adequately "fit" the data.

Students	Ethnic	Sex	Interview	Hours	GPA
1	1	2	11.0	4.0	5.50
2	1	2	10.0	5.0	4.10
3	1	2	12.0	73.0	5.00
4	1	2	11.5	68.0	4.22
5	1	2	10.8	82.0	5.00
6	1	1	12.5	72.5	5.00
7	1	1	9.5	64.0	4.60
8	1	1	9.5	78.0	4.25
9	1	1	8.0	64.0	4.00
10	1	1	7.5	13.0	2.00
11	2	2	9.0	37.0	4.25
12	2	2	8.2	4.0	4.00
13	2	2	10.7	38.5	4.61
14	2	2	8.5	3.0	2.93
15	2	2	12.5	10.5	5.50
16	2	1	12.0	80.0	4.77
17	2	1	12.2	6.0	5.00
18	2	1	7.0	6.5	3.25
19	2	1	8.6	22.0	2.66
20	2	1	8.3	28.5	3.37
21	3	2	10.9	12.0	5.00
22	3	2	9.0	9.0	4.00
23	3	2	10.0	5.0	5.00
24	3	2	7.2	12.0	3.87
25	3	2	8.5	4.0	3.00
26	3	1	10.0	8.0	4.77
27	3	1	8.5	8.0	5.00
28	3	1	10.0	22.0	5.08
29	3	1	11.4	61.5	5.57
30	3	1	11.9	37.0	6.00

8.10 Given the following intercorrelation matrix between measures where Y = freshman mathematics grade, X_1 = Ohio State psychological examination, X_2 = English-usage examination, X_3 = algebra examination, X_4 = engineering aptitude test:

Variable	Y	X_1	X_2	X_3	X_4
X_1	0.51				
X_2	0.51	0.70			
X_3	0.61	0.53	0.61		
X_4	0.39	0.39	0.29	0.28	
Means	5.70	4.10	5.44	5.37	4.95
Standard deviation	2.42	1.92	1.84	2.26	2.14

Do a complete multiple regression on these data and summarize your conclusions. Assume that $N = 100$ students.

8.11 In a study of the effect of 15 possible variables on the current annual gift of a corporation to a college (X_5) the variables are

$$X_6 = \text{number of students enrolled in current year}$$
$$X_7 = \text{number of families represented}$$
$$X_8 = \text{number of years gifts have been given}$$
$$X_9 = \text{average previous year's gift}$$
$$X_{10} = \text{related company gift giving}$$
$$X_{12} = \text{school ranking score}$$
$$X_{13} = \text{previous industry average}$$
$$X_{14} = \text{last annual gift}$$
$$X_{15} = \text{continuity of giving}$$
$$X_{400} = \text{initialized year code}$$
$$X_{401} = X_{15} \text{ times } X_8 \text{ or } X_{15}X_8$$
$$X_{402} = X_{15}X_9$$
$$X_{403} = X_9X_{10}$$
$$X_{404} = X_9X_{12}$$
$$X_{405} = X_{10}X_{13}$$

This means, standard deviations, and intercorrelations are given on p. 183. (Note that Y is X_5.)

a. On the basis of the above printout what is the first variable to be entered in a stepwise regression? Explain how you know.
b. On the basis of just the information given find the best fitting prediction equation for predicting X_5 from this first variable.
c. How good is this prediction?
d. What is the standard error of estimate?

8.12 At step 4 in Problem 8.11 the printout is as appears on p. 184.

Explain what the next step should be and discuss the results up to this point. That is, what is the equation and how good is it at step 4?

VARIABLE	MEAN	STANDARD DEV	CASES
VAR005	928.6732	1366.7388	428
VAR006	2.5935	3.5189	428
VAR007	1.4860	1.9459	428
VAR008	2.3107	1.6859	428
VAR009	758.8312	1256.5571	428
VAR010	31.0234	12.7108	428
VAR012	2.1098	2.3469	428
VAR013	1055.3668	551.0675	428

VARIABLE	MEAN	STANDARD DEV	CASES
VAR014	870.9416	1457.7314	428
VAR015	1.5374	.7688	428
VAR400	2.6869	1.0886	428
VAR401	4.3925	3.4560	428
VAR402	1452.6631	2508.6356	428
VAR403	26116.8654	46509.4447	428
VAR404	2864.2585	7008.5368	428
VAR405	35430.7944	30254.2850	428

CORRELATION COEFFICIENTS.
A VALUE OF 99.00000 IS PRINTED IF A COEFFICIENT CANNOT BE COMPUTED.

	VAR005	VAR006	VAR007	VAR008	VAR009	VAR010	VAR012	VAR013	VAR014	VAR015	VAR400	VAR401	VAR402	VAR403	VAR404
VAR006	.81905														
VAR007	.82577	.95509													
VAR008	.32498	.24833	.30721												
VAR009	.90660	.75420	.76877	.35370											
VAR010	.15696	.06880	.08826	.37342	.16162										
VAR012	.45557	.41633	.45340	.38852	.42938	.22601									
VAR013	.35856	.28676	.31122	.25947	.37505	.38490	.22787								
VAR014	.94601	.76364	.78550	.40917	.94737	.19040	.46417	.36034							
VAR015	.23269	.17443	.19602	.64959	.29678	.29683	.25536	.14065	.30286						
VAR400	.00423	-.09933	-.07504	.33896	-.04935	.35188	.02449	-.02835	.03081	.17351					
VAR401	.33833	.26331	.31841	.97663	.35514	.38364	.38476	.27688	.41257	.68900	.34460				
VAR402	.91290	.75593	.77198	.35907	.99461	.16903	.43196	.37904	.94574	.31593	-.03161	.37409			
VAR403	.78538	.65135	.68025	.38523	.88714	.40633	.41576	.47599	.83095	.28850	.01813	.38862	.88608		
VAR404	.88086	.71984	.75139	.33351	.91755	.12960	.55398	.34913	.91084	.22296	-.01851	.33929	.92208	.79807	
VAR405	.30962	.21382	.23816	.36283	.33875	.75783	.26445	.82746	.33214	.22363	.12471	.37455	.34442	.57640	.29343

183

DEPENDENT VARIABLE.. VAR005 GIFT AMOUNT-CURRENT YEAR
VARIABLE(S) ENTERED ON STEP NUMBER 4... VAR402

MULTIPLE R	.95961	ANALYSIS OF VARIANCE DF	SUM OF SQUARES	MEAN SQUARE
R SQUARE	.92085	REGRESSION 4.	734496748.42831	183624187.10708
ADJUSTED R SQUARE	.92011	RESIDUAL 423.	63128607.66488	149240.20725
STD DEVIATION	386.31620	COEFF OF VARIABILITY 41.6 PCT		

F 1230.39354 SIGNIFICANCE .000

VARIABLES IN THE EQUATION

VARIABLE	B	STD ERROR B	F SIGNIFICANCE	BETA ELASTICITY
VAR014	.67427766	.41821414E-01	259.94419 0	.7191686 .63236
VAR006	83.964369	8.3754680	100.50135 0	.2161777 .23448
VAR008	-44.547260	12.262990	13.196212 .000	-.0549505 -.11084
VAR402	.48527309E-01	.23309294E-01	4.3342523 .038	.0890714 .07591
(CONSTANT)	156.10232	33.501401	21.711646 .000	

VARIABLES NOT IN THE EQUATION

VARIABLE	PARTIAL	TOLERANCE	F SIGNIFICANCE
VAR007	.03201	.07823	.43296776 .511
VAR009	-.24317	.01024	26.522381 .000
VAR010	.04090	.85174	.70715462 .401
VAR012	.06119	.72233	1.5860747 .209
VAR013	.06956	.83808	2.0518152 .153
VAR015	-.07302	.55286	2.2622789 .133
VAR400	.09747	.83043	4.0473548 .045
VAR401	.08495	.04471	3.0676690 .081
VAR403	-.08399	.20790	2.9981534 .084
VAR404	.06182	.13483	1.619306 .204
VAR405	.05444	.81126	1.2542856 .263

DEPENDENT VARIABLE.. VAR005 GIFT AMOUNT-CURRENT YEAR
VARIABLE(S) ENTERED ON STEP NUMBER 6.. VAR015

			DF	SUM OF SQUARES	MEAN SQUARE	F	SIGNIFICANCE
MULTIPLE R	.96260	ANALYSIS OF VARIANCE				885.72234	.000
R SQUARE	.92660	REGRESSION	6.	739075972.00258	123179328.66710		
ADJUSTED R SQUARE	.92555	RESIDUAL	421.	58549384.09061	139072.17124		
STD DEVIATION	372.92381	COEFF OF VARIABILITY	40.2 PCT				

VARIABLES IN THE EQUATION

VARIABLE	B	STD ERROR B	F SIGNIFICANCE	BETA	ELASTICITY
VAR014	.70883853	.41544582E-01	291.11585 0	.7560304	.66477
VAR006	81.738979	8.1061090	101.67953 0	.2104482	.22827
VAR008	-27.217424	15.022736	3.2824337 .071	-.0335736	-.06772
VAR402	43168396	.71849885E-01	36.097711 0	.7923516	.67526
VAR009	-.79551599	.14402992	30.506499 0	-.7313843	-.65003
VAR015	-79.011302	32.030175	6.0849891 .014	-.0444463	-.13080
(CONSTANT)	260.26386	43.589110	35.651001 0		

VARIABLES NOT IN THE EQUATION

VARIABLE	PARTIAL	TOLERANCE	F SIGNIFICANCE
VAR007	.00702	.07714	.20696127E-01 .886
VAR010	.03429	.84517	.49451217 .482
VAR012	.05735	.72198	1.3860130 .240
VAR013	.05954	.83159	1.4943132 .222
VAR400	.04105	.79030	.70892366 .400
VAR401	-.05131	.02612	1.1087083 .293
VAR403	-.06089	.20156	1.5627801 .212
VAR404	.02388	.12820	.23959657 .625
VAR405	.04308	.80747	.78080119 .377

8.13 For Problem 8.11 the printout at step 6 is as appears on p. 185.

Explain the next step and analyze the results as completely as you can.

8.14 In a study on the effect of several variables on posttest achievement scores in biology in a rural high school involving 48 students the following table was presented. It included variables that had a significant effect on achievement at the 10 percent level of significance.

Step	Variable Entered	Multiple R	R^2	R^2 Change
1	Figural elaboration pretest	0.46587	0.21704	0.21704
2	Achievement pretest	0.83022	0.68927	0.47223
3	Piagetian pretest	0.86536	0.74885	0.05958
4	Attitudes pretest	0.90606	0.82094	0.07209

Discuss these results and test whether or not the last two variables were necessary.

8.15 To study the effect of planting rate on yield in an agricultural problem, 13 data points were taken with the following results.

X 12   12   16   16   16   20   20
Y 130.5 129.6 142.5 140.3 143.4 144.8 144.1

X 23   23   23   23   28   28
Y 145.1 147.8 146.6 148.4 134.8 135.1

a. Plot a scattergram and note what type of curve you might expect of Y versus X.
b. Taking $X^2 = X_2$ and $X = X_1$, use multiple regression to find the equation of the best fitting quadratic.
c. Compare this quadratic with the best linear fit.

8.16 The printout of a computer program yielded the following statistics.

$$B_0 = 140.27 \qquad r_{y1}^2 = 0.0907$$
$$B_1 = 1.24 \qquad R_{y.12}^2 = 0.3285$$
$$B_2 = 2.65 \qquad R_{y.123}^2 = 0.4211$$
$$B_3 = -0.30 \qquad R_{y.1234}^2 = 0.4548$$
$$B_4 = -1.27 \qquad S_y = 16.0$$
$$N = 45$$

a. Explain the meaning of the R squares given above.
b. Test and determine which regression model is adequate for predicting Y.
c. Compute the standard error of estimate for your model in (b).
d. Explain whether or not the program was a stepwise (step-up) program.

3^f Factorial Experiments

9.1 INTRODUCTION

As 2^f factorial experiments represent an interesting special case of factorial experimentation, so also do 3^f factorial experiments. The 3^f factorials consider f factors at three levels; thus there are 2 df between the levels of each of these factors. If the three levels are quantitative and equispaced, the methods of Chapter 7 may be used to extract linear and quadratic effects and to test these for significance. The 3^f factorials also play an important role in· more complicated design problems, which will be discussed in subsequent chapters. For this chapter, it will be assumed that the design is a completely randomized design and that the levels of the factors considered are fixed levels. Such levels may be either qualitative or quantitative.

9.2 3^2 FACTORIAL

If just two factors are crossed in an experiment and each of the two is set at three levels, there are $3 \times 3 = 9$ treatment combinations. Since each factor is at three levels, the notation of Chapter 6 will no longer suffice. There is now a low, intermediate, and high level for each factor, which may be designated as 0, 1, and 2. A model for this arrangement would be

$$Y_{ij} = \mu + A_i + B_j + AB_{ij} + \varepsilon_{ij}$$

where $i = 1, 2, 3, j = 1, 2, 3$, and the error term is confounded with the AB interaction unless there are some replications in the nine cells, in which case,

$$Y_{ijk} = \mu + A_i + B_j + AB_{ij} + \varepsilon_{k(ij)}$$

and $k = 1, 2, \ldots, n$ for n replications.

To introduce some notation for treatment combinations when three levels are involved, consider the data layout in Figure 9.1. In Figure 9.1 two digits are used to describe each of the nine treatment combinations. The

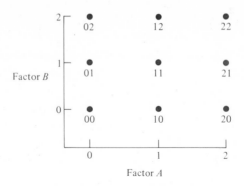

Figure 9.1 3^2 data layout.

first digit indicates the level of factor A, and the second digit the level of factor B. Thus 12 means A at its intermediate level and B at its highest level. This notation can easily be extended to more factors and as many levels as are necessary. It could have been used for 2^f factorials as 00, 10, 01, and 11, corresponding respectively to (1), a, b, and ab. The only reason for not using this digital notation on 2^f factorials is that so much of the literature includes this (1), a, b, . . . notation. By proper choice of coefficients on these treatment combinations, the linear and quadratic effects of both A and B can be determined, as well as their interactions, such as $A_L \times B_L$, $A_L \times B_Q$, $A_Q \times B_L$, and $A_Q \times B_Q$. The methods of analysis will be illustrated in a simple hypothetical example.

Example 9.1   Suppose the responses in Table 9.1 were recorded for the treatment combinations indicated in the upper-left-hand corner of each cell.

Table 9.1 3^2 Factorial with Responses and Totals

Factor B	Factor A			
	0	1	2	$T_{.j}$
0	00 1	10 -2	20 3	2
1	01 0	11 4	21 1	5
2	02 2	12 -1	22 2	3
$T_{i.}$	3	1	6	10

Table 9.2 ANOVA for 3^2 Factorial of Table 9.1

Source	df	SS	MS
A_i	2	4.22	2.11
B_j	2	1.56	0.78
AB_{ij}	4	23.11	5.77
Totals	8	28.89	

Analyzing the data of Table 9.1 by the general methods of Chapter 5 gives

$$\text{SS}_{\text{total}} = 1^2 + 0^2 + 2^2 + \cdots + 2^2 - \frac{(10)^2}{9} = 28.89$$

$$\text{SS}_A = \frac{3^2 + 1^2 + 6^2}{3} - \frac{(10)^2}{9} = 4.22$$

$$\text{SS}_B = \frac{2^2 + 5^2 + 3^2}{3} - \frac{(10)^2}{9} = 1.56$$

$$\text{SS}_{\text{error}} = 28.89 - 4.22 - 1.56 = 23.11$$

and the ANOVA is given in Table 9.2.

A further breakdown of this analysis is now possible by recalling that coefficients of $-1, 0, +1$ applied to the responses at low, intermediate, and high levels of a factor will measure its linear effect, whereas coefficients of $+1, -2, +1$ applied to these same responses will measure the quadratic effect of this factor. As in the case of 2^f factorials, products of coefficients will give the proper coefficients for various interactions. This can best be shown by Table 9.3, which indicates the coefficients for each effect to be used with the nine treatment combinations.

Table 9.3 Coefficients for a 3^2 Factorial with Quantitative Levels

	Treatment Combination									
Factor	00	01	02	10	11	12	20	21	22	$\sum c_i^2$
A_L	−1	−1	−1	0	0	0	+1	+1	+1	6
A_Q	+1	+1	+1	−2	−2	−2	+1	+1	+1	18
B_L	−1	0	+1	−1	0	+1	−1	0	+1	6
B_Q	+1	−2	+1	+1	−2	+1	+1	−2	+1	18
A_LB_L	+1	0	−1	0	0	0	−1	0	+1	4
A_LB_Q	−1	+2	−1	0	0	0	+1	−2	+1	12
A_QB_L	−1	0	+1	+2	0	−2	−1	0	+1	12
A_QB_Q	+1	−2	+1	−2	+4	−2	+1	−2	+1	36
Y_{ij}	1	0	2	−2	4	−1	3	1	2	

From Table 9.3 it can be seen that A_L compares all highest levels of $A(+1)$ with all lowest levels of $A(-1)$. A_Q compares the extreme levels with twice the intermediate levels. Both of these effects are taken across *all* levels of B. Now B_L compares the highest versus the lowest level of B at the 0 level of A, then at level 1 of A, then at level 2 of A, reading from left to right across B_L. Similarly, B_Q compares the extreme levels of B with twice the intermediate level at all three levels of A. The coefficients for interaction are found by multiplying corresponding main-effect coefficients. An examination of these coefficients in the light of what interactions there are should make the coefficients seem quite plausible. The sums of squares of the coefficients are given at the right of Table 9.3.

Applying these coefficients to the responses for each treatment combination gives

$$
\begin{aligned}
A_L &= -1(1) - 1(0) - 1(2) + 0(-2) + 0(4) \\
&\quad + 0(-1) + 1(3) + 1(1) + 1(2) = 3 \\
A_Q &= +1(1) + 1(0) + 1(2) - 2(-2) - 2(4) \\
&\quad - 2(-1) + 1(3) + 1(1) + 1(2) = 7 \\
B_L &= -1(1) + 0(0) + 1(2) - 1(-2) + 0(4) \\
&\quad + 1(-1) - 1(3) + 0(1) + 1(2) = 1 \\
B_Q &= +1(1) - 2(0) + 1(2) + 1(-2) - 2(4) \\
&\quad + 1(-1) + 1(3) - 2(1) + 1(2) = -5 \\
A_L B_L &= +1(1) + 0(0) - 1(2) + 0(-2) + 0(4) \\
&\quad + 0(-1) - 1(3) + 0(1) + 1(2) = -2 \\
A_L B_Q &= -1(1) + 2(0) - 1(2) + 0(-2) + 0(4) \\
&\quad + 0(-1) + 1(3) - 2(1) + 1(2) = 0 \\
A_Q B_L &= -1(1) + 0(0) + 1(2) + 2(-2) + 0(4) \\
&\quad - 2(-1) - 1(3) + 0(1) + 1(2) = -2 \\
A_Q B_Q &= +1(1) - 2(0) + 1(2) - 2(-2) + 4(4) \\
&\quad - 2(-1) + 1(3) - 2(1) + 1(2) = 28
\end{aligned}
$$

The corresponding sums of squares become

$$
SS_{A_L} = \frac{(3)^2}{6} = 1.50 \qquad SS_{A_L B_L} = \frac{(-2)^2}{4} = 1.00
$$

$$
SS_{A_Q} = \frac{(7)^2}{18} = 2.72 \qquad SS_{A_L B_Q} = \frac{0^2}{12} = 0.00
$$

$$
SS_{B_L} = \frac{(1)^2}{6} = 0.17 \qquad SS_{A_Q B_L} = \frac{(-2)^2}{12} = 0.33
$$

$$
SS_{B_Q} = \frac{(-5)^2}{18} = 1.39 \qquad SS_{A_Q B_Q} = \frac{(28)^2}{36} = 21.78
$$

Summarizing, we obtain Table 9.4.

Table 9.4   ANOVA Breakdown
for 3^2 Factorial

Source	df	SS	
A_i	2	4.22	
A_L	1		1.50
A_Q	1		2.72
B_j	2	1.56	
B_L	1		0.17
B_Q	1		1.39
AB_{ij}	4	23.11	
$A_L B_L$	1		1.00
$A_L B_Q$	1		0.00
$A_Q B_L$	1		0.33
$A_Q B_Q$	1		21.78
Totals	8	28.89	

The results of this analysis will not be tested, as there is no separate measure of error, and the interaction effect is obviously large compared with other effects. As the numbers used here are purely hypothetical, the purpose has been only to show how such data can be analyzed and how the notation can be used.

It may be instructive to examine graphically the meaning of this high $A_Q B_Q$ interaction since it dwarfs all main effects and other interactions in the example.

If the $A_L B_L$ interaction is graphed using only the extreme levels of each factor, the results are as shown in Figure 9.2. Although the lines do cross, this is a very small interaction. Note also that the average change in response for the lowest level of A to the highest level of A (shown by $\times$) is very small, indicating a very small A_L effect as the data show.

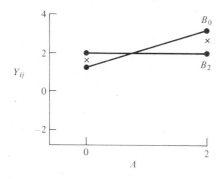

Figure 9.2 $A_L B_L$ interaction.

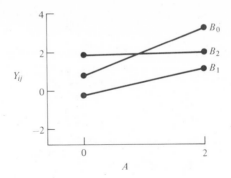

Figure 9.3 $A_L B_Q$ interaction.

Plotting $A_L B_Q$ means using all three levels of B, giving Figure 9.3, which again shows little interaction.

For $A_Q B_L$ the results are as shown in Figure 9.4, which once again shows little interaction. This graph does indicate that the quadratic effect of A is more pronounced than its linear effect. This is borne out by the data.

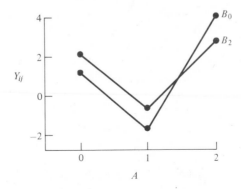

Figure 9.4 $A_Q B_L$ interaction.

Replotting Figure 9.3 with B as abscissa gives Figure 9.5, which shows the same lack of serious interaction as in Figure 9.3, but does show a slight quadratic bend in factor B over the linear effect.

So far no startling results have been seen. Now plot the middle level of B on Figure 9.4 to obtain Figure 9.6. Note the way this "curve" reverses its trend compared with the other two. This shows that not until both factors are considered at all three of their levels does this $A_Q B_Q$ interaction show up. It can also be seen by adding the middle level of A on Figure 9.5 (see Figure 9.7).

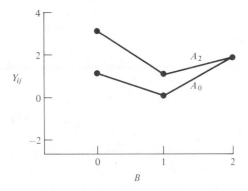

Figure 9.5 $A_L B_Q$ interaction again.

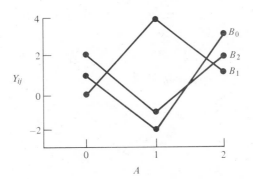

Figure 9.6 $A_Q B_Q$ interaction.

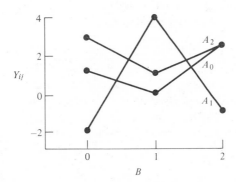

Figure 9.7 $A_Q B_Q$ interaction again.

Table 9.5   Diagonal Computations

Factor B	Factor A			Factor A		
	0	1	2	0	1	2
0	1	−2	3	1	−2	3
1	0	4	1	0	4	1
2	2	−1	2	2	−1	2

Before leaving this problem, reconsider the data of Table 9.1. Add the data by diagonals rather than by rows or columns. First consider the diagonals downward from left to right where the main diagonal is $1 + 4 + 2 = 7$, the next one to the right is $-2 + 1 + 2 = 1$, and the last $+2$ is found by repeating the table again to the right of the present one as in Table 9.5.

Similarly, the next downward diagonal gives $3 + 0 + (-1) = 2$. The sum of squares between these three diagonal terms is then

$$\frac{(7)^2 + (1)^2 + (2)^2}{3} - \frac{(10)^2}{9} = 6.89$$

If the diagonals are now considered downward and to the left, their totals are

$$3 + 4 + 2 = 9$$
$$1 + 1 - 1 = 1$$
$$-2 + 0 + 2 = 0$$

and their sum of squares is

$$\frac{(9)^2 + (1)^2 + (0)^2}{3} - \frac{(10)^2}{9} = 16.22$$

These two somewhat artificial sums of squares of 6.89 and 16.22 are seen to add up to the interaction sum of squares

$$6.89 + 16.22 = 23.11$$

These two components of interaction have no physical significance, but simply illustrate another way to extract two orthogonal components of interaction. Testing each of these separately for significance has no meaning, but this arbitrary breakdown is very useful in more complex designs. Some authors refer to these two components as the I and J components of interaction

$$I(AB) = \quad 6.89 \quad \text{2 df}$$
$$J(AB) = 16.22 \quad \text{2 df}$$
$$\text{total } A \times B = 23.11 \quad \text{4 df}$$

Each such component carries 2 df. These are sometimes referred to as the AB^2 and AB components of $A \times B$ interaction. In this notation, effects can be multiplied together using a modulus of 3, since this is a 3^f factorial. A *modulus* of 3 means that the resultant number is equal to the remainder when the number in the usual base of 10 is divided by 3. Thus $4 = 1$ in modulus 3, as 1 is the remainder when 4 is divided by 3. The following associations also hold:

$$\begin{array}{ccccc} \text{numbers:} & 0 & 3 = 0 & 6 = 0 & 9 = 0 \\ & 1 & 4 = 1 & 7 = 1 & 10 = 1 \\ & 2 & 5 = 2 & 8 = 2 & 11 = 2 \\ & \vdots \end{array}$$

When using the form $A^p B^q$, it is postulated that the only exponent allowed on the first letter in the expression is a 1. To make it a 1, the expression can be squared and reduced, modulus 3. For example,

$$A^2 B = (A^2 B)^2 = A^4 B^2 = AB^2$$

Hence AB and AB^2 are the only components of the $A \times B$ interaction with 2 df each. Here the two types of notation are related as follows:

$$I(AB) = AB^2$$
$$J(AB) = AB$$

To summarize this simple experiment, all effects can be expressed with 2 df each, as in Table 9.6.

Table 9.6 3^2 Factorial by 2-df Analysis

Source	df	SS
A_i	2	4.22
B_j	2	1.56
$I(AB) = AB^2$	2	6.89
$J(AB) = AB$	2	16.22
Totals	8	28.89

It will be found very useful to break such an experiment down into 2-df effects when more complex designs are considered. Here, this breakdown is presented merely to show another way to partition the interaction effect.

Let us consider a practical example.

Example 9.2 The effect of two factors—prechamber volume ratio (V) and injection timing (T)—on the parts per million of noxious gas emitted from an engine was to be studied. The volume ratio was set at three equispaced

Table 9.7   Gas Emission Data of
Example 9.2

Time	Volume (V)		
(T)	Low	Medium	High
Short	6.27	8.08	7.34
	5.43	8.04	7.87
Medium	6.94	7.48	8.61
	6.51	7.52	8.32
Long	7.22	8.65	9.02
	7.05	8.97	9.07

levels and the time was also set at three equispaced levels. Two engines
were built for each of the nine treatment combinations and the amount of
gas emitted was recorded for the 18 engines as shown in Table 9.7.

The ANOVA for Table 9.7, assuming a completely randomized design,
is given in Table 9.8.

Table 9.8   ANOVA for Gas Emission Data

Source	df	SS	MS	F	Prob.
Volume (V_i)	2	11.44	5.72	71.5	0.000
Time (T_j)	2	4.17	2.08	26.0	0.000
$V \times T$ interaction (VT_{ij})	4	1.39	0.35	4.4	0.031
Error	9	0.70	0.08		

The ANOVA of Table 9.8 shows strong main effects and an interaction
significant at the 5 percent significance level. As both factors are quantitative
and equispaced we can treat these data as we did the hypothetical data of
Table 9.3 to give Table 9.9.

Upon taking each contrast and squaring and dividing by 2 (as $n = 2$)
and the sum of the squares of the coefficients, one gets Table 9.10.

From Table 9.10 we see that volume ratio produced both a strong
linear and quadratic effect on the amount of gas emitted. The timing shows
a strong linear effect and a significant interaction is shown as a quadratic
by quadratic interaction. One must then plot all nine treatment means to
see the significant effects as shown in Figure 9.8.

Note that without the T_2 line in Figure 9.8 there would be little obvious
interaction. Careful study of this illustration should confirm the reasonableness of the results given in Table 9.10.

Table 9.9   Quantitative Level Breakdown of Example 9.2

Factor	00	01	02	10	11	12	20	21	22	Contrast
V_L	-1	-1	-1	0	0	0	$+1$	$+1$	$+1$	10.81
V_Q	$+1$	$+1$	$+1$	-2	-2	-2	$+1$	$+1$	$+1$	-7.83
T_L	-1	0	$+1$	-1	0	$+1$	-1	0	$+1$	6.95
T_Q	$+1$	-2	$+1$	$+1$	-2	$+1$	$+1$	-2	$+1$	2.25
$V_L T_L$	$+1$	0	-1	0	0	0	-1	0	$+1$	0.31
$V_L T_Q$	-1	$+2$	-1	0	0	0	$+1$	-2	$+1$	0.37
$V_Q T_L$	-1	0	$+1$	$+2$	0	-2	-1	0	$+1$	2.45
$V_Q T_Q$	$+1$	-2	$+1$	-2	$+4$	-2	$+1$	-2	$+1$	-8.97
$T_{ij.}$	11.70	13.45	14.27	16.12	15.00	17.62	15.21	16.93	18.09	

Table 9.10   ANOVA Breakdown for Example 9.2

Source	df	SS		MS	F	Prob.
V_i	2	11.44		5.72		
V_L	1		9.74	9.74	121.75***	0.000
V_Q	1		1.70	1.70	21.25**	0.001
T_j	2	4.17		2.08		
T_L	1		4.03	4.03	50.38***	0.000
T_Q	1		0.14	0.14	1.75	0.219
VT_{ij}	4	1.38		0.35		
$V_L T_L$	1		0.01	0.01	<1	
$V_L T_Q$	1		0.00	0.00	<1	
$V_Q T_L$	1		0.25	0.25	3.12	0.111
$V_Q T_Q$	1		1.12	1.12	14.00**	0.005
Error	9	0.70		0.08		

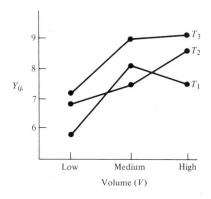

Figure 9.8   Treatment means of Example 9.2.

As an exercise let us also pull out the VT and VT^2 components of interaction from the totals on the diagonals of Table 9.7.

$$SS_{VT} = \frac{(44.48)^2 + (47.66)^2 + (46.25)^2}{6} - \frac{(138.39)^2}{18} = 0.85$$

$$SS_{VT^2} = \frac{(44.79)^2 + (47.32)^2 + (46.28)^2}{6} - \frac{(138.39)^2}{18} = 0.54$$

which is approximately (1.39 versus 1.38) the total interaction.

Since the chief concern in this example is to reduce the amount of noxious gas to a minimum one could examine all nine treatment means in a Newman–Keuls sense and look for conditions that might result in a minimum mean.

The means are

$$5.85, 6.72, 7.13, 7.50, 7.60, 8.06, 8.47, 8.81, 9.04$$

$$S_{\bar{Y}_{ij\cdot}} = \sqrt{\frac{0.08}{2}} = 0.2$$

From Appendix Table E.1 for $n_2 = 9$ df:

$p =$	2	3	4	5	6	7	8	9
	3.20	3.95	4.42	4.76	5.02	5.24	5.43	5.60

and LSRs $= 0.64, 0.79, 0.88, 0.95, 1.00, 1.05, 1.09, 1.12$.

From these the lines can be drawn as shown above. Clearly, the minimum emission is when one has low prechamber volume ratio and the shortest injection time as the mean of 5.85 ppm is significantly lower than any other of the nine means.

9.3 3^3 FACTORIAL

If an experimenter has three factors, each at three levels, or a $3 \times 3 \times 3 = 3^3$ factorial, there are several ways to break down the effects of factors A, B, and C and their associated interactions. If the order of experimentation is completely randomized, the model for such an experiment is

$$Y_{ijk} = \mu + A_i + B_j + AB_{ij} + C_k + AC_{ik} + BC_{jk} + ABC_{ijk} + \varepsilon_{ijk}$$

with the last two terms confounded unless there is replication within the cells. In this model $i = 1, 2, 3$, $j = 1, 2, 3$, and $k = 1, 2, 3$, making 27 treatment combinations. These 27 treatment combinations may be as shown in Table 9.11.

Table 9.11 3^3 Factorial Treatment Combinations

Factor B	Factor C	Factor A 0	1	2
0	0	000	100	200
	1	001	101	201
	2	002	102	202
1	0	010	110	210
	1	011	111	211
	2	012	112	212
2	0	020	120	220
	1	021	121	221
	2	022	122	222

Association of the proper coefficients on these 27 treatment combinations would allow the Table 9.12 breakdown of an ANOVA if all effects were set at quantitative levels.

In an actual problem these three-way interactions would be hard to explain, and quite often the ABC interaction is left with its 8 df for use as an error term to test the main effects A, B, C, and the two-way interactions.

Table 9.12 3^3 Factorial Analysis for Linear and Quadratic Effects

Source	df	Source	df
A_i	2	$A_Q C_L$	1
A_L	1	$A_Q C_Q$	1
A_Q	1	BC_{jk}	4
B_j	2	$B_L C_L$	1
B_L	1	$B_L C_Q$	1
B_Q	1	$B_Q C_L$	1
AB_{ij}	4	$B_Q C_Q$	1
$A_L B_L$	1	ABC_{ijk}	8
$A_L B_Q$	1	$A_L B_L C_L$	1
$A_Q B_L$	1	$A_L B_L C_Q$	1
$A_Q B_Q$	1	$A_L B_Q C_L$	1
C_k	2	$A_L B_Q C_Q$	1
C_L	1	$A_Q B_L C_L$	1
C_Q	1	$A_Q B_L C_Q$	1
AC_{ik}	4	$A_Q B_Q C_L$	1
$A_L C_L$	1	$A_Q B_Q C_Q$	1
$A_L C_Q$	1	Total	26

Table 9.13 3^3 Factorial in 2-df Analyses

Source	df	
A	2	
B	2	
AB	2	} 4
AB^2	2	
C	2	
AC	2	} 4
AC^2	2	
BC	2	} 4
BC^2	2	
ABC	2	
ABC^2	2	} 8
AB^2C	2	
AB^2C^2	2	
Total	26	

Another possible partitioning of these effects is in terms of 2-df effects using I and J components on AB, AC, and BC interactions. These could be designated as AB, AB^2, AC, AC^2, and BC, BC^2, each with 2 df. However, the three-way interaction with its 8 df may need a further breakdown. Sometimes ABC is broken into four 2-df components called $X(ABC)$, $Y(ABC)$, $Z(ABC)$, and $W(ABC)$, or, using the notation of the last section: AB^2C, ABC^2, ABC, and AB^2C^2. Here again no first letter is squared, and $A^2BC = (A^2BC)^2 = A^4B^2C^2 = AB^2C^2$ modulus 3. Such a partitioning would yield Table 9.13.

Example 9.3 A problem involving the effect of three factors, each at three levels, was proposed by Professor Burr of Purdue University. Here the measured variable was yield and the factors that might affect this response were days, operators, and concentrations of solvent. Three days, three operators, and three concentrations were chosen. Days and operators were qualitative effects, concentrations were quantitative and set at 0.5, 1.0, 2.0. Although these are not equispaced, the logarithms of these three levels are equispaced, and the logarithms can then be used if a curve fitting is warranted. For the purposes of this chapter, all levels of all factors will be considered as fixed and the design will be considered as completely randomized. It was decided to take three replications of each of the $3^3 = 27$ treatment combinations. The data, after coding by subtracting 20.0, are as presented in Table 9.14.

Table 9.14   Example Data on 3^3 Factorial with Three Replications

	Day D_i								
	5/14			5/15			5/16		
	Operator O_j								
Concentration C_k	A	B	C	A	B	C	A	B	C
0.5	1.0	0.2	0.2	1.0	1.0	1.2	1.7	0.2	0.5
	1.2	0.5	0.0	0.0	0.0	0.0	1.2	0.7	1.0
	1.7	0.7	−0.3	0.5	0.0	0.5	1.2	1.0	1.7
1.0	5.0	3.2	3.5	4.0	3.2	3.7	4.5	3.7	3.7
	4.7	3.7	3.5	3.5	3.0	4.0	5.0	4.0	4.5
	4.2	3.5	3.2	3.5	4.0	4.2	4.7	4.2	3.7
2.0	7.5	6.0	7.2	6.5	5.2	7.0	6.7	7.5	6.2
	6.5	6.2	6.5	6.0	5.7	6.7	7.5	6.0	6.5
	7.7	6.2	6.7	6.2	6.5	6.8	7.0	6.0	7.0

If these data are analyzed on a purely qualitative basis, the methods of Chapter 5 can be used. The resulting ANOVA is shown in Table 9.15.

The model for this example is merely $Y_{ijkm} = \mu$ plus the sum of the terms in the Source column in Table 9.15. From this analysis the concentration effect is tremendous, and the days, operators, and day × operator interaction are all significant at the 1 percent level of significance.

Since concentrations are at quantitative levels, the linear and quadratic effects of concentrations may be computed, as well as the interactions between linear effect of concentration and days, quadratic effect of concentration and days, linear effect of concentration and operators, and quadratic effect of concentration and operators. It is not usually worthwhile to extract

Table 9.15   First ANOVA for Example 9.3

Source	df	SS	MS
D_i	2	3.48	1.74**
O_j	2	6.14	3.07**
DO_{ij}	4	4.07	1.02**
C_k	2	468.99	234.49***
DC_{ik}	4	0.59	0.15
OC_{jk}	4	0.89	0.22
DOC_{ijk}	8	1.09	0.14
$\varepsilon_{m(ijk)}$	54	9.98	0.18
Totals	80	495.23	

Table 9.16   Cell Totals for $D \times C$ and $O \times C$ Interactions

	(a)				(b)		
	Concentration				Concentration		
Day	0.5	1.0	2.0	Operator	0.5	1.0	2.0
5/14	5.2	34.5	60.5	A	9.5	39.1	61.6
5/15	4.2	33.1	56.6	B	4.3	32.5	55.3
5/16	9.2	38.0	60.4	C	4.8	34.0	60.6
Totals	18.6	105.6	177.5 301.7	Totals	18.6	105.6	177.5 301.7

three-way interaction in this way. To calculate these quantitative effects it is usually helpful to construct some two-way tables for the interactions that are being computed. Two of these are shown as Table 9.16(a) and Table 9.16(b).

From Table 9.16(a), applying the linear and quadratic coefficients to the concentration totals, we have

Sums of Squares

$$C_L = -1(18.6) + 0(105.6) + 1(177.5) = 158.9$$

$$SS_{C_L} = \frac{(158.9)^2}{27(2)} = 467.58$$

$$C_Q = +1(18.6) - 2(105.6) + 1(177.5) = -15.1$$

$$SS_{C_Q} = \frac{(-15.1)^2}{27(6)} = 1.41$$

$$SS_C = 468.99$$

For the $D \times C$ interactions, consider each level of days separately. At

$$5/14\colon C_L = -1(5.2) + 0(34.5) + 1(60.5) = 55.3$$
$$5/15\colon C_L = -1(4.2) + 0(33.1) + 1(56.6) = 52.4$$
$$5/16\colon C_L = -1(9.2) + 0(38.0) + 1(60.4) = 51.2$$

The $D \times C_L$ $SS_{\text{interaction}}$ is then

$$\frac{(55.3)^2 + (52.4)^2 + (51.2)^2}{9(2)} - \frac{(158.9)^2}{27(2)} = 0.49$$

For quadratic effects, at

$$5/14\colon C_Q = +1(5.2) - 2(34.5) + 1(60.5) = -3.3$$
$$5/15\colon C_Q = +1(4.2) - 2(33.1) + 1(56.6) = -5.4$$
$$5/16\colon C_Q = +1(9.2) - 2(38.0) + 1(60.4) = -6.4$$

The $D \times C_Q$ $\mathrm{SS_{interaction}}$ is then

$$\frac{(-3.3)^2 + (-5.4)^2 + (-6.4)^2}{9(6)} - \frac{(-15.1)^2}{27(6)} = 0.09$$

and

$$\mathrm{SS}_{D \times C} = \mathrm{SS}_{D \times C_L} + \mathrm{SS}_{D \times C_Q} = 0.49 + 0.09 = 0.58$$

If the same procedure is now applied to the data of Table 9.16(b), we have

$$\mathrm{SS}_{O \times C_L} = \frac{(52.1)^2 + (51.0)^2 + (55.8)^2}{9(2)} - \frac{(158.9)^2}{27(2)} = 0.70$$

$$\mathrm{SS}_{O \times C_Q} = \frac{(-7.1)^2 + (-5.4)^2 + (-2.6)^2}{9(6)} - \frac{(-15.1)^2}{27(6)} = 0.19$$

and

$$\mathrm{SS}_{O \times C} = \mathrm{SS}_{O \times C_L} + \mathrm{SS}_{O \times C_Q} = 0.70 + 0.19 = 0.89$$

The resulting ANOVA can now be shown in Table 9.17.

Table 9.17   Second ANOVA for Example 9.3

Source	df	SS	MS
D_i	2	3.48	1.74**
O_j	2	6.14	3.07**
DO_{ij}	4	4.07	1.02**
C_L	1	467.58	467.58***
C_Q	1	1.41	1.41**
$D \times C_L$	2	0.49	0.24
$D \times C_Q$	2	0.09	0.04
$O \times C_L$	2	0.70	0.35
$O \times C_Q$	2	0.19	0.09
DOC_{ijk}	8	1.09	0.14
$\varepsilon_{m(ijk)}$	54	9.98	0.18
Totals	80	495.22	

This second analysis shows that the linear effect and the quadratic effect of concentration are extremely significant. Two plots of Figure 9.8 may help in picturing what is really happening in this experiment.

Figure 9.9(a) shows the effect of operators, days, and $D \times O$ interaction. Figure 9.9(b) indicates that the linear effect of concentration far outweighs the quadratic effect and there is no significant interaction. If a straight line or three straight lines were fit to these data, the logs of the concentrations would be used, as the logs are equispaced.

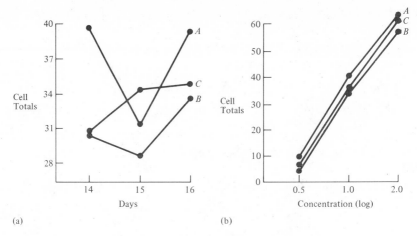

Figure 9.9 Plots of 3^3 example of Table 9.14.

Although this would usually conclude the analysis of this problem, each interaction will be broken down into its diagonal, or I and J, components in order to illustrate the technique. To compute the two diagonal components of the two-factor interactions, the two parts of Table 9.16 can be used, along with a similar table for the $D \times O$ cells (see Table 9.18).

From Table 9.18 the diagonal components of the $D \times O$ interaction are

$I(D \times O)$

$$= \frac{(39.5 + 28.6 + 34.8)^2 + (30.2 + 34.1 + 39.5)^2 + (30.5 + 33.3 + 31.2)^2}{27}$$

$$- \frac{(301.7)^2}{81} = 1.74$$

Call it DO^2.

$$J(D \times O) = \frac{(30.5 + 28.6 + 39.5)^2 + (96.2)^2 + (106.9)^2}{27} - \frac{(301.7)^2}{81}$$

$$= 2.33$$

Table 9.18   Cell Totals for $D \times O$ Interactions

	Operator		
Day	A	B	C
5/14	39.5	30.2	30.5
5/15	31.2	28.6	34.1
5/16	39.5	33.3	34.8

Table 9.19 Cell Totals for $D \times O$ Interaction at Each Level of Concentration

O_i	D_i at C_1			D_i at C_2			D_i at C_3		
A	3.9	1.5	4.1	13.9	11.0	14.2	21.7	18.7	21.2
B	1.4	1.0	1.9	10.4	10.2	11.9	18.4	17.4	19.5
C	−0.1	1.7	3.2	10.2	11.9	11.9	20.4	20.5	19.7

Call it *DO*. These total $1.74 + 2.33 = 4.07$, the $D \times O$ interaction sum of squares.

Applying the same technique to the two parts of Table 9.16 gives

$$I(DC) = DC^2 = \frac{(98.7)^2 + (102.7)^2 + (100.3)^2}{27} - \frac{(301.7)^2}{81} = 0.30$$

$$J(DC) = DC = \frac{(102.8)^2 + (99.8)^2 + (99.1)^2}{27} - \frac{(301.7)^2}{81} = 0.29$$

$$D \times C = 0.59$$

$$I(OC) = OC^2 = \frac{(102.6)^2 + (99.9)^2 + (99.2)^2}{27} - \frac{(301.7)^2}{81} = 0.24$$

$$J(OC) = OC = \frac{(98.9)^2 + (104.0)^2 + (98.8)^2}{27} - \frac{(301.7)^2}{81} = 0.65$$

$$O \times C = 0.89$$

To break down the 8 df of the $D \times O \times C$ interaction, form an $O \times D$ table showing each of the three levels of concentration C as in Table 9.19.

For each of these concentration levels, find the I and J effect totals, for example, at C_1,

$$I \text{ components are } 8.1, 3.3, 7.2$$

$$J \text{ components are } 5.0, 6.1, 7.5$$

Now form a table with these I and J components at each level of C (see Table 9.20).

Table 9.20 Diagonal Totals for Each Level of Concentration

C_k	$I(DO)$			$J(DO)$		
	i_0	i_1	i_2	j_0	j_1	j_2
0.5	8.1	3.3	7.2	5.0	6.1	7.5
1.0	36.0	33.1	36.5	34.6	33.3	37.7
2.0	58.8	58.6	60.1	59.0	56.8	61.7

Treat each half of Table 9.20 as a simple interaction and compute the I and J components. Thus,

$$DO^2C^2 = I[C \times I(DO)] = \frac{(101.3)^2 + (98.6)^2 + (101.8)^2}{27} - \frac{(301.7)^2}{81} = 0.22$$

$$DO^2C = J[C \times I(DO)] = \frac{(99.1)^2 + (99.4)^2 + (103.2)^2}{27} - \frac{(301.7)^2}{81} = 0.39$$

$$DOC^2 = I[C \times J(DO)] = \frac{(100.0)^2 + (102.8)^2 + (98.9)^2}{27} - \frac{(301.7)^2}{81} = 0.30$$

$$DOC = J[C \times J(DO)] = \frac{(99.8)^2 + (102.4)^2 + (99.5)^2}{27} - \frac{(301.7)^2}{81} = 0.19$$

$$\text{total } D \times O \times C = 1.10$$

compared with 1.09 in Table 9.17. This last breakdown into four parts could also have been accomplished by considering the $C \times O$ interaction at three levels of D_i, or the $C \times D$ interaction at three levels of O_j.

The resulting analysis is summarized in Table 9.21. This analysis is in substantial agreement with Tables 9.15 and 9.17. No new tests would be performed on the data in Table 9.21, as they represent only an arbitrary breakdown of the interactions into 2-df components. The purpose of such a breakdown is discussed in subsequent chapters. For testing hypotheses on interaction, these components are added together again.

Table 9.21 Third ANOVA
for 3^3 Experiment

Source	df	SS
D_i	2	3.48
O_j	2	6.14
DO	2	2.33
DO^2	2	1.74
C_k	2	468.99
DC	2	0.29
DC^2	2	0.30
OC	2	0.65
OC^2	2	0.24
DOC	2	0.19
DOC^2	2	0.30
DO^2C	2	0.39
DO^2C^2	2	0.22
$\varepsilon_{m(ijk)}$	54	9.98
Totals	80	495.24

9.4   SUMMARY

The summary at the end of Chapter 6 may now be extended for Part II.

Experiment	Design	Analysis

II. Two or more factors
  A. Factorial
    (crossed)		

	1. Completely randomized $Y_{ijk} = \mu + A_i + B_j$ $\qquad + AB_{ij} + \varepsilon_{k(ij)}, \ldots$ for more factors	1.
	a. General case	a. ANOVA with interactions
	b. 2^f case	b. Yates method or general ANOVA; use (1), a, b, ab, ...
	c. 3^f case	c. General ANOVA; use 00, 10, 20, 01, 11, ... and $A \times B = AB + AB^2$, ... for interaction

PROBLEMS

9.1   Pull-off force in pounds on glued boxes at three temperatures and three humidities with two observations per treatment combination in a completely randomized experiment gives:

	Temperature A		
Humidity B	Cold	Ambient	Hot
50%	0.8 2.8	1.5 3.2	2.5 4.2
70%	1.0 1.6	1.6 1.8	1.8 1.0
90%	2.0 2.2	1.5 0.8	2.5 4.0

Do a complete analysis of this problem by the general methods of Chapter 5.

9.2   Assuming the temperatures in Problem 9.1 are equispaced, extract linear and quadratic effects of both temperature and humidity as well as all components of interaction.

9.3   From Problem 9.1 extract the AB and AB^2 components of interaction.

9.4 Develop a "Yates method" for this 3^2 experiment and check the results with those above.

9.5 When the data were collected in Example 9.2 on ppm of noxious gas, data were also recorded on percent smoke emitted by the 18 engines. Results were as follows:

Time	Volume (V)		
(T)	Low	Medium	High
Short	0.3	0.1	0.4
	0.4	0.4	0.4
Medium	0.1	0.1	0.4
	0.4	0.2	1.2
Long	0.8	0.7	2.5
	2.0	1.6	3.6

Do an ANOVA by the methods of Chapter 5.

9.6 From Problem 9.5 extract linear and quadratic components of both main effects and the interaction.

9.7 From Problem 9.5 extract the VT and the VT^2 components of interaction.

9.8 Plot any results that stand out in Problem 9.6.

9.9 Run further tests to recommend the best volume and time combinations for minimizing smoke in Problem 9.5.

9.10 Compare the results obtained in Problem 9.8 with the results of Example 9.2 in the text and make some overall recommendations about these engine designs.

9.11 A behavior variable on concrete pavements was measured for three surface thicknesses: 3 inches, 4 inches, and 5 inches; three base thicknesses: 0 inch, 3

	Surface Thickness (inches)								
	3			4			5		
Subbase Thickness (inches)	Base Thickness (inches)			Base Thickness (inches)			Base Thickness (inches)		
	0	3	6	0	3	6	0	3	6
4	2.8	4.3	5.7	4.1	5.4	6.7	6.0	6.3	7.1
	2.6	4.5	5.3	4.4	5.5	6.9	6.2	6.5	6.9
8	4.1	5.7	6.9	5.3	6.5	7.7	6.1	7.2	8.1
	4.4	5.8	7.1	5.1	6.7	7.4	5.8	7.1	8.4
12	5.5	7.0	8.1	6.5	7.7	8.8	7.0	8.0	9.1
	5.3	6.8	8.3	6.7	7.5	9.1	7.2	8.3	9.0

inches, and 6 inches, and three subbase thicknesses: 4 inches, 8 inches, and 12 inches. Two observations were made under each of the 27 pavement conditions and complete randomization performed. The results were as shown on page 208.

Do a complete ANOVA of this experiment by the methods of Chapter 5.

9.12 Since all three factors are quantitative and equispaced, determine linear and quadratic effects for each factor and all interaction breakdowns. Test for significance in Problem 9.11.

9.13 Break down the interactions of Problem 9.11 into 2-df components such as AB, AB^2, ABC, AB^2C, and so on.

9.14 Use a Yates method to solve Problem 9.11 and check the results.

9.15 Plot any significant results of Problem 9.11.

9.16 From Example 9.3 fit the proper degree regression equation for the quantitative variable and express it in terms of the original variable.

9.17 The following data are on the wet-film thickness (in mils) of lacquer. The factors studied were: type of resin (two types), gate-blade setting in mils (three settings), and weight fraction of nonvolatile material in the lacquer (three fractions).

	Resin Type					
	1			2		
Gate	Weight Fraction			Weight Fraction		
Setting	0.20	0.25	0.30	0.20	0.25	0.30
2	1.6	1.5	1.5	1.5	1.4	1.6
	1.5	1.3	1.3	1.4	1.3	1.4
4	2.7	2.5	2.4	2.4	2.6	2.2
	2.7	2.5	2.3	2.3	2.4	2.1
6	4.0	3.6	3.5	4.0	3.7	3.4
	3.9	3.8	3.4	4.0	3.6	3.3

Do an ANOVA of the above data and pull out any quantitative effects in terms of their components.

9.18 Using $\alpha = 0.01$, write prediction equations where possible and check them (Problem 9.17).

9.19 If the gate setting is 5 mils and the weight fraction 0.20, what is the best prediction of wet-film thickness? If interaction is ignored, what is the prediction? How much difference does it make? (Problem 9.17.)

9.20 Plot results indicated in Problem 9.18.

10
Fixed, Random, and Mixed Models

10.1 INTRODUCTION

In Chapter 1 it was pointed out that, in the planning stages of an experiment, the experimenter must decide whether the levels of factors considered are to be set at fixed values or are to be chosen at random from many possible levels. In the intervening chapters it has always been assumed that the factor levels were fixed. In practice it may be desirable to choose the levels of some factors at random, depending on the objectives of the experiment. Are the results to be judged for these levels alone or are they to be extended to more levels of which those in the experiment are but a random sample? In the case of some factors such as temperature, time, or pressure, it is usually desirable to pick fixed levels, often near the extremes and at some inter- mediate points, because a random choice might not cover the range in which the experimenter is interested. In such cases of fixed quantitative levels, we often feel safe in interpolating between the fixed levels chosen. Other factors such as operators, days, or batches may often be only a small sample of all possible operators, days, or batches. In such cases the particular operator, day, or batch may not be very important but only whether or not operators, days, or batches in general increase the variability of the experiment.

It is not reasonable to decide after the data have been collected whether the levels are to be considered fixed or random. This decision must be made prior to the running of the experiment, and if random levels are to be used, they must be chosen from all possible levels by a random process. In the case of random levels, it will be assumed that the levels are chosen from an infinite population of possible levels. Bennett and Franklin [4] discuss a case in which the levels chosen are from a finite set of possible levels.

When all levels are fixed, the mathematical model of the experiment is called a *fixed model*. When all levels are chosen at random, the model is called a *random model*. When several factors are involved, some at fixed levels and others at random levels, the model is called a *mixed model*.

10.2   SINGLE-FACTOR MODELS

In the case of a single-factor experiment the factor may be referred to as a *treatment effect*, as in Chapter 3; and if the design is completely randomized, the model is

$$Y_{ij} = \mu + \tau_j + \varepsilon_{ij} \tag{10.1}$$

Whether the treatment levels are fixed or random, it is assumed in this model that μ is a fixed constant and the errors are normally and independently distributed with a zero mean and the same variance, that is, ε_{ij} are NID $(0, \sigma_\varepsilon^2)$. The decision as to whether the levels of the treatments are fixed or random will affect the assumptions about the treatment term τ_j. The different assumptions and other differences will be compared in parallel columns.

Fixed Model	*Random Model*
1. Assumptions: τ_j's are fixed constants. $$\sum_{j=1}^{k} \tau_j = \sum_{j=1}^{k} (\mu_{.j} - \mu) = 0$$ (These add to zero as they are the only treatment means being considered.)	1. Assumptions: τ_j's are random variables and are NID $(0, \sigma_\tau^2)$ (Here σ_τ^2 represents the variance among the τ_j's or among the true treatment means $\mu_{.j}$. The τ_j average to zero when averaged over all possible levels, but for the k levels of the experiment they usually will not average 0.)

Figure 10.1(a) shows three fixed means whose average is μ as these are the only means of concern and $\sum_j \tau_j = \sum_j (\mu_j - \mu) = 0$. Figure 10.1(b) shows three random means whose average is obviously not μ as these are but three means chosen at random from many possible means. These means and their corresponding τ_j's are assumed to form a normal distribution with a standard deviation of σ_τ.

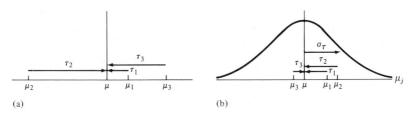

(a)   (b)

Figure 10.1   Assumed means in (a) fixed and (b) random models.

2. Analysis: Procedures as given in Chapter 3 for computing SS.	2. Analysis: Same as for fixed model.

3. EMS:

Source	df	EMS
τ_j	$k - 1$	$\sigma_\varepsilon^2 + n\phi_\tau$
ε_{ij}	$k(n - 1)$	σ_ε^2

3. EMS:

Source	df	EMS
τ_j	$k - 1$	$\sigma_\varepsilon^2 + n\sigma_\tau^2$
ε_{ij}	$k(n - 1)$	σ_ε^2

4. Hypothesis tested:

$$H_0: \tau_j = 0 \quad \text{(for all } j\text{)}$$

4. Hypothesis tested:

$$H_0: \sigma_\tau^2 = 0$$

The expected mean square (EMS) column turns out to be extremely important in more complex experiments as an aid in deciding how to set up an F test for significance. The EMS for any term in the model is the long-range average of the calculated mean square when the Y_{ij} from the model is substituted in algebraic form into the mean square computation. The derivation of these EMS values is often complicated, but those for the single-factor model were derived in Chapter 3 and those for two-factor models are derived in a later section of this chapter.

For the fixed model, if the hypothesis is true that $\tau_j = 0$ for all j, that is, all the k fixed treatment means are equal, then $\sum_j \tau_j^2 = 0$ and the EMS for τ_j and ε_{ij} are both σ_ε^2. Hence the observed mean squares for treatments and error mean square are both estimates of the error variance, and they can be compared by means of an F test. If this F test shows a significantly high value, it must mean that $n \sum_j \tau_j^2 / (k - 1) = n\phi_\tau$ is not zero and the hypothesis is to be rejected.

For the random model, if the hypothesis is true that $\sigma_\tau^2 = 0$, that is, the variance among all treatment means is zero, then again each mean square is an estimate of the error variance. Again, an F test between the two mean squares is appropriate.

From the two tables in step 3 above, it is seen that for a single-factor experiment there is no difference in the test to be made after the analysis, and the only difference is in the generality of the conclusions. If H_0 is rejected, there is probably a difference between the k fixed treatment means for the fixed model; for the random model there is a difference between all treatments of which the k examined are but a random sample.

10.3 TWO-FACTOR MODELS

For two factors A and B the model in the general case is

$$Y_{ijk} = \mu + A_i + B_j + AB_{ij} + \varepsilon_{k(ij)}$$

with

$$i = 1, 2, \ldots, a \quad j = 1, 2, \ldots, b \quad k = 1, 2, \ldots, n$$

provided the design is completely randomized. In this model, it is again assumed that μ is a fixed constant and $\varepsilon_{k(ij)}$'s are NID $(0, \sigma_\varepsilon^2)$. If both A and B are at fixed levels, the model is a fixed model. If both are at random levels, the model is a random model, and if one is at fixed levels and the other at random levels, the model is a mixed model. Comparing each of these models gives:

Fixed	*Random*	*Mixed*
1. Assumptions: A_i's are fixed constants and $$\sum_{i=1}^{a} A_i = 0$$	1. Assumptions: A_i's are NID $(0, \sigma_A^2)$	1. Assumptions: A_i's are fixed and $$\sum_{i}^{a} A_i = 0$$
B_j's are fixed constants and $$\sum_{j=1}^{b} B_j = 0$$	B_j's are NID $(0, \sigma_B^2)$	B_j's are NID $(0, \sigma_B^2)$
AB_{ij}'s are fixed constants and $$\sum_{i} AB_{ij} = 0$$ $$\sum_{j} AB_{ij} = 0$$	AB_{ij}'s are NID $(0, \sigma_{AB}^2)$	AB_{ij}'s are NID $(0, \sigma_{AB}^2)$ but $$\sum_{i}^{a} AB_{ij} = 0$$ $$\sum_{j}^{b} AB_{ij} \neq 0$$ (for A fixed, B random)
2. Analysis: Procedures of Chapter 5 for sums of squares	2. Analysis: Same	2. Analysis: Same
3. EMS:	3. EMS:	3. EMS:

Source	df	EMS (Fixed)	EMS (Random)	EMS (Mixed)
A_i	$a - 1$	$\sigma_\varepsilon^2 + nb\phi_A$	$\sigma_\varepsilon^2 + n\sigma_{AB}^2 + nb\sigma_A^2$	$\sigma_\varepsilon^2 + n\sigma_{AB}^2 + nb\phi_A$
B_j	$b - 1$	$\sigma_\varepsilon^2 + na\phi_B$	$\sigma_\varepsilon^2 + n\sigma_{AB}^2 + na\sigma_B^2$	$\sigma_\varepsilon^2 + na\sigma_B^2$
AB_{ij}	$(a - 1)(b - 1)$	$\sigma_\varepsilon^2 + n\phi_{AB}$	$\sigma_\varepsilon^2 + n\sigma_{AB}^2$	$\sigma_\varepsilon^2 + n\sigma_{AB}^2$
$\varepsilon_{k(ij)}$	$ab(n - 1)$	σ_ε^2	σ_ε^2	σ_ε^2

4. Hypotheses tested: 4. Hypotheses tested: 4. Hypotheses tested:

$H_1:\quad A_i = 0$ for all i ⠀⠀⠀ $H_1:\ \sigma_A^2 = 0$ ⠀⠀⠀⠀⠀ $H_1:\quad A_i = 0$ for all i

$H_2:\quad B_j = 0$ for all j ⠀⠀⠀ $H_2:\ \sigma_B^2 = 0$ ⠀⠀⠀⠀⠀ $H_2:\ \sigma_B^2 = 0$

$H_3: AB_{ij} = 0$ for all ⠀⠀⠀⠀ $H_3: \sigma_{AB}^2 = 0$ ⠀⠀⠀⠀ $H_3: \sigma_{AB}^2 = 0$
⠀⠀⠀⠀⠀⠀ i and j

In the assumptions for the mixed model the fact that summing the interaction term over the fixed factor ($\sum_i$) is zero but summing it over the random factor ($\sum_j$) is not zero affects the expected mean squares, as seen in item 3 on page 213.

For the fixed model the mean squares for A, B, and AB are each compared with the error mean square to test the respective hypotheses, as should be clear from an examination of the EMS column when the hypotheses are true. For the random model the third hypothesis of no interaction is tested by comparing the mean square for interaction with the mean square for error, but the first and second hypotheses are each tested by comparing the mean square for the main effect (A_i or B_j) with the mean square for the interaction as seen by their expected mean square values. For a mixed model the interaction hypothesis is tested by comparing the interaction mean square with the error mean square. The random effect B_j is also tested by comparing its mean square with the error mean square. The fixed effect (A_i), however, is tested by comparing its mean square with the interaction mean square.

From these observations on a two-factor experiment, the importance of the EMS column is evident, as this column can be used to see how the tests of hypotheses should be run. It is also important to note that these EMS expressions can be determined prior to the running of the experiment. This will indicate whether or not a good test of a hypothesis exists. In some cases the proper test indicated by the EMS column will have insufficient degrees of freedom to be sufficiently sensitive, in which case the investigator might wish to change the experiment. This would involve such changes as a choice of more levels of some factors, or changing from random to fixed levels of some factors.

10.4   EMS RULES

The two examples above have shown the importance of the EMS column in determining what tests of significance are to be run after the analysis is completed. Because of the importance of this EMS column in these and more complex models, it is often useful to have some simple method of determining these values from the model for the given experiment. A set of rules can be stated that will determine the EMS column very rapidly, without recourse to their derivation. The rules will be illustrated on the two-factor mixed model of Section 10.3. To determine the EMS column for any model:

1. Write the variable terms in the model as row headings in a two-way table.

A_i
B_j
AB_{ij}
$\varepsilon_{k(ij)}$

2. Write the subscripts in the model as column headings; over each subscript write F if the factor levels are fixed, R if random. Also write the number of observations each subscript is to cover.

	a	b	n
	F	R	R
	i	j	k
A_i			
B_j			
AB_{ij}			
$\varepsilon_{k(ij)}$			

3. For each row (each term in the model) copy the number of observations under each subscript, providing the subscript does not appear in the row heading.

	a	b	n
	F	R	R
	i	j	k
A_i		b	n
B_j	a		n
AB_{ij}			n
$\varepsilon_{k(ij)}$			

4. For any bracketed subscripts in the model, place a 1 under those subscripts that are inside the brackets.

	a	b	n
	F	R	R
	i	j	k
A_i		b	n
B_j	a		n
AB_{ij}			n
$\varepsilon_{k(ij)}$	1	1	

5. Fill the remaining cells with a 0 or a 1, depending upon whether the subscript represents a fixed F or a random R factor.

	a	b	n
	F	R	R
	i	j	k
A_i	0	b	n
B_j	a	1	n
AB_{ij}	0	1	n
$\varepsilon_{k(ij)}$	1	1	1

6. To find the expected mean square for any term in the model:

   a. Cover the entries in the column (or columns) that contain non-bracketed subscript letters in this term in the model (for example, for A_i, cover column i; for $\varepsilon_{k(ij)}$, cover column k).

   b. Multiply the remaining numbers in each row. Each of these products is the coefficient for its corresponding term in the model, provided the subscript on the term is also a subscript on the term whose expected mean square is being determined. The sum of these coefficients multiplied by the variance of their corresponding terms (ϕ_τ or σ_τ^2) is the EMS of the term being considered (for example, for A_i, cover column i). The products of the remaining coefficients are bn, n, n, and 1, but the first n is not used, as there is no i in its term (B_j). The resulting EMS is then $bn\phi_A + n\sigma_{AB}^2 + 1 \cdot \sigma_\varepsilon^2$. For all terms, these rules give:

	a	b	n	
	F	R	R	
	i	j	k	EMS
A_i	0	b	n	$\sigma_\varepsilon^2 + n\sigma_{AB}^2 + nb\phi_A$
B_j	a	1	n	$\sigma_\varepsilon^2 + na\sigma_B^2$
AB_{ij}	0	1	n	$\sigma_\varepsilon^2 + n\sigma_{AB}^2$
$\varepsilon_{k(ij)}$	1	1	1	σ_ε^2

These results are seen to be in agreement with the EMS values for the mixed model in Section 10.3. Here ϕ_A is, of course, a fixed type of variance

$$\phi_A = \frac{\sum_i A_i^2}{a - 1}$$

Although the rules seem rather involved, they become very easy to use with a bit of practice. Two examples will illustrate the concept.

Example 10.1 The viscosity of a slurry is to be determined by four randomly selected laboratory technicians. Material from each of five mixing machines is bottled and divided in such a way as to provide two samples for each technician to test for viscosity. These are the only mixing machines of interest and the samples can be presented to the technicians in a completely randomized order.

The model here assumes four random technicians and five fixed mixing machines, and each technician measures samples of each machine twice. The model is shown as the vertical column of Table 10.1 and the remainder of the table shows how the EMS column is determined.

Table 10.1 EMS for Example 10.1

Source	df	4 R i	5 F j	2 R k	EMS
T_i	3	1	5	2	$\sigma_\varepsilon^2 + 10\sigma_T^2$
M_j	4	4	0	2	$\sigma_\varepsilon^2 + 2\sigma_{TM}^2 + 8\phi_M$
TM_{ij}	12	1	0	2	$\sigma_\varepsilon^2 + 2\sigma_{TM}^2$
$\varepsilon_{k(ij)}$	20	1	1	1	σ_ε^2

The proper F tests are quite obvious from Table 10.1 and all tests have adequate degrees of freedom for a reasonable test.

Example 10.2 An industrial engineering student wished to determine the effect of five different clearances on the time required to position and assemble mating parts. As all such experiments involve operators, it was natural to consider a random sample of operators to perform the experiment. He also decided the part should be assembled directly in front of the operator and at arm's length from the operator. He tried four different angles, from $0°$ directly in front of the operator through $30°$, $60°$, and $90°$ from this position. Thus four factors were involved, any one of which might affect the time required to position and assemble the part. The experimenter decided to replicate each setup six times and to randomize completely the order of experimentation. Here operators O_i were at random levels (six being chosen), angles A_j at four fixed levels ($0°$, $30°$, $60°$, $90°$), clearances C_k at five fixed levels and locations L_m fixed either in front of or at arm's length from the operator. This is a $6 \times 4 \times 5 \times 2$ factorial experiment with six replications, run in a completely randomized design. The expected mean square values can be determined from the rules given in Section 10.4 as shown in Table 10.2.

Table 10.2   EMS for Clearance Problem

Source	df	6 R i	4 F j	5 F k	2 F m	6 R q	EMS
O_i	5	1	4	5	2	6	$\sigma_\varepsilon^2 + 240\sigma_O^2$
A_j	3	6	0	5	2	6	$\sigma_\varepsilon^2 + 60\sigma_{OA}^2 + 360\phi_A$
OA_{ij}	15	1	0	5	2	6	$\sigma_\varepsilon^2 + 60\sigma_{OA}^2$
C_k	4	6	4	0	2	6	$\sigma_\varepsilon^2 + 48\sigma_{OC}^2 + 288\phi_C$
OC_{ik}	20	1	4	0	2	6	$\sigma_\varepsilon^2 + 48\sigma_{OC}^2$
AC_{jk}	12	6	0	0	2	6	$\sigma_\varepsilon^2 + 12\sigma_{OAC}^2 + 72\phi_{AC}$
OAC_{ijk}	60	1	0	0	2	6	$\sigma_\varepsilon^2 + 12\sigma_{OAC}^2$
L_m	1	6	4	5	0	6	$\sigma_\varepsilon^2 + 120\sigma_{OL}^2 + 720\phi_L$
OL_{im}	5	1	4	5	0	6	$\sigma_\varepsilon^2 + 120\sigma_{OL}^2$
AL_{jm}	3	6	0	5	0	6	$\sigma_\varepsilon^2 + 30\sigma_{OAL}^2 + 180\phi_{AL}$
OAL_{ijm}	15	1	0	5	0	6	$\sigma_\varepsilon^2 + 30\sigma_{OAL}^2$
CL_{km}	4	6	4	0	0	6	$\sigma_\varepsilon^2 + 24\sigma_{OCL}^2 + 144\phi_{CL}$
OCL_{ikm}	20	1	4	0	0	6	$\sigma_\varepsilon^2 + 24\sigma_{OCL}^2$
ACL_{jkm}	12	6	0	0	0	6	$\sigma_\varepsilon^2 + 6\sigma_{OACL}^2 + 36\phi_{ACL}$
$OACL_{ijkm}$	60	1	0	0	0	6	$\sigma_\varepsilon^2 + 6\sigma_{OACL}^2$
$\varepsilon_{q(ijkm)}$	1200	1	1	1	1	1	σ_ε^2

From this table it is easily seen that all interactions involving operators and the operator main effect are tested against the error mean square at the bottom of the table. All interactions and main effects involving fixed factors are tested by the mean square just below them in the table.

The rules given in this section are general enough to be applied to the most complex designs, as will be seen in later chapters.

10.5   EMS DERIVATIONS

Single-Factor Experiment

The EMS expressions for a single-factor experiment were derived in Chapter 3, Section 3.6.

Two-Factor Experiment

Using the definitions of expected values given in Chapter 2 and the procedures of Section 3.6 the EMS expressions may be derived for the two-factor experiment.

For a two-factor experiment, the model is

$$Y_{ijk} = \mu + A_i + B_j + AB_{ij} + \varepsilon_{k(ij)} \tag{10.2}$$

with

$$i = 1, 2, \ldots, a \qquad j = 1, 2, \ldots, b \qquad k = 1, 2, \ldots, n$$

The sum of squares for factor A is

$$\text{SS}_A = \sum_{i=1}^{a} nb(\bar{Y}_{i..} - \bar{Y}_{...})^2$$

From the model in Equation (10.2)

$$\bar{Y}_{i..} = \sum_{j}^{b} \sum_{k}^{n} Y_{ijk}/bn = \sum_{j}^{b} \sum_{k}^{n} (\mu + A_i + B_j + AB_{ij} + \varepsilon_{k(ij)})/bn$$

$$\bar{Y}_{i..} = \mu + A_i + \sum_{j}^{b} B_j/b + \sum_{j}^{b} AB_{ij}/b + \sum_{j}^{b} \sum_{k}^{n} \varepsilon_{k(ij)})/bn$$

$$\bar{Y}_{...} = \sum_{i}^{a} \sum_{j}^{b} \sum_{k}^{n} Y_{ijk}/abn$$

$$= \sum_{i}^{a} \sum_{j}^{b} \sum_{k}^{n} (\mu + A_i + B_j + AB_{ij} + \varepsilon_{k(ij)})/abn$$

$$\bar{Y}_{...} = \mu + \sum_{i}^{a} A_i/a + \sum_{j}^{b} B_j/b + \sum_{i}^{a} \sum_{j}^{b} AB_{ij}/ab$$

$$+ \sum_{i}^{a} \sum_{j}^{b} \sum_{k}^{n} \varepsilon_{k(ij)}/nab$$

Subtracting gives

$$\bar{Y}_{i..} - \bar{Y}_{...} = \left(A_i - \sum_{i}^{a} A_i/a \right) + \left(\sum_{j}^{b} AB_{ij}/b - \sum_{i}^{a} \sum_{j}^{b} AB_i/ab \right)$$

$$+ \left(\sum_{j}^{b} \sum_{k}^{n} \varepsilon_{k(ij)}/bn - \sum_{i}^{a} \sum_{j}^{b} \sum_{k}^{n} \varepsilon_{k(ij)}/abn \right)$$

Note that the B effect cancels out of the A sum of squares as it should, since A and B are orthogonal effects in a factorial experiment. Squaring and adding gives

$$\text{SS}_A = nb \sum_{i=1}^{a} \left(A_i - \sum_{i}^{a} A_i/a \right)^2$$

$$+ nb \sum_{i=1}^{a} \left(\sum_{j}^{b} AB_{ij}/b - \sum_{i}^{a} \sum_{j}^{b} AB_{ij}/ab \right)^2$$

$$+ nb \sum_{i=1}^{a} \left(\sum_{j}^{b} \sum_{k}^{n} \varepsilon_{k(ij)}/bn - \sum_{i}^{a} \sum_{j}^{b} \sum_{k}^{n} \varepsilon_{k(ij)}/abn \right)^2$$

$$+ \text{(cross-product terms)}$$

Taking the expected value for a fixed model where

$$\sum_i A_i = 0 \qquad \sum_{i\ or\ j} AB_{ij} = 0$$

the result is

$$E(SS_A) = nb \sum_i^a A_i^2 + 0 + \frac{nb}{n^2b^2}(abn - bn)\sigma_\varepsilon^2$$

$$E(MS_A) = E[SS_A/(a-1)] = nb \sum_{i=1}^a A_i^2/(a-1) + \sigma_\varepsilon^2 = nb\phi_A + \sigma_\varepsilon^2$$

which agrees with Section 10.3 for the fixed model.

If now the levels of A and B are random,

$$\sum_i A_i \neq 0 \qquad \sum_{i\ or\ j} AB_{ij} \neq 0$$

$$E(SS_A) = nb(a-1)\sigma_A^2 + \frac{nb}{b^2}(ab - b)\sigma_{AB}^2 + \frac{nb}{n^2b^2}(abn - bn)\sigma_\varepsilon^2$$

$$E(MS_A) = nb\sigma_A^2 + n\sigma_{AB}^2 + \sigma_\varepsilon^2$$

as stated in Section 10.3 for a random model.

If the model is mixed with A fixed and B random,

$$\sum_i^a A_i = 0 \qquad \sum_i^a AB_{ij} = 0$$

but

$$\sum_j^b AB_{ij} \neq 0$$

then

$$E(SS_A) = nb \sum_i^a A_i^2 + \frac{nb}{b^2}(ab - b)\sigma_{AB}^2 + \frac{nb}{n^2b^2}(nab - nb)\sigma_\varepsilon^2$$

$$E(MS_A) = nb \sum_i^a A_i^2/(a-1) + n\sigma_{AB}^2 + \sigma_\varepsilon^2 = nb\phi_A + n\sigma_{AB}^2 + \sigma_\varepsilon^2$$

which agrees with the value stated in Section 10.3 for a mixed model.

Using these expected value methods, one can derive all EMS values given in Section 10.3. It might be noted that if A were random in the mixed model

$$\sum_j^a AB_{ij} = 0$$

then the interaction term would not appear in the factor A sum of squares. This is true of B in the mixed model of Section 10.3.

These few derivations should be sufficient to show the general method of derivation and to demonstrate the advantages of the simple rules in Section 10.4 in determining these EMS values.

10.6   THE PSEUDO-F TEST

Occasionally the EMS column for a given experiment indicates that there is no exact F test for one or more factors in the design model. Consider the following example.

Example 10.3   Two days in a given month were randomly selected in which to run an experiment. Three operators were also selected at random from a large pool of available operators. The experiment consisted of measuring the dry-film thickness of varnish in mils for three different gate settings: 2, 4, and 6 mils. Two determinations were made by each operator each day and at each of the three gate settings. Results are shown in Table 10.3.

Table 10.3   Dry-Film Thickness Experiment

Gate Setting	Day					
	1			2		
	Operator			Operator		
	A	*B*	*C*	*A*	*B*	*C*
2	0.38	0.39	0.45	0.40	0.39	0.41
	0.40	0.41	0.40	0.40	0.43	0.40
4	0.63	0.72	0.78	0.68	0.77	0.85
	0.59	0.70	0.79	0.66	0.76	0.84
6	0.76	0.95	1.03	0.86	0.86	1.01
	0.78	0.96	1.06	0.82	0.85	0.98

Assuming that days and operators are random effects, gate settings are fixed, and the design is completely randomized, the analysis yields Table 10.4.

Table 10.4   Analysis of Dry-Film Thickness Experiment

Source	df	SS	MS	EMS
Days D	1	0.0010	0.0010	$\sigma_\varepsilon^2 + 6\sigma_{DO}^2 + 18\sigma_D^2$
Operators O	2	0.1121	0.0560	$\sigma_\varepsilon^2 + 6\sigma_{DO}^2 + 12\sigma_O^2$
$D \times O$ interaction	2	0.0060	0.0030**	$\sigma_\varepsilon^2 + 6\sigma_{DO}^2$
Gate setting G	2	1.5732	0.7866**	$\sigma_\varepsilon^2 + 2\sigma_{DOG}^2 + 4\sigma_{OG}^2$ $+ 6\sigma_{DG}^2 + 12\phi_G$
$D \times G$ interaction	2	0.0113	0.0056	$\sigma_\varepsilon^2 + 2\sigma_{DOG}^2 + 6\sigma_{DG}^2$
$O \times G$ interaction	4	0.0428	0.0107	$\sigma_\varepsilon^2 + 2\sigma_{DOG}^2 + 4\sigma_{OG}^2$
$D \times O \times G$ interaction	4	0.0099	0.0025**	$\sigma_\varepsilon^2 + 2\sigma_{DOG}^2$
Error	18	0.0059	0.0003	σ_ε^2
Totals	35	1.7622		

** Two asterisks indicate significance at the 1-percent level.

All F tests are clear from the EMS column except for the test on gate setting, which is probably the most important factor in the experiment. Only the three-way interaction shows significance. It is obvious from the results that gate setting is the most important factor, but how can it be tested? If the $D \times G$ interaction is assumed to be zero, then the gate setting can be tested against the $O \times G$ interaction term. On the other hand, if the $O \times G$ interaction is assumed to be zero, gate setting can be tested against the $D \times G$ interaction term. Although neither of these interactions is significant at the 5 percent level, both are numerically larger than the $D \times O \times G$ interaction against which they are tested. In this case any test on G is contingent upon these tests on interaction. One method for testing hypotheses in such situations was developed by Satterthwaite and is given in Bennett and Franklin [4, pp. 367–368].

The scheme consists of constructing a mean square as a linear combination of the mean squares in the experiment, where the EMS for this mean squares includes the same terms as in the EMS of the term being tested, except for the variance of that term. For example, to test the gate-setting effect G in Table 10.4 a mean square is to be constructed whose expected value is

$$\sigma_\varepsilon^2 + 2\sigma_{DOG}^2 + 4\sigma_{OG}^2 + 6\sigma_{DG}^2$$

This can be found by the linear combination

$$MS = MS_{DG} + MS_{OG} - MS_{DOG}$$

as its expected value is

$$E(MS) = \sigma_\varepsilon^2 + 2\sigma_{DOG}^2 + 4\sigma_{OG}^2 + \sigma_\varepsilon^2 + 2\sigma_{DOG}^2 + 6\sigma_{DG}^2 - \sigma_\varepsilon^2 - 2\sigma_{DOG}^2$$
$$= \sigma_\varepsilon^2 + 2\sigma_{DOG}^2 + 4\sigma_{OG}^2 + 6\sigma_{DG}^2$$

An F test can now be constructed using the mean square for gate setting as the numerator and this mean square as the denominator. Such a test is called a pseudo-F or F' test. The real problem here is to determine the degrees of freedom for the denominator mean square. According to Bennett and Franklin, if

$$MS = a_1(MS)_1 + a_2(MS)_2 + \cdots$$

and $(MS)_1$ is based on v_1 df, $(MS)_2$ is based on v_2 df, and so on, then the degrees of freedom for MS are

$$v = \frac{(MS)^2}{a_1^2[(MS)_1^2/v_1] + a_2^2[(MS)_2^2/v_2] + \cdots}$$

In the case of testing for the gate-setting effect above, $a_1 = 1$, $a_2 = 1$, and $a_3 = -1$ and the degrees of freedom are $v_1 = 4$, $v_2 = 2$, and $v_3 = 4$. Here

$MS = 0.0107 + 0.0056 - 0.0025 = 0.0138$ and its df is

$$v = \frac{(0.0138)^2}{(1)^2[(0.0107)^2/4] + (1)^2[(0.0056)^2/2] + (-1)^2[(0.0025)^2/4]}$$

$$= \frac{1.9044 \times 10^{-4}}{0.4586 \times 10^{-4}} = 4.2$$

Hence the F' test is

$$F' = \frac{MS_G}{MS} = \frac{0.7866}{0.0138} = 57.0$$

with 2 and 4.2 df, which is significant at the 1 percent level of significance based on F with 2 and 4 or 2 and 5 df.

10.7 REMARKS

The examples in this chapter should be sufficient to show the importance of the EMS column in deciding just what mean squares should be compared in an F test of a given hypothesis. This EMS column is also useful (usually in random models) to solve for components of variance as illustrated in Chapter 3, Section 3.6.

One special case is of interest. In a two-factor factorial when there is but one observation per cell ($k = 1$), the EMS columns of Section 10.3 reduce to those in Table 10.5.

Table 10.5   EMS for One Observation Per Cell

Source	EMS (Fixed)	EMS (Random)	EMS (Mixed)
A_i	$\sigma_\varepsilon^2 + b\phi_A$	$\sigma_\varepsilon^2 + \sigma_{AB}^2 + b\sigma_A^2$	$\sigma_\varepsilon^2 + \sigma_{AB}^2 + b\phi_A$
B_j	$\sigma_\varepsilon^2 + a\phi_B$	$\sigma_\varepsilon^2 + \sigma_{AB}^2 + a\sigma_B^2$	$\sigma_\varepsilon^2 + a\sigma_B^2$
AB_{ij} or ε_{ij}	$\sigma_\varepsilon^2 + \phi_{AB}$	$\sigma_\varepsilon^2 + \sigma_{AB}^2$	$\sigma_\varepsilon^2 + \sigma_{AB}^2$

A glance at these EMS values will show that there is no test for the main effects A and B in a fixed model, as interaction and error are hopelessly confounded. The only test possible is to assume that there is no interaction; then $\phi_{AB} = 0$, and the main effects are tested against the error. If a no-interaction assumption is not reasonable from information outside the experiment, the investigator should not run one observation per cell but should replicate the data in a fixed model.

For a random model both main effects can be tested whether interaction is present or not. For a mixed model there is a test for the fixed effect A but no test for the random effect B. This may not be a serious drawback, since the fixed effect is often the most important; the B effect is included chiefly for reduction of the error term. Such a situation is seen in a randomized block design where treatments are fixed, but blocks may be chosen at random.

In the discussion of the single-factor experiment it was assumed that there were equal sample sizes n for each treatment. If this is not the case, it can be shown that the expected treatment mean square is

$$\sigma_\varepsilon^2 + n_0 \sigma_\tau^2 \quad \text{or} \quad \sigma_\varepsilon^2 + n_0 \phi_\tau$$

and

$$n_0 = \frac{N^2 - \sum_{j=1}^{k} n_j^2}{(k-1)N}$$

where

$$N = \sum_{j=1}^{k} n_j$$

The test for treatment effect is to compare the treatment mean square with the error mean square; the use of n_0 is primarily for computing components of variance.

PROBLEMS

10.1   An experiment is run on the effects of three randomly selected operators and five fixed aluminizers on the aluminum thickness of a TV tube. Two readings are made for each operator–aluminizer combination. The following ANOVA table is compiled.

Source	df	SS	MS
Operators	2	107,540	53,770
Aluminizers	4	139,805	34,951
$O \times A$ interaction	8	84,785	10,598
Error	15	230,900	15,393
Totals	29	563,030	

Assuming complete randomization, determine the EMS column for this problem and make the indicated significance tests.

10.2   Consider a three-factor experiment where factor A is at a levels, factor B at b levels, and factor C at c levels. The experiment is to be run in a completely randomized manner with n observations for each treatment combination. Assuming factor A is run at a random levels and both B and C at fixed levels, determine the EMS column and indicate what tests would be made after the analysis.

10.3   Repeat Problem 10.2 with A and B at random levels, but C at fixed levels.

10.4   Repeat Problem 10.2 with all three factors at random levels.

10.5   Consider the completely randomized design of a four-factor experiment similar to Example 10.2. Assuming factors A and B are at fixed levels and C and D are at random levels, set up the EMS column and indicate the tests to be made.

10.6   Assuming three factors are at random levels and one is at fixed levels in Problem 10.5, work out the EMS column and the tests to be run.

10.7   A physical education experiment will be conducted to investigate the effects of four types of exercise on heart rate. Five subjects are chosen at random from a physical education class. A subject does each exercise twice. The 40 measurements are done in a completely randomized order with each subject allowed at least 10 minutes' rest between exercises.
a. Give the model. State the parameters of its random variables.
b. Work out the expected mean squares of all the effects. Also give the degrees of freedom for the effects.
c. Tell how to form the F's for the tests that can be made.

10.8   Four elementary schools are chosen at random in a large city system and two methods of instruction are tried in the third, fourth, and fifth grades,
a. Outline an ANOVA layout to test the effect of schools, methods, and grades on gain in reading scores assuming ten students per class.
b. Work out the EMS column and indicate what tests are possible in this situation.

10.9   As part of an experiment on testing viscosity four operators were chosen at random and four instruments were randomly selected from many instruments available. Each operator was to test a sample of product twice on each instrument. Results showed:

Source	df	MS
O_i	3	1498
I_j	3	1816
OI_{ij}	9	218
$\varepsilon_{k(ij)}$	16	67

Determine the EMS column and find the percentage of the total variance in these readings that is attributable to each term in the model, if it is significant.

10.10 Six different formulas or mixes of concrete are to be purchased from five competing suppliers. Tests of crushing strength are to be made on two blocks constructed according to each formula-supplier combination. One block will be poured, hardened, and tested for each combination before the second block is formed, thus making two replications of the whole experiment. Assuming mixes and suppliers fixed and replications random, set up a mathematical model for this situation, outline its ANOVA table, and show what F tests are appropriate.

10.11 Determine the EMS column for Problem 3.6 and solve for components of variance.

10.12 Derive the expression for n_0 in the EMS column of a single-factor completely randomized experiment where the n's are unequal.

10.13 Verify the numerical results of Table 10.4 using the methods of Chapter 6.

10.14 Verify the EMS column of Table 10.4 by the method of Section 10.4.

10.15 Set up F' tests for Problem 10.4 and explain how the df would be determined.

10.16 Set up F' tests for Problem 10.5 and explain how the df would be determined.

11

Nested and Nested-Factorial Experiments

11.1 INTRODUCTION

Example 11.1 In a recent in-plant training course the members of the class were assigned a final problem. Each class member was to go into the plant and set up an experiment using the techniques that had been discussed in class. One engineer wanted to study the strain readings of glass cathode supports from five different machines. Each machine had four "heads" on which the glass was formed, and she decided to take four samples from each head. She treated this experiment as a 5 × 4 factorial with four replications per cell. Complete randomization of the testing for strain readings presented no problem. Her model was

$$Y_{ijk} = \mu + M_i + H_j + MH_{ij} + \varepsilon_{k(ij)}$$

with

$$i = 1, 2, \ldots, 5 \qquad j = 1, \ldots, 4 \qquad k = 1, \ldots, 4$$

Her data and analysis appear in Table 11.1. In this model she assumed that both machines and heads were fixed, and used the 10 percent significance level. The results indicated no significant machine or head effect on strain readings, but there was a significant interaction at the 10 percent level of significance.

The question was raised as to whether the four heads were actually removed from machine *A* and mounted on machine *B*, then on *C*, and so on. Of course, the answer was "no" as each machine had its own four heads. Thus machines and heads did not form a factorial experiment, as the heads on each machine were unique for that particular machine. In such a case the experiment is called a *nested experiment*: levels of one factor are nested within, or are subsamples of, levels of another factor. Such experiments are also sometimes called *hierarchical* experiments.

Table 11.1   Data and ANOVA for Strain-Reading Problem

Head	Machine				
	A	B	C	D	E
1	6	10	0	11	1
	2	9	0	0	4
	0	7	5	6	7
	8	12	5	4	9
2	13	2	10	5	6
	3	1	11	10	7
	9	1	6	8	0
	8	10	7	3	3
3	1	4	8	1	3
	10	1	5	8	0
	0	7	0	9	2
	6	9	7	4	2
4	7	0	7	0	3
	4	3	2	8	7
	7	4	5	6	4
	9	1	4	5	0

Source	df	SS	MS	EMS	F	$F_{0.90}$
M_i	4	45.08	11.27	$\sigma_\varepsilon^2 + 16\phi_M$	1.05	2.04
H_j	3	46.45	15.48	$\sigma_\varepsilon^2 + 20\phi_H$	1.45	2.18
MH_{ij}	12	236.42	19.70	$\sigma_\varepsilon^2 + 4\phi_{MH}$	1.84	1.66
$\varepsilon_{k(ij)}$	60	642.00	10.70			
Totals	79	969.95				

11.2   NESTED EXPERIMENTS

The preceding example can now be reanalyzed by treating it as a nested experiment, since heads are nested within machines. Such a factor may be represented in the model as $H_{j(i)}$, where j covers all levels 1, 2, . . . within the ith level of M_i. The number of levels of the nested factor need not be the same for all levels of the other factor. They are all equal in this problem, that is, $j = 1, 2, 3, 4$ for all i. The errors, in turn, are nested within the levels of i and j; $\varepsilon_{k(ij)}$ and $k = 1, 2, 3, 4$ for all i and j.

To emphasize the fact that the heads on each machine are different heads, the data layout in Table 11.2 shows heads 1, 2, 3, and 4 on machine A; heads 5, 6, 7, and 8 on machine B; and so on.

Table 11.2   Data for Strain-Reading Problem in a Nested Arrangement

Machine	A				B				C				D				E			
Head	1	2	3	4	5	6	7	8	9	10	11	12	13	14	15	16	17	18	19	20
	6	13	1	7	10	2	4	0	0	10	8	7	11	5	1	0	1	6	3	3
	2	3	10	4	9	1	1	3	0	11	5	2	0	10	8	8	4	7	0	7
	0	9	0	7	7	1	7	4	5	6	0	5	6	8	9	6	7	0	2	4
	8	8	6	9	12	10	9	1	5	7	7	4	4	3	4	5	9	3	2	0
Head totals	16	33	17	27	38	14	21	8	10	34	20	18	21	26	22	19	21	16	7	14
Machine totals		93				81				82				88				58		

As the heads that are mounted on the machine can be chosen from many possible heads, we might consider the four heads as a random sample of heads that might be used on a given machine. If such heads were selected at random for the machines, the model would be

$$Y_{ijk} = \mu + M_i + H_{j(i)} + \varepsilon_{k(ij)}$$

with

$$i = 1, \ldots, 5 \qquad j = 1, \ldots, 4 \qquad k = 1, \ldots, 4$$

This nested model has no interaction present, as the heads are not crossed with the five machines. If heads are considered random and machines fixed, the proper EMS values can be determined by the rules given in Chapter 10, as shown in Table 11.3.

This breakdown shows that the head effect is to be tested against the error, and the machine effect is to be tested against the heads-within-machines effect.

To analyze the data for a nested design, first determine the total sum of squares

$$SS_{total} = 6^2 + 2^2 + \cdots + 4^2 + 0^2 - \frac{(402)^2}{80} = 969.95$$

Table 11.3   EMS for Nested Experiment

	5	4	4	
	F	R	R	
Source	i	j	k	EMS
M_i	0	4	4	$\sigma_\varepsilon^2 + 4\sigma_H^2 + 16\phi_M$
$H_{j(i)}$	1	1	4	$\sigma_\varepsilon^2 + 4\sigma_H^2$
$\varepsilon_{k(ij)}$	1	1	1	σ_ε^2

and for machines

$$SS_M = \frac{(93)^2 + (81)^2 + (82)^2 + (88)^2 + (58)^2}{16} - \frac{(402)^2}{80} = 45.08$$

To determine the sum of squares between heads within machines, consider each machine separately.

Machine A:

$$SS_H = \frac{(16)^2 + (33)^2 + (17)^2 + (27)^2}{4} - \frac{(93)^2}{16} = 50.19$$

Machine B:

$$SS_H = \frac{(38)^2 + (14)^2 + (21)^2 + (8)^2}{4} - \frac{(81)^2}{16} = 126.18$$

Machine C:

$$SS_H = \frac{(10)^2 + (34)^2 + (20)^2 + (18)^2}{4} - \frac{(82)^2}{16} = 74.75$$

Machine D:

$$SS_H = \frac{(21)^2 + (26)^2 + (22)^2 + (19)^2}{4} - \frac{(88)^2}{16} = 6.50$$

Machine E:

$$SS_H = \frac{(21)^2 + (16)^2 + (7)^2 + (14)^2}{4} - \frac{(58)^2}{16} = 25.25$$

$$\text{total } SS_H = 282.87$$

The error sum of squares by subtraction is then

$$969.95 - 45.08 - 282.87 = 642.00$$

The degrees of freedom between heads within machine A are $4 - 1 = 3$; for all five machines the degrees of freedom will be $5 \times 3 = 15$. The analysis follows in Table 11.4.

From this analysis, machines appear to have no significant effect on strain readings, but there is a slightly significant (10 percent level) effect of heads within machines on the strain readings. Note that what the experimenter took as head effect (3 df) and interaction effect (12 df) is really heads-within-machines effect (15 df). These results might suggest a more careful adjustment between heads on the same machine. Since there are significant differences between heads within machines, the heads-within-machines sum of squares of 282.87 might well be analyzed further. Returning

Table 11.4 ANOVA for Nested Strain-Reading Problem

Source	df	SS	MS	EMS	F	$F_{0.90}$
M_i	4	45.08	11.27	$\sigma_\varepsilon^2 + 4\sigma_H^2 + 16\phi_M$	<1	2.36
$H_{j(i)}$	15	282.87	18.85	$\sigma_\varepsilon^2 + 4\sigma_H^2$	1.76	1.60
$\varepsilon_{k(ij)}$	60	642.00	10.70	σ_ε^2		
Totals	79	969.95				

to the method by which this term was computed, one can look at its partitioning into five parts with 3 df per machine, and since all parts are orthogonal, a further breakdown and tests of Table 11.4 can be made as in Table 11.5.

Table 11.5 Detailed ANOVA of Example 11.1

Source	df	SS		MS	F	$F_{0.90}$
M_i	4	45.08		11.27	<1	2.36
$H_{j(i)}$	15	282.87				
$H_{j(A)}$	3		50.19	16.73	1.56	2.18
$H_{j(B)}$	3		126.18	42.06	3.93*	2.18
$H_{j(C)}$	3		74.75	24.92	2.33*	2.18
$H_{j(D)}$	3		6.50	2.17	<1	2.18
$H_{j(E)}$	3		25.25	8.42	<1	2.18
$\varepsilon_{k(ij)}$	60	642.00		10.70		

* Significant at 10 percent level.

From Table 11.5, using a 10 percent significance level, we see a significant difference between heads on machine B and machine C. One might then examine the head means within each of these machines in a Newman–Keuls sense.

For machine B, means are: 2.00 3.50 5.25 9.50
For heads: 8 6 7 5

$$s_{\bar{Y}} = \sqrt{\frac{10.70}{4}} = 1.64 \text{ with 60 df for } n_2.$$

From Appendix Table E.1 (although one should have a 10 percent table)

p: 2 3 4
2.83 3.40 3.74

LSR: 4.64 5.58 6.13

So groups are 2.00 3.50 5.25 9.50

One should then examine head 5 as its mean is larger than two of the other three.

For machine C means are 2.50, 4.50, 5.00, 8.50 for heads 9, 12, 11, 10, and since the largest gap is $8.5 - 2.5 = 6$, which is less than 6.13, no significant differences are detected. This is so because the differences are significant at the 10 percent level and not at the 5 percent level on which Table E.1 is based. One might still examine the heads on machine C for lack of consistency. In a nested model it is also seen that the 5 SS between heads within each machine are "pooled," or added, to give 282.87. This assumes that these sums of squares (which are proportional to variances) within each machine are of about the same magnitude. This might be questioned, as these sums of squares are 50.19, 126.18, 74.75, 6.50, and 25.25. If these five are really different, it appears as if the greatest variability is in machine B. This should indicate the need for work on each machine with an aim toward more homogeneous strain readings between heads on each machine. This example shows the importance of recognizing the difference between a nested experiment and a factorial experiment.

11.3 ANOVA RATIONALE

To see that the sums of squares computed in Section 11.2 were correct, consider the nested model

$$Y_{ijk} = \mu + A_i + B_{j(i)} + \varepsilon_{k(ij)}$$

or

$$Y_{ijk} \equiv \mu + (\mu_{i..} - \mu) + (\mu_{ij.} - \mu_{i..}) + (Y_{ijk} - \mu_{ij.})$$

which is an identity.

Using the best estimates of these population means from the sample data, the sample model is

$$Y_{ijk} \equiv \bar{Y}_{...} + (\bar{Y}_{i..} - \bar{Y}_{...}) + (\bar{Y}_{ij.} - \bar{Y}_{i..}) + (Y_{ijk} - \bar{Y}_{ij.})$$

with

$$i = 1, 2, \ldots, a \qquad j = 1, 2, \ldots, b \qquad k = 1, 2, \ldots, n$$

Transposing $\bar{Y}_{...}$ to the left of this expression, squaring both sides, and adding over $i, j,$ and k gives

$$\sum_i^a \sum_j^b \sum_k^n (Y_{ijk} - \bar{Y}_{...})^2 = \sum_{i=1}^a nb(\bar{Y}_{i..} - \bar{Y}_{...})^2 + \sum_i^a \sum_j^b n(\bar{Y}_{ij.} - \bar{Y}_{i..})^2$$

$$+ \sum_i^a \sum_j^b \sum_k^n (Y_{ijk} - \bar{Y}_{ij.})^2$$

as the sums of cross products equal zero. This expresses the idea that the total sum of squares is equal to the sum of squares between levels of A, plus the sum of squares between levels of B within each level of A, plus the sum of the squares of the errors. The degrees of freedom are

$$(abn - 1) \equiv (a - 1) + a(b - 1) + ab(n - 1)$$

Dividing each independent sum of squares by its corresponding degrees of freedom gives estimates of population variance as usual. For computing purposes, the sum of squares as given above should be expanded in terms of totals, with the general results shown in Table 11.6.

Table 11.6 General ANOVA for a Nested Experiment

Source	df	SS	MS
A_i	$a - 1$	$\sum_i^a \dfrac{T_{i..}^2}{nb} - \dfrac{T_{...}^2}{nab}$	$\dfrac{SS_A}{a - 1}$
$B_{j(i)}$	$a(b - 1)$	$\sum_i^a \sum_j^b \dfrac{T_{ij.}^2}{n} - \sum_i^a \dfrac{T_{i..}^2}{nb}$	$\dfrac{SS_B}{a(b - 1)}$
$\varepsilon_{k(ij)}$	$ab(n - 1)$	$\sum_i^a \sum_j^b \sum_k^n Y_{ijk}^2 - \sum_i^a \sum_j^b \dfrac{T_{ij.}^2}{n}$	$\dfrac{SS_\varepsilon}{ab(n - 1)}$
Totals	$abn - 1$	$\sum_i^a \sum_j^b \sum_k^n Y_{ijk}^2 - \dfrac{T_{...}^2}{nab}$	

This is essentially the form followed in the problem of the last section. Note that the

$$SS_{B_{j(i)}} = \sum_i^a \sum_j^b \frac{T_{ij.}^2}{n} - \sum_i^a \frac{T_{i..}^2}{nb}$$

$$= \sum_i^a \left(\sum_j^b \frac{T_{ij.}^2}{n} - \frac{T_{i..}^2}{nb} \right)$$

which shows the way in which the sum of squares was calculated in the last section: by getting the sum of squares between levels of B for each level of A and then pooling over all levels of A.

11.4 NESTED-FACTORIAL EXPERIMENTS

In many experiments where several factors are involved, some may be factorial or crossed with others; some may be nested within levels of the others. When both factorial and nested factors appear in the same experiment, it is known as a *nested-factorial experiment*. The analysis of such an experiment is simply an extension of the methods of Chapter 5 and this chapter. Care must be exercised, however, in computing some of the interactions. Levels of both factorial and nested factors may be either fixed or random. The methods of Chapter 10 can be used to determine the EMS values and the proper tests to be run.

Example 11.2 The nested-factorial experiment is best explained by an example. An investigator wished to improve the number of rounds per minute that could be fired from a Navy gun. He devised a new loading method, which he hoped would increase the number of rounds per minute when compared with the existing method of loading. To test this hypothesis he needed teams of men to operate the equipment. As the general physique of a man might affect the speed with which he could handle the loading of the gun, he chose teams of men in three general groupings—slight, average, and heavy or rugged men. The classification of such men was on the basis of an Armed Services classification table. He chose three teams at random to represent each of the three physique groupings. Each team was presented with the two methods of gun loading in a random order and each team used each method twice. The model for this experiment was

$$Y_{ijkm} = \mu + M_i + G_j + MG_{ij} + T_{k(j)} + MT_{ik(j)} + \varepsilon_{m(ijk)}$$

where

$$M_i = \text{methods}, i = 1, 2$$
$$G_j = \text{groups}, j = 1, 2, 3$$
$$T_{k(j)} = \text{teams within groups}, k = 1, 2, 3 \text{ for all } j$$
$$\varepsilon_{m(ijk)} = \text{random error}, m = 1, 2 \text{ for all } i, j, k$$

The EMS values are shown in Table 11.7, which indicates the proper F tests to run.

Table 11.7 EMS for Gun-Loading Problem

	2	3	3	2	
	F	*F*	*R*	*R*	
Source	*i*	*j*	*k*	*m*	EMS
M_i	0	3	3	2	$\sigma_\varepsilon^2 + 2\sigma_{MT}^2 + 18\phi_M$
G_j	2	0	3	2	$\sigma_\varepsilon^2 + 4\sigma_T^2 + 12\phi_G$
MG_{ij}	0	0	3	2	$\sigma_\varepsilon^2 + 2\sigma_{MT}^2 + 6\phi_{MG}$
$T_{k(j)}$	2	1	1	2	$\sigma_\varepsilon^2 + 4\sigma_T^2$
$MT_{ik(j)}$	0	1	1	2	$\sigma_\varepsilon^2 + 2\sigma_{MT}^2$
$\varepsilon_{m(ijk)}$	1	1	1	1	σ_ε^2

Table 11.8   Data and ANOVA for Gun-Loading Problem

Group	I			II			III		
Team	1	2	3	4	5	6	7	8	9
Method I	20.2	26.2	23.8	22.0	22.6	22.9	23.1	22.9	21.8
	24.1	26.9	24.9	23.5	24.6	25.0	22.9	23.7	23.5
Method II	14.2	18.0	12.5	14.1	14.0	13.7	14.1	12.2	12.7
	16.2	19.1	15.4	16.1	18.1	16.0	16.1	13.8	15.1

Source	df	SS	MS	EMS
M_i	1	651.95	651.95	$\sigma_\varepsilon^2 + 2\sigma_{MT}^2 + 18\phi_M$
G_j	2	16.05	8.02	$\sigma_\varepsilon^2 + 4\sigma_T^2 + 12\phi_G$
MG_{ij}	2	1.19	0.60	$\sigma_\varepsilon^2 + 2\sigma_{MT}^2 + 6\phi_{MG}$
$T_{k(j)}$	6	39.26	6.54	$\sigma_\varepsilon^2 + 4\sigma_T^2$
$MT_{ik(j)}$	6	10.72	1.79	$\sigma_\varepsilon^2 + 2\sigma_{MT}^2$
$\varepsilon_{m(ijk)}$	18	41.59	2.31	σ_ε^2
Totals	35	760.76		

The data and analysis of this experiment appear in Table 11.8.

It would be well for the reader to verify that the sums of squares of this table are correct. The only term that is somewhat different in this model than those previously handled is $MT_{ik(j)}$, that is, the interaction between methods and teams within groups. The safest way to compute this term is to compute the $M \times T$ interaction sums of squares within each of the three groups separately and then pool these sums of squares. (See Tables 11.9 through 11.11.)

Table 11.9   Data on Gun-Loading Problem for Group I

	Team			Method Totals
	1	2	3	
Method I	20.2	26.2	23.8	
	24.1	26.9	24.9	
	44.3	53.1	48.7	146.1
Method II	14.2	18.0	12.5	
	16.2	19.1	15.4	
	30.4	37.1	27.9	95.4
Team totals	74.7	90.2	76.6	241.5

Table 11.10   Data on Gun-Loading Problem for Group II

	Team			Method Totals
	4	5	6	
Method I	22.0	22.6	22.9	
	23.5	24.6	25.0	
	45.5	47.2	47.9	140.6
Method II	14.1	14.0	13.7	
	16.1	18.1	16.0	
	30.2	32.1	29.7	92.0
Team totals	75.7	79.3	77.6	232.6

Table 11.11   Data on Gun-Loading Problem for Group III

	Team			Method Totals
	7	8	9	
Method I	23.1	22.9	21.8	
	22.9	23.7	23.5	
	46.0	46.6	45.3	137.9
Method II	14.1	12.2	12.7	
	16.1	13.8	15.1	
	30.2	26.0	27.8	84.0
Team totals	76.2	72.6	73.1	221.9

To compute the $M \times T$ interaction for group I, we have

$$SS_{cell} = \frac{(44.3)^2 + (53.1)^2 + (48.7)^2 + (30.4)^2 + (37.1)^2 + (27.9)^2}{2}$$

$$- \frac{(241.5)^2}{12} = 5116.39 - 4860.21 = 256.18$$

$$SS_{method} = \frac{(146.1)^2 + (95.4)^2}{6} - 4860.21 = 214.19$$

$$SS_{team} = \frac{(74.7)^2 + (90.2)^2 + (76.6)^2}{4} - 4860.21 = 35.74$$

$$SS_{M \times T \text{ interaction}} = 256.18 - 214.19 - 35.74 = 6.25$$

For Group II

$$SS_{cell} = \frac{(45.5)^2 + (47.2)^2 + (47.9)^2 + (30.2)^2 + (32.1)^2 + (29.7)^2}{2}$$

$$- \frac{(232.6)^2}{12} = 199.96$$

$$SS_{method} = \frac{(140.6)^2 + (92.0)^2}{6} - 4508.56 = 196.83$$

$$SS_{team} = \frac{(75.7)^2 + (79.3)^2 + (77.6)^2}{4} - 4508.56 = 1.62$$

$$SS_{M \times T \text{ interaction}} = 199.96 - 196.83 - 1.62 = 1.51$$

For Group III

$$SS_{cell} = \frac{(46.0)^2 + (46.6)^2 + (45.3)^2 + (30.2)^2 + (26.0)^2 + (27.8)^2}{2}$$

$$- \frac{(221.9)^2}{12} = 246.96$$

$$SS_{method} = \frac{(137.9)^2 + (84.0)^2}{6} - 4103.30 = 242.10$$

$$SS_{team} = \frac{(76.2)^2 + (72.6)^2 + (73.1)^2}{4} - 4103.30 = 1.90$$

$$SS_{M \times T \text{ interaction}} = 246.96 - 242.10 - 1.90 = 2.96$$

Pooling for all three groups gives

$$SS_{M \times T} = 6.25 + 1.51 + 2.96 = 10.72$$

which is recorded in Table 11.8.

The results of this experiment show a very significant method effect (the new method averaged 23.58 rounds per minute, and the old method averaged only 15.08 rounds per minute). The results also show a significant difference between teams within groups at the 5 percent significance level. This points up individual differences in the men. No other effects or interactions were significant.

A further analysis of the significant differences between teams within the three groups shows the sums of squares to be 35.74, 1.62, and 1.90 respectively. With the error mean square in Table 11.8 of 2.31 it is obvious that the significant difference between teams within the three groups is concentrated in group I. The mean number of rounds per minute for group I's three teams (across methods) are

$$
\begin{array}{cccc}
\text{Team} & 1 & 3 & 2 \\
\text{Mean} & 18.68 & 19.15 & 22.55
\end{array}
$$

A Newman–Keuls will show that team 2 is exceptionally faster than the other two teams.

11.5 COMPUTER PROGRAMS

A nested or nested-factorial experiment, of course, can be analyzed by a computer program. Many such program packages do not have one for a nested experiment. If this is the case, one can feed the data into a factorial program as shown in Chapter 5 and then combine appropriate terms to give the proper sums of squares for the nested terms.

Recall in Example 11.1 concerning machines and heads within machines that if it were (erroneously) analyzed as a factorial, the results would be as given in Table 11.1. However, what appears as an interaction is really part of the between-heads, within-machines term. Combining H_j and MH_{ij}'s SS gives 46.45 + 236.42 = 282.87 with 3 + 12 = 15 df as given in Table 11.4. Such a procedure can be extended to the nested factorial.

For the example on gun loading (Example 11.2) the data were entered into an SPSS program taking teams as 1, 2, and 3 in each group and the resulting ANOVA is as shown in Table 11.12. In Table 11.12 if teams (T)

Table 11.12 Computer Printout of Example 11.2

SOURCE OF VARIATION	SUM OF SQUARES	DF	MEAN SQUARE	F	SIGNIFICANCE OF F
MAIN EFFECTS	680.774	5	136.155	58.927	.001
G	16.052	2	8.026	3.474	.053
T	12.772	2	6.386	2.764	.090
M	651.951	1	651.951	282.162	.001
2-WAY INTERACTIONS	33.234	8	4.154	1.798	.143
G T	26.487	4	6.622	2.866	.053
G M	1.187	2	.594	.257	.776
T M	5.561	2	2.780	1.203	.323
3-WAY INTERACTIONS	5.161	4	1.290	.558	.696
G T M	5.161	4	1.290	.558	.696
EXPLAINED	719.170	17	42.304	18.309	.001
RESIDUAL	41.590	18	2.311		
TOTAL	760.760	35	21.736		

and the $G \times T$ interaction SS are combined: $12.772 + 26.487 = 39.259$ as the teams within groups, which checks with Table 11.8. To get the methods by teams within groups $[MT_{ik(j)}]$ interaction, combine $M \times T$ with $M \times T \times G$ or SS: $5.561 + 5.161 = 10.722$, which is the same as the value in Table 11.8.

The computer printout will also give means computed in all combinations of effects for further analysis if needed.

11.6 REPEATED-MEASURES DESIGN AND NESTED-FACTORIAL EXPERIMENTS

Many statisticians who work with psychologists and educators on the design of their experiments treat repeated-measures designs as a unique topic in these fields of application of statistics. It is the purpose of this section to show that these designs are but a special case of factorial and nested-factorial experiments. Numerical examples and mathematical models are used to illustrate the correspondence between these designs.

One of the simplest of these cases is when a pretest and posttest are given to the same group of subjects after a certain time lapse during which some special instruction may have been administered.

Example 11.3 A recent study at Purdue University was made on a measurement of physical strength (in pounds) given to seven subjects before and after a specified training period. Results are shown in Table 11.13.

Table 11.13 Pretest and Posttest Measures for Example 11.3

Subjects	Pretest	Posttest
1	100	115
2	110	125
3	90	105
4	110	130
5	125	140
6	130	140
7	105	125

Since the same seven subjects were given each test, we have two repeated measures on each subject. Winer [24, p. 266] would handle this as a between-subjects, within-subjects repeated-measures design and report results as shown in Table 11.14.

Table 11.14   ANOVA for Table 11.13

Source of Variation	df	SS	MS	F
Between subjects (S_i)	6	2084.71	347.45	
Within subjects	7	901.00	—	
Tests (T_j)	1		864.29	864.29	145
Residual	6		35.71	5.96	
Totals	13	2985.71		

This analysis implies that the subjects model is

$$Y_{ij} = \mu + S_i + \varepsilon_{j(i)}$$

with df: 6   7

And then the within-subjects' data are further broken down into tests and residual. This residual is actually the subject by test interaction so the model becomes

$$Y_{ij} = \mu + S_i + T_j + ST_{ij}$$

with df: 6   1   6

This model contains no error term since only one pretest score and one post-test score are available on each subject.

Now if this problem were considered in more general terms instead of the so-called repeated-measures format, one would consider this data layout as a two-factor factorial experiment with one observation per treatment. Subjects should be chosen at random and tests should be fixed. Using the algorithm given in Chapter 10, the resulting ANOVA layout would be as shown in Table 11.15.

Here $k = 1$ and a separate error $\varepsilon_{k(ij)}$ is not retrievable. A glance at the EMS column indicates that the only proper F test is to compare the test mean square with the $S \times T$ interaction mean square, which was the test as shown in Table 11.14.

Table 11.15   EMS Determination for Table 11.13

| | | 7 | 2 | 1 | |
| | | R | F | R | |
Source	df	i	j	k	EMS
S_i	6	1	2	1	$\sigma_\varepsilon^2 + 2\sigma_S^2$
T_j	1	7	0	1	$\sigma_\varepsilon^2 + \sigma_{ST}^2 + 7\phi_T$
ST_{ij}	6	1	0	1	$\sigma_\varepsilon^2 + \sigma_{ST}^2$
$\varepsilon_{k(ij)}$	0	1	1	1	σ_ε^2 (not retrievable)

Of course, another approach to this example is to take differences between post- and pretest scores on each individual and test the hypothesis that the true mean of the difference is zero. Using this scheme, a t of 12.05 is found and it is quite well known that this t when squared equals the F of our ANOVA table:

$$t^2 = (12.05)^2 = 145 = F$$

Example 11.4   A slightly more complex problem involves three factors where the subjects are nested within groups. These groups could be classes or experimental conditions and then again each subject is subjected to repeated measures. The study cited on physical strength was extended so that three groups of subjects were involved, two being subjected to special experimental training and the third acting as a control with no special training. Again each subject was given a pre- and posttest where the measurement here was velocity of a baseball throw in meters per second. The resulting data are given in Table 11.16. If these data are treated as a

Table 11.16   Pretest and Posttest Throwing Velocities of Three Groups of Subjects in Meters/Second

Group	Subject	Pretest	Posttest
I	1	26.25	29.50
	2	24.33	27.62
	3	22.52	25.71
	4	29.33	31.55
	5	28.90	31.35
	6	25.13	29.07
	7	29.33	31.15
II	8	27.47	28.74
	9	25.19	26.11
	10	23.53	25.45
	11	24.57	25.58
	12	26.88	27.70
	13	27.86	28.82
	14	28.09	28.99
III	15	22.27	22.52
	16	21.55	21.79
	17	23.31	23.53
	18	30.03	30.21
	19	28.17	28.65
	20	28.09	28.33
	21	27.55	27.86

Table 11.17   ANOVA of Table 11.16

Source	df	SS	MS	EMS	F
Between subjects	20	271.05	—		
Groups (G_i)	2	28.14	14.07	$\sigma_e^2 + 2\sigma_S^2 + 14\phi_G$	1.04
Subjects within groups ($S_{j(i)}$)	18		242.91	13.50	$\sigma_e^2 + 2\sigma_S^2$	
Within subjects	21	35.73	—		
Tests (T_k)	1		21.26	21.26	$\sigma_e^2 + \sigma_{TS}^2 + 21\phi_T$	183***
$G \times T$	2		12.38	6.19	$\sigma_e^2 + \sigma_{TS}^2 + 7\phi_{GT}$	53**
$T \times S_{kj(i)}$	18		2.09	0.116	$\sigma_e^2 + \sigma_{TS}^2$	

repeated-measures design, the results are usually presented as in Winer [24, p. 520], here shown as Table 11.17.

If the layout of Table 11.16 is treated as a nested-factorial experiment where subjects are nested within groups and then tests are factorial on both groups and subjects, the model can be written as follows:

$$Y_{ijk} = \mu + G_i + S_{j(i)} + T_k + GT_{ik} + TS_{kj(i)} + \varepsilon_{m(ijk)}$$

with df: 2   18   1   2   18   0

including a nonretrievable error term. Use of this model and the algorithm cited above gives Table 11.18.

Table 11.18   EMS Determination for Table 11.16

		F	R	F	R	
		3	7	2	1	
Source	df	i	j	k	m	EMS
G_i	2	0	7	2	1	$\sigma_e^2 + 2\sigma_S^2 + 14\phi_G$
$S_{j(i)}$	18	1	1	2	1	$\sigma_e^2 + 2\sigma_S^2$
T_k	1	3	7	0	1	$\sigma_e^2 + \sigma_{TS}^2 + 21\phi_T$
GT_{ik}	2	0	7	0	1	$\sigma_e^2 + \sigma_{TS}^2 + 7\phi_{GT}$
$TS_{kj(i)}$	18	1	1	0	1	$\sigma_e^2 + \sigma_{TS}^2$
$\varepsilon_{m(ijk)}$	0	1	1	1	1	σ_e^2 (not retrievable)

This scheme generates the EMS column that was simply reported in Table 11.17, and that column indicates the proper tests to be made. The numerical results are, of course, the same.

The nested-factorial experiment covers many situations in addition to the repeated-measures design. The factor in the nest may be farms within townships, classes within schools, heads within machines, samples within batches, and so on. It treats all situations where three factors are involved where one is nested and another is factorial. Usually in a repeated-measures design the subjects are chosen at random and the other two factors are considered as fixed. By setting up the model and using the EMS algorithm one may have factors that are random as well as fixed. For example, classes could very well be chosen at random from a given grade level and then the students within the classes chosen at random for the experiment.

As more factors are added in an experiment, one only has to expand the mathematical model and determine the proper tests to make based on the EMS column. One does not need to think in terms of repeat measures. To illustrate further, Winer [24, p. 540] considers two cases of three-factor experiments with repeated measures. Case I has two of the factors (fixed) crossing (or repeated on) all subjects and in case II subjects are nested within two factors (fixed) and the third factor crosses (or is repeated on) all subjects. Both of these cases, and more complex ones, can easily be handled as nested-factorial experiments by writing the proper model. Handling them in this way does not require that the three factors all be fixed. It simply requires that the subjects be treated as another factor nested within one or more of the other factors.

Models for case I and case II show the relationship between the two approaches.

Case I:

$$Y_{ijkm}$$

$$= \mu + \overbrace{A_i + S_{j(i)}}^{\substack{\text{between} \\ \text{subjects}}}$$

$$\overbrace{+ B_k + AB_{ik} + BS_{kj(i)} + C_m + AC_{im} + CS_{mkj(i)} + BC_{km} + ABC_{ikm} + BCS_{mkj(i)}}^{\text{within subjects}}$$

Case II:

$$Y_{ijkm} = \mu + \overbrace{A_i + B_j + AB_{ij} + S_{k(ij)}}^{\text{between subjects}}$$

$$+ \overbrace{C_m + AC_{im} + BC_{jm} + ABC_{ijm} + CS_{mk(ij)}}^{\text{within subjects}}$$

No error terms have been included since they are not retrievable.

These examples, it is hoped, are sufficient to remove any mystery surrounding the repeated-measures designs by showing that they fit into more general models familiar to statisticians in all fields.

11.7   SUMMARY

The summary at the end of Chapter 6 may now be extended for Part II.

Experiment	Design	Analysis
II. Two or more factors		
A. Factorial (crossed)		
	1. Completely randomized $Y_{ijk} = \mu + A_i + B_j$ $\qquad + AB_{ij} + \varepsilon_{k(ij)}, \ldots$ for more factors	1.
	a. General case	a. ANOVA with interactions
	b. 2^n case	b. Yates' method or general ANOVA; use (1), a, b, ab, $\ldots$
	c. 3^n case	c. General ANOVA; use $00, 10, 20, 01, 11, \ldots,$ and $A \times B = AB + AB^2, \ldots$ for interaction
B. Nested (hierarchical)		
	1. Completely randomized $Y_{ijk} = \mu + A_i$ $\qquad + B_{j(i)} + \varepsilon_{k(ij)}$	1. Nested ANOVA
C. Nested factorial		
	1. Completely randomized $Y_{ijkm} = \mu + A_i + B_{j(i)}$ $\qquad + C_k + AC_{ik}$ $\qquad + BC_{jk(i)} + \varepsilon_{m(ijk)}$	1. Nested-factorial ANOVA

PROBLEMS

11.1   Porosity readings on condenser paper were recorded for paper from four rolls taken at random from each of three lots. The results were as follows. Analyze these data, assuming lots are fixed and rolls random.

Lot		I				II				III		
Roll	1	2	3	4	5	6	7	8	9	10	11	12
	1.5	1.5	2.7	3.0	1.9	2.3	1.8	1.9	2.5	3.2	1.4	7.8
	1.7	1.6	1.9	2.4	1.5	2.4	2.9	3.5	2.9	5.5	1.5	5.2
	1.6	1.7	2.0	2.6	2.1	2.4	4.7	2.8	3.3	7.1	3.4	5.0

11.2 In Problem 11.1 how would the results change (if they do) if the lots were chosen at random?

11.3 Set up the EMS column and indicate the proper tests to make if A is a fixed factor at five levels, B is nested within A at four random levels for each level of A, C is nested within B at three random levels, and two observations are made in each "cell."

11.4 Repeat Problem 11.3 for A and B crossed or factorial and C nested within the A, B cells.

11.5 Two types of machines are used to wind coils. One type is hand operated; two machines of this type are available. The other type is power operated; two machines of this type are available. Three coils are wound on each machine from two different wiring stocks. Each coil is then measured for the outside diameter of the wire at a middle position on the coil. The results were as follows. (Units are 10^{-5} inch.)

Machine Type	Hand		Power	
Machine Number	2	3	5	8
Stock 1	3279	3527	1904	2464
	3262	3136	2166	2595
	3246	3253	2058	2303
Stock 2	3294	3440	2188	2429
	2974	3356	2105	2410
	3157	3240	2379	2685

Set up the model for this problem and determine what tests can be run.

11.6 Do a complete analysis of Problem 11.5.

11.7 If the outside diameter readings in Problem 11.5 were taken at three fixed positions on the coil (tip, middle, and end), what model would now be appropriate and what tests could be run?

11.8 An experiment is to be designed to compare faculty morale in three types of junior high school organizations: grades 6–8, grades 7–9, and grades 7–8. Two schools are randomly selected to represent each of these three types of organizations and data are to be collected from several teachers in each school. Someone suggests that the morale of male and female teachers might differ so it is agreed to choose random samples of five male and five female teachers from each school so that sex differences in morale may be checked out too.

a. Set up a mathematical model for this experiment, outline into ANOVA, and indicate what tests can be made.

b. Make any recommendations that seem reasonable based on your answers to part (a), assuming that the data have not been collected as yet.

11.9   In an attempt to study the effectiveness of two corn varieties (V_1 and V_2), two corn-producing counties (C) in Iowa were chosen at random. Then four farms (F) were randomly selected within each county. Seed from both varieties were sent to these farms and planted in random plots. When harvested, the number of bushels of corn per acre was recorded for each variety on each farm.

a. Write a mathematical model for this situation.

b. Determine the EMS for varieties in this experiment.

c. Assuming only counties (C) showed significantly different average yields, and $MS_C = 130$ and $MS_F = 10$ (farms or error), find the percentage of the variance in this experiment that can be attributed to county differences.

11.10 An educator proposes a new teaching method and wishes to compare the achievement of students using his method with that of students using a traditional method. Twenty students are randomly placed into two groups with ten students per group. Tests are given to all 20 students at the beginning of a semester, at the end of the semester, and ten weeks after the end of the semester. The educator wishes to see whether there is a difference in the average achievement between the two methods at each of the three time periods.

a. Write a mathematical model for this situation.

b. Set up an ANOVA table and show the F tests that can be made.

c. If the educator's method is really better than the traditional method, what would you expect if you graphed the results at each time period? Show by sketch.

d. If method A—the new method—gave a mean achievement score across all time periods of 60.0, explain in some detail how you would set confidence limits on this method mean.

11.11 In Example 11.4 in the text the ANOVA of Table 11.17 suggests that further analysis is needed. Using the data of Table 11.16, make further analyses and state your conclusions. (A graph might also be helpful.)

11.12 In studying the intellectual attitude toward science as measured on the Bratt attitude scale, a science educator gave a pretest and posttest to an experimental group exposed to a new curriculum and a control group that was not so exposed. There were 15 students in each group. Set up a mathematical model for this situation and determine what tests are appropriate.

11.13 In Problem 11.12 a strong interaction was found between the groups and the tests with means as follows: pretest control = 57.00, pretest experimental = 54.96, posttest control = 54.66, posttest experimental = 65.66. Assuming the sum of squares for the error term used to test for this interaction was 492.61, make further tests and state some conclusions based on this information.

11.14 A researcher wished to test the figural fluency of three groups of Egyptian students: those taught by the Purdue Creative Thinking Program (PCTP) in a restricted atmosphere, those taught by PCTP in a permissive atmosphere, and a control group taught in a traditional manner with no creativity training. Four classes were randomly assigned to each of the three methods. Before training began, all pupils were administered a Group Embedded Figures Test and classified as either field-dependent of field-independent students. After six weeks of training results on the averages of students within these classes on this one subtest (figural fluency) were found to be

Source	df	SS	MS	F	Prob.
Methods (M_i)		124.60			
Classes within method ($C_{j(i)}$)		35.10			
Cognitive style (F_k)		2.76			
$M \times F$ interaction (MF_{ik})		25.76			
$F \times C$ interaction ($FC_{kj(i)}$)		20.97			

Complete the table and state your conclusions.

11.15 Fifty-four randomly selected students are assigned to one of three curricula (C) of science lessons. Over a period of a semester six lessons (L) were presented to all students in these three groups according to their prescribed curricula. The first two curricula were experimental A—taught question-asking skills with evaluation feedback, and experimental B—taught with question-asking skills without the evaluation feedback. The third curriculum was used as a control with no instruction in question-asking skill or feedback. Results on the proportion of high-level questions asked on a test gave

Source	df	SS	MS	F	Prob.
C_i	2	0.556	0.278	4.14	0.0211
$S_{k(i)}$	51	3.417	0.067		
L_j	5	0.250	0.050	2.25	0.0493
CL_{ij}	10	0.430	0.043	1.92	0.0431
$LS_{jk(i)}$	255	5.610	0.022		

Discuss these results and explain how the F tests were made.

11.16 For Problem 11.15 it was known in advance that the experimenter wished to compare the two experimental curricula results with the control, and also to compare the two experimental curricula results with each other. Set up proper a priori tests for this purpose and test the results based on the following means and the table given in Problem 11.15.

Mean proportion for curriculum A: 0.2224
Mean proportion for curriculum B: 0.1929
Mean proportion for curriculum C: 0.1236

11.17 In a study made of the characteristics associated with guidance competence versus counseling competence, 144 students were divided into nine groups of 16 each. These nine groups represented all combinations of three levels of guidance ranking (high, medium, low) and three levels of counseling ranking (high, medium, low). All subjects were then given nine subtests. Assuming the rankings as two fixed factors, the subtests as fixed, and the subjects within the nine groups as random, set up a data layout and mathematical model for this experiment.

11.18 Determine the ANOVA layout and the EMS column for Problem 11.17 and indicate the proper tests that can be made.

11.19 Three days of sampling where each sample was subjected to two types of size graders gave the following results, coded by subtracting 4 percent moisture and multiplying by 10.

Day	1		2		3	
Grader	A	B	A	B	A	B
Sample 1	4	11	5	11	0	6
2	6	7	17	13	−1	−2
3	6	10	8	15	2	5
4	13	11	3	14	8	2
5	7	10	14	20	8	6
6	7	11	11	19	4	10
7	14	16	6	11	5	18
8	12	10	11	17	10	13
9	9	12	16	4	16	17
10	6	9	−1	9	8	15
11	8	13	3	14	7	11

Assuming graders fixed, days random, and samples within days random, set up a mathematical model for this experiment and determine the EMS column.

11.20 Complete the ANOVA for Problem 11.19 and comment on the results.

11.21 In a filling process two random runs are made. For each run six hoppers are used, but they may not be the same hoppers on the second run. Assume that hoppers are a random sample of possible hoppers. Data are taken from the bottom, middle, and top of each hopper in order to check on a possible position effect on filling. Two observations are made within positions and runs on all hoppers. Set up a mathematical model for this problem and the associated EMS column.

11.22 The data below are from the experiment described in Problem 11.21.

Head	Run 1						Run 2					
	Bottom		Middle		Top		Bottom		Middle		Top	
1	19	0	31	25	25	13	6	6	0	6	0	0
2	13	0	19	38	31	25	0	13	0	31	6	0
3	13	0	31	19	13	0	28	6	25	19	13	28
4	13	0	19	31	25	13	28	0	6	0	28	25
5	19	19	19	38	19	25	16	13	13	19	0	0
6	6	0	0	0	0	0	13	6	6	19	0	19

Do a complete analysis of these data.

11.23 Some research on the abrasive resistance of filled epoxy plastics was carried out by molding blocks of plastic using five fillers: iron oxide, iron filings, copper, alumina, and graphite. Two concentration ratios were used for the ratio of filler to epoxy resin: $\frac{1}{2}$:1 and 1:1. Three sample blocks were made up of each of the ten combinations above. These blocks were then subjected to a reciprocating motion where gritcloth was used as the abrasive material in all tests. After 10,000 cycles, each sample block was measured in three places at each of three fixed positions on the plates: I, II, and III. These blocks had been measured before cycling so that the difference in thickness was used as the measured variable. Measurements were made to the nearest 0.0001 inch. Assuming complete randomization of the order of testing of the 30 blocks, set up a mathematical model for this situation and set up the ANOVA table with an EMS column. We are interested in the effect of concentration ratio, fillers, and positions on the block. The three observations at each position will be considered as error, and there may well be differences between the average of the three sample blocks within each treatment combination.

11.24 Data for Problem 11.23 were found to be as shown in the following tables. Complete an ANOVA for these data and state your conclusions.

11.25 From the results of Problem 11.24 what filler–concentration combination would you recommend if you were interested in a minimum amount of wear? The 1:1 ratio is cheaper than $\frac{1}{2}$:1.

	Alumina						Graphite					
Concentration	$\frac{1}{2}$:1			1:1			$\frac{1}{2}$:1			1:1		
Position	I	II	III	I	II	III	I	II	III	I	II	III
Sample 1	7.0	5.4	6.4	7.1	5.0	6.4	8.8	5.5	9.0	18.0	14.4	18.5
	7.3	5.1	5.7	6.9	5.7	6.6	8.3	5.9	11.0	17.6	12.4	19.2
	6.6	5.0	7.7	6.5	4.0	6.2	6.9	4.7	11.3	18.4	12.6	19.4
	20.9	15.5	19.8	20.5	14.7	19.2	24.0	16.1	31.3	54.0	39.4	57.1
Sample 2	6.8	5.1	4.6	6.8	4.0	7.5	7.7	6.8	10.1	15.6	11.6	17.6
	6.9	5.5	5.8	7.2	5.2	7.8	6.5	7.0	11.2	14.6	12.8	18.7
	6.2	4.8	5.8	4.8	4.4	6.0	5.9	7.2	10.8	15.1	13.4	19.3
	19.9	15.4	16.2	18.8	13.6	21.3	20.1	21.0	32.1	45.3	37.8	55.6
Sample 3	7.3	6.3	5.1	6.2	4.4	5.7	7.2	7.0	7.7	13.3	11.7	15.5
	7.6	5.3	6.5	6.6	4.8	6.6	6.8	6.9	9.3	13.8	13.4	18.0
	5.6	4.8	7.1	4.4	3.8	5.6	6.2	5.2	9.8	13.6	12.1	19.6
	20.5	16.4	18.7	17.2	13.0	17.9	20.2	19.1	26.8	40.7	37.2	53.1
Totals		163.3			156.2			210.7			420.2	

	Iron Filings						Iron Oxide						Copper					
Concentration	$\frac{1}{2}$:1			1:1			$\frac{1}{2}$:1			1:1			$\frac{1}{2}$:1			1:1		
Position	I	II	III	I	II	III	I	II	III	I	II	III	I	II	III	I	II	III
Sample 1	2.1	1.1	1.3	3.6	1.6	2.3	3.0	1.1	3.2	1.8	1.3	2.4	2.7	1.4	2.8	2.8	1.5	2.4
	2.1	1.1	1.7	3.8	1.0	2.5	3.8	1.8	4.1	2.1	1.0	1.9	2.9	2.2	3.8	2.6	1.1	2.1
	1.0	0.9	1.7	2.9	1.6	2.4	3.0	1.1	3.3	1.9	1.6	1.8	3.0	1.8	3.4	1.9	1.4	2.0
	5.2	3.1	4.7	10.3	4.2	7.2	9.8	4.0	10.6	5.8	3.9	6.1	8.6	5.4	10.0	7.3	4.0	6.5
Sample 2	1.7	1.1	1.7	2.1	1.1	1.7	3.1	1.6	2.2	2.1	1.0	1.5	2.5	1.8	2.4	2.1	1.0	1.5
	1.8	1.0	2.0	2.6	1.1	1.0	3.0	1.7	3.0	2.0	0.8	2.0	3.0	2.6	3.3	2.5	1.0	1.6
	1.3	0.8	2.0	1.6	0.9	1.4	2.7	1.2	2.7	2.3	1.0	2.0	2.0	2.1	2.4	1.5	1.2	1.7
	4.8	2.9	5.7	6.3	3.1	4.1	8.8	4.5	7.9	6.4	2.8	5.5	7.5	6.5	8.1	6.1	3.2	4.8
Sample 3	3.2	0.8	1.4	2.3	1.1	1.8	2.1	1.5	2.0	1.6	1.2	1.6	3.3	2.1	3.4	2.6	1.3	2.6
	2.8	0.8	2.0	2.6	1.2	2.0	2.9	1.9	2.5	1.6	1.1	1.4	3.2	2.0	4.1	2.9	1.5	3.3
	2.0	0.5	1.9	2.1	0.7	2.4	1.6	2.5	2.4	1.9	1.2	2.1	3.2	2.0	3.8	2.8	1.2	2.7
	8.0	2.1	5.3	7.0	3.0	6.2	6.6	5.9	6.9	5.1	3.5	5.1	9.7	6.1	11.3	8.3	4.0	8.6
Totals	41.8			51.4			65.0			44.2			73.2			52.8		

12

Experiments of Two or More Factors—
Restrictions on Randomization

12.1 INTRODUCTION

In the discussion of factorial and nested experiments it was assumed that the whole experiment was performed in a completely randomized manner. In practice, however, it may not be feasible to run the entire experiment several times in one day, or by one experimenter, and so forth. It then becomes necessary to restrict this complete randomization and block the experiment in the same manner that a single-factor experiment was blocked in Chapter 4. Instead of running several replications of the experiment all at one time, it may be possible to run one complete replication on one day, a second complete replication on another day, a third replication on a third day, and so on. In this case each replication is a block, and the design is a randomized block design with a complete factorial or nested experiment randomized within each block.

Occasionally a second restriction on randomization is necessary, in which case a Latin square design might be used, with the treatments in the square representing a complete factorial or nested experiment.

12.2 FACTORIAL EXPERIMENT IN A RANDOMIZED BLOCK DESIGN

Just as the factorial arrangements were extracted from the treatment effect in Equations (5.1) and (5.2), so can these arrangements be extracted from the treatment effect in the randomized complete block design of Chapter 4 (Section 4.2). For the completely randomized design, you will recall that

$$Y_{ij} = \mu + \tau_{ij} + \varepsilon_{k(ij)}$$

can be subdivided into

$$Y_{ijk} = \mu + A_i + B_j + AB_{ij} + \varepsilon_{k(ij)}$$

where $A_i + B_j + AB_{ij} = \tau_{ij}$ of the first model.

When a complete experiment is replicated several times and R_k represents the blocks or replications, the randomized block model is

$$Y_{ijkm} = \mu + R_k + \tau_{ij} + \varepsilon_{m(ijk)}$$

where R_k represents the blocks or replications and τ_{ij} represents the treatments. If the treatments are formed from a two-factor factorial experiment, then

$$\tau_{ij} = A_i + B_j + AB_{ij}$$

and the complete model is

$$Y_{ijk} = \mu + R_k + A_i + B_j + AB_{ij} + \varepsilon_{ijk} \qquad (12.1)$$

Here it is assumed that there is no interaction between replications (blocks) and treatments, which is the usual assumption underlying a randomized block design. Any such interaction is confounded in the error term ε_{ijk}.

To get a more concrete picture of this type of design, consider three levels of factor A, two levels of B, and four replications. The treatment combinations may be written as A_1B_1, A_1B_2, A_2B_1, A_2B_2, A_3B_1, and A_3B_2. Such a design assumes that all six of these treatment combinations can be run on a given day, if days are the replications or blocks. A layout in which these six treatment combinations are randomized within each replication might be

Replication I: A_1B_2 A_3B_1 A_3B_2 A_2B_1 A_1B_1 A_2B_2

Replication II: A_2B_2 A_1B_1 A_3B_2 A_2B_1 A_1B_2 A_3B_1

Replication III: A_2B_1 A_3B_2 A_1B_2 A_3B_1 A_2B_2 A_1B_1

Replication IV: A_1B_1 A_3B_1 A_3B_2 A_2B_1 A_1B_2 A_2B_2

An analysis breakdown would be that shown in Table 12.1.

To see how the interaction of replications and treatments is used as error, consider the expanded model and the corresponding EMS terms exhibited

Table 12.1   Two-Factor Experiment in a Randomized Block Design

Source	df	
Replications R_k	3	
Treatments τ_{ij}	5	
$\quad A_i$		2
$\quad B_j$		1
$\quad AB_{ij}$		2
Error ε_{ijk}	15	
Total	23	

in Table 12.2. Here the factors are fixed and replications are considered as random.

Table 12.2   EMS Terms for Two-Factor Experiment in a Randomized Block

Source	df	3 F i	2 F j	4 R k	1 R m	EMS
R_k	3	3	2	1	1	$\sigma_\varepsilon^2 + 6\sigma_R^2$
A_i	2	0	2	4	1	$\sigma_\varepsilon^2 + 2\sigma_{RA}^2 + 8\phi_A$
RA_{ik}	6	0	2	1	1	$\sigma_\varepsilon^2 + 2\sigma_{RA}^2$
B_j	1	3	0	4	1	$\sigma_\varepsilon^2 + 3\sigma_{RB}^2 + 12\phi_B$
RB_{jk}	3	3	0	1	1	$\sigma_\varepsilon^2 + 3\sigma_{RB}^2$
AB_{ij}	2	0	0	4	1	$\sigma_\varepsilon^2 + \sigma_{RAB}^2 + 4\phi_{AB}$
RAB_{ijk}	6	0	0	1	1	$\sigma_\varepsilon^2 + \sigma_{RAB}^2$
$\varepsilon_{m(ijk)}$	0	1	1	1	1	σ_ε^2 (not retrievable)
Total	23					

Since the degrees of freedom are quite low for the tests indicated above, and since there is no separate estimate of error variance, it is customary to assume that $\sigma_{RA}^2 = \sigma_{RB}^2 = \sigma_{RAB}^2 = 0$ and to pool these three terms to serve as the error variance. The reduced model is then as shown in Table 12.3. Here all effects are tested against the 15-df error term. Although we usually pool the interactions of replications with treatment effects for the error term, this is not always necessary. It will depend upon the degrees of freedom available for testing various hypotheses in Table 12.2, and also whether or not any repeat measurements can be taken within a replication. If the latter is possible, a separate estimate of σ_ε^2 is available and all tests in Table 12.2 may be made.

Table 12.3   EMS Terms for Two-Factor Experiment, Randomized Block Design, Reduced Model

Source	df	EMS
R_k	3	$\sigma_\varepsilon^2 + 6\sigma_R^2$
A_i	2	$\sigma_\varepsilon^2 + 8\phi_A$
B_j	1	$\sigma_\varepsilon^2 + 12\phi_B$
AB_{ij}	2	$\sigma_\varepsilon^2 + 4\phi_{AB}$
ε_{ijk}	15	σ_ε^2
Total	23	

Example 12.1   At Purdue a researcher was interested in determining the effect of both nozzle types and operators on the rate of fluid flow in cubic centimeters through these nozzles. The researcher considered three fixed nozzle types and five randomly chosen operators. Each of these 15 combinations was to be run in a random order on each of three days. These three days will be considered as three replications of the complete 3×5 factorial experiment. After subtracting 96.0 cm^3 from each reading and multiplying all readings by 10, the data layout and observed readings (in cubic centimeters) are as recorded in Table 12.4.

Table 12.4   Nozzle Example Data

Operator		1			2			3			4			5	
Nozzle	A	B	C	A	B	C	A	B	C	A	B	C	A	B	C
Replication															
I		6	13	10	26	4	−35	11	17	11	21	−5	12	25	15	−4
II		6	6	10	12	4	0	4	10	−10	14	2	−2	18	8	10
III	−15	13	−11	5	11	−14	4	17	−17	7	−5	−16	25	1	24

The model for this example is

$$Y_{ijk} = \mu + R_k + N_i + O_j + NO_{ij} + \varepsilon_{ijk}$$

with

$$k = 1, 2, 3 \qquad i = 1, 2, 3 \qquad j = 1, 2, \dots, 5$$

The R_k levels are random, O_j levels are random, and N_i levels are fixed. If m repeat measurements are considered in the model, but $m = 1$, the EMS values are those appearing in Table 12.5.

Table 12.5   EMS for Nozzle Example

| | | 3 | 5 | 3 | 1 | |
| | | F | R | R | R | |
Source	df	i	j	k	m	EMS
R_k	2	3	5	1	1	$\sigma_\varepsilon^2 + 15\sigma_R^2$
N_i	2	0	5	3	1	$\sigma_\varepsilon^2 + 3\sigma_{NO}^2 + 15\phi_N$
O_j	4	3	1	3	1	$\sigma_\varepsilon^2 + 9\sigma_O^2$
NO_{ij}	8	0	1	3	1	$\sigma_\varepsilon^2 + 3\sigma_{NO}^2$
$\varepsilon_{m(ijk)}$	28	1	1	1	1	σ_ε^2
Total	44					

These EMS values indicate that all main effects and the interaction can be tested with reasonable precision (df). The sums of squares are computed in the usual manner from the subtotals in Table 12.6.

Table 12.6   Subtotals in Nozzle Example

Replication	Total	Nozzle	Operator 1	2	3	4	5	Nozzle Total
I	127	A	−3	43	19	42	68	169
II	92	B	32	19	44	−8	24	111
III	29	C	9	−49	−16	−6	30	−32
Totals	248		38	13	47	28	122	248

Using these subtotals, the usual sums of squares can easily be computed and the results displayed as in Table 12.7.

Table 12.7   ANOVA for Nozzle Example

Source	df	SS	MS	EMS	F	$F_{0.05}$
R_k	2	328.84	164.42	$\sigma_\varepsilon^2 + 15\sigma_R^2$	1.70	3.34
N_i	2	1426.97	713.48	$\sigma_\varepsilon^2 + 3\sigma_{NO}^2 + 15\phi_N$	3.13	4.46
O_j	4	798.79	199.70	$\sigma_\varepsilon^2 + 9\sigma_O^2$	2.06	2.71
NO_{ij}	8	1821.48	227.68	$\sigma_\varepsilon^2 + 3\sigma_{NO}^2$	2.35*	2.29
ε_{ijk}	28	2709.16	96.76	σ_ε^2		
Totals	44	7085.24				

The results of tests of hypotheses on this model show only a significant nozzle × operator interaction at the 5 percent level of significance. The plot in Figure 12.1 of all totals for the right-hand data of Table 12.6 shows this interaction graphically.

If desirable, the model of this example could be expanded to include an estimate of replication by operator and replication by nozzle interactions, using the three-way interaction ($N \times O \times$ Replication) as the error. The results of such a breakdown are recorded in Table 12.8.

The F test values are shown in Table 12.8 except for the test on nozzles labeled with an asterisk. Since there is no direct test, an F' test is used as described in Chapter 10. Here

$$MS = MS_{NR} + MS_{NO} - MS_{NRO}$$

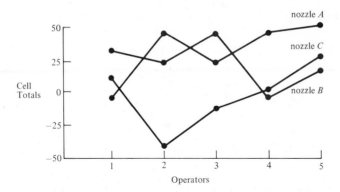

Figure 12.1 Interaction plot of nozzle × operator interaction in nozzle example.

Table 12.8 ANOVA for Nozzle Example with Replication Interactions

Source	df	SS	MS	EMS	F
R_k	2	328.84	164.42	$\sigma_\varepsilon^2 + 3\sigma_{RO}^2 + 15\sigma_R^2$	1.86
N_i	2	1426.97	713.48	$\sigma_\varepsilon^2 + 3\sigma_{NO}^2 + 5\sigma_{NR}^2 + 15\phi_N$	*
NR_{ik}	4	272.23	68.06	$\sigma_\varepsilon^2 + 5\sigma_{NR}^2$	< 1
O_j	4	798.79	199.70	$\sigma_\varepsilon^2 + 3\sigma_{RO}^2 + 9\sigma_O^2$	2.26
RO_{jk}	8	705.60	88.20	$\sigma_\varepsilon^2 + 3\sigma_{RO}^2$	< 1
NO_{ij}	8	1821.48	227.68	$\sigma_\varepsilon^2 + 3\sigma_{NO}^2$	2.10
ε_{ijk} or NRO_{ijk}	16	1731.33	108.21	σ_ε^2	
Totals	44	7085.24			

	Operator						Nozzle			
Replication	1	2	3	4	5	Total	A	B	C	Total
I	29	−5	39	28	36	127	89	44	−6	127
II	22	16	4	14	36	92	54	30	8	92
III	−13	2	4	−14	50	29	26	37	−34	29
Totals	38	13	47	28	122	248	169	111	−32	248

Here the coefficients are 1, 1, and −1, and the corresponding mean squares and degrees of freedom are

$$\text{MS}_{NR} = 68.06 \quad \text{with } v_1 = 4 \text{ df}$$
$$\text{MS}_{NO} = 227.68 \quad \text{with } v_2 = 8 \text{ df}$$
$$\text{MS}_{NRO} = 108.21 \quad \text{with } v_3 = 16 \text{ df}$$

The F' test is then

$$F' = \frac{MS_N}{MS_{NR} + MS_{NO} - MS_{NRO}}$$

$$= \frac{713.48}{68.06 + 227.68 - 108.21}$$

$$= \frac{713.48}{187.53} = 3.80$$

and

$$v = \frac{(187.53)^2}{(1)^2[(68.06)^2/4] + (1)^2[(227.68)^2/8] + (-1)^2[(108.21)^2/16]} = 4.2$$

With 2 and 4 df the 5 percent F value is 6.94, and with 2 and 5 df it is 5.79; hence the $F' = 3.80$ is not significant at the 5 percent level. Hence all F tests indicate no significant effects. The $N \times O$ interaction does not show significance as it did in Table 12.7, because the test is less sensitive due to the reduction in degrees of freedom from 28 to 16 in the error mean square. Since both the $N \times R$ and $R \times O$ interaction F tests are less than one, we are probably justified in assuming that σ_{NR}^2 and σ_{RO}^2 are zero and in pooling these with σ_{NRO}^2 for a 28-df error mean square as in Table 12.7.

12.3 FACTORIAL EXPERIMENT IN A LATIN SQUARE DESIGN

If there are two restrictions on the randomization, and an equal number of restriction levels and treatment levels are used, Chapter 4 has shown that a Latin square design may be used. The model for this design is

$$Y_{ijk} = \mu + R_i + \tau_j + \gamma_k + \varepsilon_{ijk}$$

where τ_j represents the treatments, and both R_i and γ_k represent restrictions on the randomization. If a 4×4 Latin square is used for a given experiment, such as the one in Section 4.5, the four treatments A, B, C, and D could represent the four treatment combinations of a 2×2 factorial (Table 12.9). Here

$$\tau_j = A_m + B_q + AB_{mq}$$

Table 12.9 4×4 Latin Square (Same as Table 4.12)

Position	Car			
	I	II	III	IV
1	C	D	A	B
2	B	C	D	A
3	A	B	C	D
4	D	A	B	C

with its 3 df. Thus a factorial experiment $(A \times B)$ is run in a Latin square design. For the problem of Section 4.5, instead of four tire brands, the four treatments could consist of two brands (A_1 and A_2) and two types (B_1 and B_2). The types might be black- and white-walled tires. The complete model would then be

$$Y_{ikmq} = \mu + R_i + \gamma_k + A_m + B_q + AB_{mq} + \varepsilon_{ikmq}$$

and the degree-of-freedom breakdown would be that shown in Table 12.10.

This idea could be extended to running a factorial in a Graeco-Latin square, and so on.

Table 12.10 Degree-of-Freedom Breakdown for Table 12.9

Source	df	
R_i	3	
γ_k	3	
A_m	1	
B_q	1	3 df for τ_j
AB_{mq}	1	
ε_{ikmq}	6	
Total	15	

12.4 REMARKS

In the preceding sections factorial experiments were considered where the design of the experiment was a randomized block or a Latin square. It is also possible to run a nested experiment or a nested-factorial experiment in a randomized block or Latin square design. In the case of a nested experiment, the treatment effect might be broken down as

$$\tau_m = A_i + B_{j(i)}$$

and when this experiment has one restriction on the randomization, the model is

$$Y_{ijk} = \mu + R_k + \underbrace{A_i + B_{j(i)}}_{\tau_m} + \varepsilon_{ijk}$$

If a nested factorial is repeated on several different days, the model would be

$$Y_{ijkm} = \mu + R_k + \underbrace{A_i' + B_{j(i)} + C_m + AC_{im} + BC_{mj(i)}}_{\tau_m} + \varepsilon_{ijkm}$$

where the whole nested factorial is run in a randomized block design. The analysis of such designs follows from the methods of Chapter 11.

12.5 SUMMARY

The summary at the end of Chapter 11 may now be extended.

Experiment	Design	Analysis

II. Two or more factors

A. Factorial (crossed)

    1. Completely randomized   1.
$$Y_{ijk} = \mu + A_i + B_j + AB_{ij} + \varepsilon_{k(ij)}, \cdots$$
    for more factors

    a. General case   a. ANOVA with interactions

    b. 2^f case   b. Yates' method or general ANOVA

    c. 3^f case   c. General ANOVA

    2. Randomized block   2.

    a. Complete   a. Factorial ANOVA with replications R_k
$$Y_{ijk} = \mu + R_k + A_i + B_j + AB_{ij} + \varepsilon_{ijk}$$

    3. Latin square   3.

    a. Complete   a. Factorial ANOVA with replications and positions
$$Y_{ijkm} = \mu + R_k + \gamma_m + A_i + B_j + AB_{ij} + \varepsilon_{ijkm}$$

B. Nested (hierarchical)

    1. Completely randomized   1. Nested ANOVA
$$Y_{ijk} = \mu + A_i + B_{j(i)} + \varepsilon_{k(ij)}$$

    2. Randomized block   2. Nested ANOVA with blocks R_k

    a. Complete   a. Nested ANOVA with blocks R_k
$$Y_{ijk} = \mu + R_k + A_i + B_{j(i)} + \varepsilon_{ijk}$$

    3. Latin square   3.

    a. Complete   a. Nested ANOVA with blocks and positions
$$Y_{ijkm} = \mu + R_k + \gamma_m + A_i + B_{j(i)} + \varepsilon_{ijkm}$$

C. Nested factorial

    1. Completely randomized   1. Nested-factorial ANOVA
$$Y_{ijkm} = \mu + A_i + B_{j(i)} + C_k + AC_{ik} + BC_{kj(i)} + \varepsilon_{m(ijk)}$$

Experiment	Design	Analysis

	2. Randomized block
	  a. Complete
	  $Y_{ijkm}=\mu+R_k+A_i$
	    $+B_{j(i)}+C_m+AC_{im}$
	    $+BC_{mj(i)}+\varepsilon_{ijkm}$
	3. Latin square
	  a. Complete
	  $Y_{ijkmq}=\mu+R_k+\gamma_m$
	    $+A_i+B_{j(i)}+C_q$
	    $+AC_{iq}+BC_{qj(i)}$
	    $+\varepsilon_{ijkmq}$
		2.
		  a. ANOVA with blocks R_k
		
		3.
		
		  a. ANOVA with blocks
		    and positions

PROBLEMS

12.1 Data on the glass rating of tubes taken from two fixed stations and three shifts are recorded each week for three weeks. Considering the weeks as blocks, the six treatment combinations (two stations by three shifts) were tested in a random order each week, with results as follows:

		Station 1			Station 2		
Shift		1	2	3	1	2	3
Week 1		3	3	3	6	3	6
		6	4	6	8	9	8
		6	7	7	11	11	13
Week 2		14	8	11	4	15	4
		16	8	12	6	15	7
		19	9	17	7	17	10
Week 3		2	2	2	2	2	10
		3	3	4	5	4	12
		6	4	6	7	6	13

Analyze these data as a factorial run in a randomized block design.

12.2 In the example of Section 4.5 (Table 4.13) consider tire brands A, B, C, D as four combinations of two factors, ply of tires and type of tread, where

$$A = P_1T_1 \qquad B = P_1T_2 \qquad C = P_2T_1 \qquad D = P_2T_2$$

representing the four combinations of two ply values and two tread types. Analyze the data for this revised design—a 2^2 factorial run in a Latin square design.

12.3 If in Problem 11.5 the stock factor is replaced by "days," considered random, and the rest of the experiment is performed in a random manner on each of the two days, use the same numerical results and reanalyze with this restriction.

12.4 Again, in Problem 11.5 assume that the whole experiment (24 readings) was repeated on five randomly selected days. Set up an outline of this experiment, including the EMS column, df, and so on.

12.5 Three complete replications were run in a study of aluminum thickness. Samples were taken from five machines and two types of slug—rivet and staple—in each replicate. Considering replicates random and machines and type of slug fixed, set up a model for this problem and determine what tests can be made. Comment on the adequacy of the various tests. Data are as follows:

Replication	Slug	Machine				
		1	2	3	4	5
I	Rivet	175	95	180	170	155
	Staple	165	165	175	185	130
II	Rivet	190	185	180	200	190
	Staple	170	160	175	165	190
III	Rivet	185	165	175	195	200
	Staple	190	160	200	185	200

12.6 Analyze the data of Problem 12.5 using a computer program and state your conclusions.

12.7 Data were collected for three consecutive years on the number of days after planting that the first flowering of a plant occurred. Each year five varieties of plants were planted at six different stations. Considering the repeat years random and varieties and stations fixed, outline the ANOVA for this problem.

12.8 The data for Problem 12.7 are given below.

Station/Year	Variety				
	Hazel	Coltsfoot	Anemone	Blackthorn	Mustard
Broadchalke					
1932	57	67	95	102	123
1933	46	72	90	88	101
1934	28	66	89	109	113
Total	131	205	274	299	337

	Variety				
Station/Year	Hazel	Coltsfoot	Anemone	Blackthorn	Mustard
Bratton					
1932	26	44	92	96	93
1933	38	68	89	89	110
1934	20	64	106	106	115
Total	84	176	287	291	318
Lenham					
1932	48	61	78	99	113
1933	35	60	89	87	109
1934	48	75	95	113	111
Total	131	196	262	299	333
Dorstone					
1932	50	68	85	117	124
1933	37	65	74	93	102
1934	19	61	80	107	118
Total	106	194	239	317	344
Coaley					
1932	23	74	105	103	120
1933	36	47	85	90	101
1934	18	69	85	105	111
Total	77	190	275	298	332
Ipswich					
1932	39	57	91	102	112
1933	39	61	82	93	104
1934	43	61	98	98	112
Total	121	179	271	293	328

Do a complete ANOVA on these data and discuss the results.

12.9 Puffed cereal was packaged from a special machine with six heads and the filling could come from any one of three positions—bottom, middle, or bottom of a hopper. Data were collected on a measure of filling capacity from each of these 18 possible combinations of head and position. Measurements were taken on each of two cycles within each of the combinations above. These cycles within treatments can be considered random. The whole experiment was repeated in a second run at a later date. Assuming runs are random, set up an analysis of variance table for this problem and also make up a data layout table for the collection of the data.

12.10 Data for Problem 12.9 are given below for two runs, three positions (B, M, T), six heads (H), and two cycles (C) with each combination:

	Run 1						Run 2					
	B		M		T		B		M		T	
H	C_1	C_2	C_1	C_2	C_1	C_2	C_1	C_2	C_1	C_2	C_1	C_2
1	19	0	31	25	25	13	6	6	0	6	0	0
2	13	0	19	38	31	25	0	13	0	31	6	0
3	13	0	31	19	13	0	28	6	25	19	13	28
4	13	0	19	31	25	13	28	0	6	0	28	25
5	19	19	19	38	19	25	16	13	13	19	0	0
6	6	0	0	0	0	0	13	6	6	19	0	19

   Do a complete analysis of these data. Explain whether or not you could combine terms in the table as done in the text example.

12.11 Discuss the similarities and differences between a nested experiment with two factors and a randomized block design with one factor.

12.12 Assuming the nested experiment involving machines and heads in Section 11.2 was completely repeated on three subsequent randomly selected days, outline its mathematical model and ANOVA table complete with the EMS column.

12.13 For Problem 12.12 show a complete model with *all* interactions and the reduced model. Make suggestions regarding how this experiment might be made less costly and still give about the same information.

12.14 For the nested-factorial example in Section 11.4 on gun loading set up the model assuming the whole experiment is repeated on four more randomly selected days.

12.15 Examine both the full and reduced models for Problem 12.14 and comment.

Factorial Experiment—
Split-Plot Design

13.1 INTRODUCTION

In many experiments where a factorial arrangement is desired, it may not be possible to randomize completely the order of experimentation. In the last chapter restrictions on randomization were considered, and both random-ized block and Latin square designs were discussed. There are still many practical situations in which it is not at all feasible to even randomize within a block. Under certain conditions these restrictions will lead to a split-plot design. An example will show why such designs are quite common.

Example 13.1 The data in Table 13.1 were compiled on the effect of oven temperature T and baking time B on the life Y of an electrical component [25].

Looking only at this table of data, we might think of this as a 4×3 factorial with three replications per cell and proceed to the analysis in

Table 13.1 Electrical Component Life-Test Data

Baking Time B (min)	Oven Temperature T (°F)			
	580	600	620	640
5	217	158	229	223
	188	126	160	201
	162	122	167	182
10	233	138	186	227
	201	130	170	181
	170	185	181	201
15	175	152	155	156
	195	147	161	172
	213	180	182	199

Table 13.2   ANOVA for Electrical Component
Life-Test Data as a Factorial

Source	df	SS	MS	EMS
T_i	3	12,494	4165	$\sigma_\varepsilon^2 + 9\phi_T$
B_j	2	566	283	$\sigma_\varepsilon^2 + 12\phi_B$
BT_{ij}	6	2601	434	$\sigma_\varepsilon^2 + 3\phi_{TB}$
$\varepsilon_{k(ij)}$	24	13,670	570	σ_ε^2
Totals	35	29,331		

Table 13.2. Here only the temperature has a significant effect on the life of the component at the 1 percent significance level.

We might go on now and seek linear, quadratic, and cubic effects of temperature; but before proceeding, a few questions on design should be raised. What was the order of experimentation? The data analysis above assumes a completely randomized design. This means that one of the four temperatures was chosen at random and the oven was heated to this temperature, then a baking time was chosen at random and an electrical component was inserted in the oven and baked for the time selected. After this run, the whole procedure was repeated until the data were compiled. Now, was the experiment really conducted in this manner? Of course the answer is "no." Once an oven is heated to temperature, all nine components are inserted; three are baked for 5 minutes, three for 10 minutes, and three for 15 minutes. We would argue that this is the only practical way to run the experiment. Complete randomization is too impractical as well as too expensive. Fortunately, this restriction on the complete randomization can be handled in a design called a *split-plot* design.

The four temperature levels are called *plots*. They could be called blocks, but in the previous chapter blocks and replications were used for a complete rerun of the whole experiment. (The word "plots" has been inherited from agricultural applications.) In such a setup temperature—a main effect—is confounded with these plots. If conditions change from one oven temperature to another, these changes will show up as temperature differences. Thus, in such a design, a main effect is confounded with plots. This is necessary because it is often the most practical way to order the experiment. Now once a temperature has been set up by choosing one of the four temperatures at random, three components can be placed in the oven and one component can be baked for 5 minutes, another for 10 minutes, and the third one for 15 minutes. The specific component that is to be baked for 5, 10, and 15 minutes is again decided by a random selection. These three baking-time levels may be thought of as a splitting of the plot into three parts, one part for each baking time. This defines the three parts of a main plot that are called split-plots. Note here that only three components were placed in the oven, not nine. The temperature is then changed to another level and, three more components are placed in the oven for 5, 10, and 15 minutes baking time. This same procedure is followed for all four temperatures; after this the whole experiment may be replicated. These replications may be run several days after the initial experiment; in fact, it is often desirable to collect data from two or three replications and then decide whether more replications are necessary.

The experiment conducted above was actually run as a split-plot experiment and laid out as shown in Table 13.3.

Table 13.3   Split-Plot Layout for Electrical Component Life-Test Data

Replication R	Baking Time B (min)	Oven Temperature T (°F)			
		580	600	620	640
I	5	217	158	229	223
	10	233	138	186	227
	15	175	152	155	156
II	5	188	126	160	201
	10	201	130	170	181
	15	195	147	161	172
III	5	162	122	167	182
	10	170	185	181	201
	15	213	180	182	199

In Table 13.3 the lines indicate where randomization restrictions have occurred. First a replication or block is run (double lines), then within a replication an oven temperature is chosen at random (single vertical line), and then within that temperature level baking time is randomized. Thus a set of observations with no lines separating them (such as 158, 138, 152 for temperature 600°, replication I) indicates random order of these three observations. This notion of using straight lines to show randomization restrictions was suggested in Chapter 4, Section 4.2. If data are presented as in Table 13.1, there is no way to tell how they were collected. Were they completely randomized—all 36 readings? Or were there randomization restrictions as shown in Table 13.3?

Here temperatures are confounded with plots and the $R \times T$ cells are called whole plots. Inside a whole plot, the baking times are applied to one third of the material. These plots associated with the three baking times are called split-plots. Since one main effect is confounded with plots and the other main effect is not, it is usually desirable to place in the split the main effect we are most concerned about testing, as this factor is not confounded.

We might think that factor B (baking time) is nested with the main plots but this is not the case, since the same levels of B are used in all plots. A model for this experiment would be

$$Y_{ijk} = \mu + \underbrace{R_i + T_j + RT_{ij}}_{\text{whole plot}} + \underbrace{B_k + RB_{ik} + TB_{jk} + RTB_{ijk}}_{\text{split plot}}$$

The first three variable terms in this model represent the whole plot, and the RT interaction is often referred to as the *whole-plot error*. The usual assumption is that this interaction does not exist, that this term is really an estimate of the error within the main plot. The last four terms represent the split-plot, and the RTB interaction is referred to as the *split-plot error*. Sometimes the RB term is also considered nonexistent and is combined with RTB as an error term. A separate error term might be obtained if it were feasible to repeat some observations within the split-plot. The proper EMS values can be found by considering m repeat measurements where $m = 1$ (Table 13.4).

Table 13.4   EMS for Split-Plot Electrical Component Life-Test Data

	Source	df	3 R i	4 F j	3 F k	1 R m	EMS
Whole plot	R_i	2	1	4	3	1	$\sigma_\varepsilon^2 + 12\sigma_R^2$
	T_j	3	3	0	3	1	$\sigma_\varepsilon^2 + 3\sigma_{RT}^2 + 9\phi_T$
	RT_{ij}	6	1	0	3	1	$\sigma_\varepsilon^2 + 3\sigma_{RT}^2$
Split-plot	B_k	2	3	4	0	1	$\sigma_\varepsilon^2 + 4\sigma_{RB}^2 + 12\phi_B$
	RB_{ik}	4	1	4	0	1	$\sigma_\varepsilon^2 + 4\sigma_{RB}^2$
	TB_{jk}	6	3	0	0	1	$\sigma_\varepsilon^2 + \sigma_{RTB}^2 + 3\phi_{TB}$
	RTB_{ijk}	12	1	0	0	1	$\sigma_\varepsilon^2 + \sigma_{RTB}^2$
	$\varepsilon_{m(ijk)}$	—	1	1	1	1	σ_ε^2 (not retrievable)
	Total	35					

Note in Table 13.4 that the degrees of freedom in the whole plot add to 11 as there are 12 whole plots (3 replications × 4 temperatures) and the degrees of freedom in the split-plot add to 24 as there are 2 df within each whole plot and hence 2 × 12 within all whole plots.

Since the error mean square cannot be isolated in this experiment, $\sigma_\varepsilon^2 + \sigma_{RTB}^2$ is taken as the split-plot error, and $\sigma_\varepsilon^2 + 3\sigma_{RT}^2$ is taken as the whole-plot error. The main effects and interaction of interest TB can be tested, as seen from the EMS column, although no exact tests exist for the replication effect nor for interactions involving the replications. This is not a serious disadvantage for this design, since tests on replication effects are not of interest but are isolated only to reduce the error variance. The analysis of the data from Table 13.3 follows the methods given in Chapter 5. The results are shown in Table 13.5.

Table 13.5 ANOVA for Split-Plot Electrical Component Life-Test Data

Source	df	SS	MS	EMS
R_i	2	1963	982	$\sigma_\varepsilon^2 + 12\sigma_R^2$
T_j	3	12,494	4165	$\sigma_\varepsilon^2 + 3\sigma_{RT}^2 + 9\phi_T$
RT_{ij}	6	1774	296	$\sigma_\varepsilon^2 + 3\sigma_{RT}^2$
B_k	2	566	283	$\sigma_\varepsilon^2 + 4\sigma_{RB}^2 + 12\phi_B$
RB_{ik}	4	7021	1755	$\sigma_\varepsilon^2 + 4\sigma_{RB}^2$
TB_{jk}	6	2601	434	$\sigma_\varepsilon^2 + \sigma_{RTB}^2 + 3\phi_{TB}$
RTB_{ijk}	12	2912	243	$\sigma_\varepsilon^2 + \sigma_{RTB}^2$
Totals	35	29,331		

Testing the hypothesis of no temperature effect gives

$$F_{3,6} = \frac{4165}{296} = 14.1 \text{ (significant at the 1 percent level)}$$

Testing for baking time gives

$$F_{2,4} = \frac{283}{1755} < 1 \text{ (not significant)}$$

Testing for TB interaction gives

$$F_{6,12} = \frac{434}{243} = 1.79 \text{ (not significant at the 5 percent level)}$$

No exact tests are available for testing the replication effect or replication interaction with other factors, but these effects are present only to reduce the experimental error in this split-plot design.

The results of this split-plot analysis are not too different from the results using the incorrect method of Table 13.2, but this split-plot design shows the need for careful consideration of the method of randomization before starting an experiment. This split-plot design represents a restriction on the randomization over a complete randomization in a factorial experiment.

To emphasize a bit more the difference between a completely randomized 4 × 3 factorial design and a split-plot 4 × 3 factorial design, one might let the numbers 1 through 4 on one die (red) represent temperature and the numbers 1 through 3 on another die (green) represent baking times and then use the two dice to effect the proper randomization. In the case of complete randomization both dice are tossed and the first six results are as indicated below.

Baking Time	Oven Temperature			
	580°	600°	620°	640°
5	3		4	
10		1		5
15	6		2	

The numbers 1, 2, . . . , 6 indicate the order of the experiment: first set temperature at 600, bake for 10 minutes; then set temperature at 620, bake for 15 minutes; and so on. One notes that the order is scattered throughout the 12 treatment combinations. In the case of the split-plot design a temperature is randomly chosen (roll the red die) and then the baking times are chosen at random within that temperature (roll the green die). One such randomization is shown for the first six combinations as follows:

Baking Time	Oven Temperature			
	580°	600°	620°	640°
5	3		4	
10	1		6	
15	2		5	

In this latter case it should seem more obvious that temperature is confounded with the order in which the experiment is run, as three treatments are run at 580° first, then three are run at 620° second, and so on. It is hoped that by replicating the whole experiment several times any effect of order confounded with temperature would average out. Nevertheless, in a split-plot design a main effect is confounded.

Some statisticians recommend introducing a restriction error into a model wherever such restrictions occur; see Anderson and McLean [1]. No such term has been introduced here, and so it is strongly recommended that F tests be run only within the whole plot or within the split-plot and that mean squares in the whole plot not be compared with mean squares in the split-plot regardless of the EMS column.

Since this type of design is encountered often in industrial experiments, another example will be considered.

Example 13.2 A defense-related organization was to study the pulloff force necessary to separate boxes of chaff from a tape to which they are affixed. These boxes are made of a cardboard material, approximately 3 by 3 by 1 inches, and are mounted on a strip of cloth tape that has an adhesive backing. The tape is 2 inches wide, and the boxes are placed 7 inches center to center on the strip. There are 75 boxes mounted on each strip.

The tape is pulled from the box at a 90° angle as it is wound onto a drum. During this separation process the portion of the tape still adhering to the box carries the box onto a platform. The box trips a microswitch, which energizes a plunger. The plunger then kicks the box out of the machine.

After this problem was discussed with the plant engineers, several factors were listed that might affect this pulloff force. The most important factors were temperature and humidity. It was agreed to use three fixed temperature levels, $-55°C$, $25°C$, and $55°C$, and three fixed humidity levels, 50 percent, 70 percent, and 90 percent. These gave nine basic treatment combinations. Since there might be differences in pulloff force as a result of the strip selected, it was decided to choose five different strips at random for use in this experiment. There might also be differences within a strip, so two boxes were chosen at random and cut from each strip for the test.

The test in the laboratory was accomplished by hand-holding the package, attaching a spring scale to the strip by means of a hole previously punched in the strip, and pulling the tape from the package in a direction perpendicular to the package.

In discussing the design of this experiment it seemed best to set the climatic condition (a combination of temperature and humidity) at random from one of the $3 \times 3 = 9$ conditions, and then test two boxes from each of the five strips while these conditions were maintained. Then another of the nine conditions is set, and the results again determined on two boxes from each of the five strips. This is then a restriction on the randomization, and the resulting design is a split-plot design. It was agreed to replicate the complete experiment four times. A layout for this experiment is shown in Table 13.6.

In Table 13.6, each replication is a repeat of replication I, but a new order of randomizing the nine atmospheric conditions is taken in each replication. In this design atmospheric conditions and replications form the whole plot and strips are in the split-plot.

Table 13.6   Split-Plot Design of Chaff Experiment

			Temperature T (°C)								
			−55			25			55		
			Humidity H (percent)								
	Strip S	Box	50	70	90	50	70	90	50	70	90
Replication I	1	1									
		2									
R	2	3									
		4									
	3	5									
		6									
	4	7									
		8									
	5	9									
		10									

Replication II   (repeat as above)
Replication III   (repeat as above)
Replication IV   (repeat as above)

The model for this design and its associated EMS relations are set up in Table 13.7.

From the EMS column tests can be made on the effects of replications, replications by temperature interaction, replications by humidity interaction, replications by temperature by humidity interaction, strips, and strips by all other factor interactions. Unfortunately no exact tests are available for the factors of chief importance: temperature, humidity, and temperature by humidity interaction. Of course, we could first test the hypotheses that can be tested, and if some of these are not significant at a reasonably high level (say 25 percent), assume they are nonexistent and then remove these terms from the EMS column. For example, if the TS interaction can be assumed as zero ($\sigma_{TS}^2 < 0$), the temperature mean square can be tested against the RT interaction mean square with 2 and 6 df. If this is not a reasonable assumption, a pseudo-F (F') test can be used as discussed in Chapter 10.

Unfortunately, data were available only for the first replication of this experiment, so its complete analysis cannot be given. The method of analysis is the same as given in Chapter 5, even though this experiment is rather complicated. It is presented here only to show another actual example of a split-plot design.

Table 13.7 EMS for Chaff Experiment

	Source	df	4 R i	3 F j	3 F k	5 R m	2 R q	EMS
	R_i	3	1	3	3	5	2	$\sigma_\varepsilon^2 + 18\sigma_{RS}^2 + 90\sigma_R^2$
	T_j	2	4	0	3	5	2	$\sigma_\varepsilon^2 + 6\sigma_{RTS}^2 + 24\sigma_{TS}^2 + 30\sigma_{RT}^2 + 120\phi_T$
Whole	RT_{ij}	6	1	0	3	5	2	$\sigma_\varepsilon^2 + 6\sigma_{RTS}^2 + 30\sigma_{RT}^2$
plot	H_k	2	4	3	0	5	2	$\sigma_\varepsilon^2 + 6\sigma_{RHS}^2 + 24\sigma_{HS}^2 + 30\sigma_{RH}^2 + 120\phi_H$
	RH_{ik}	6	1	3	0	5	2	$\sigma_\varepsilon^2 + 6\sigma_{RHS}^2 + 30\sigma_{RH}^2$
	TH_{jk}	4	4	0	0	5	2	$\sigma_\varepsilon^2 + 2\sigma_{RTHS}^2 + 8\sigma_{THS}^2 + 10\sigma_{RTH}^2 + 40\phi_{TH}$
	RTH_{ijk}	12	1	0	0	5	2	$\sigma_\varepsilon^2 + 2\sigma_{RTHS}^2 + 10\sigma_{RTH}^2$
Split-	S_m	4	4	3	3	1	2	$\sigma_\varepsilon^2 + 18\sigma_{RS}^2 + 72\sigma_S^2$
plot	RS_{im}	12	1	3	3	1	2	$\sigma_\varepsilon^2 + 18\sigma_{RS}^2$
	TS_{jm}	8	4	0	3	1	2	$\sigma_\varepsilon^2 + 6\sigma_{RTS}^2 + 24\sigma_{TS}^2$
	RTS_{ijm}	24	1	0	3	1	2	$\sigma_\varepsilon^2 + 6\sigma_{RTS}^2$
	HS_{km}	8	4	3	0	1	2	$\sigma_\varepsilon^2 + 6\sigma_{RHS}^2 + 24\sigma_{HS}^2$
	RHS_{ikm}	24	1	3	0	1	2	$\sigma_\varepsilon^2 + 6\sigma_{RHS}^2$
	THS_{jkm}	16	4	0	0	1	2	$\sigma_\varepsilon^2 + 2\sigma_{RTHS}^2 + 8\sigma_{THS}^2$
	$RTHS_{ijkm}$	48	1	0	0	1	2	$\sigma_\varepsilon^2 + 2\sigma_{RTHS}^2$
	$\varepsilon_{q(ijkm)}$	180	1	1	1	1	1	σ_ε^2
	Total	359						

13.2 A SPLIT-SPLIT-PLOT DESIGN

Example 13.3 In a study of the cure rate index on some samples of rubber, three laboratories, three temperatures, and three types of mix were involved. Material for the three mixes was sent to one of the three laboratories where they ran the experiment on the three mixes at the three temperatures (145°C, 155°C, and 165°C). However, once a temperature was set, all three mixes were subjected to that temperature and then another temperature was set and again all three mixes were involved, and finally the third temperature was set. Material was also sent to the second and third laboratories and similar experimental procedures were performed. There are therefore two restrictions on randomization since the laboratory is chosen, then the temperature is chosen, and then mixes can be randomized at that particular temperature and laboratory.

By complete replication of the whole experiment to achieve four replications, the three laboratories and four replications form the whole plots. Then the temperature levels may be randomized at each laboratory and in each

Table 13.8 Order of First Nine Rubber Cure Rate Index Experiments

	Temperature (°C)									
	145			155			165			
	Mix			Mix			Mix			
Laboratory	A	B	C	A	B	C	A	B	C	
1			3				4			(a)
2	8	1			5		7		9	Completely randomized
3			2						6	
1										(b)
2	4	6	1	7	9	3	2	8	5	Split-plot
3										
1										(c)
2	8	9	7	3	1	2	4	6	5	Split-split-plot
3										

Table 13.9 Rubber Cure Rate Index Data

		Temperature (°C)								
		145°			155			165		
		Mix			Mix			Mix		
Replication	Laboratory	A	B	C	A	B	C	A	B	C
I	1	18.6	14.5	21.1	9.5	7.8	11.2	5.4	5.2	6.3
	2	20.0	18.4	22.5	11.4	10.8	13.3	6.8	6.0	7.7
	3	19.7	16.3	22.7	9.3	9.1	11.3	6.7	5.7	6.6
II	1	17.0	15.8	20.8	9.4	8.3	10.0	5.3	4.9	6.4
	2	20.1	18.1	22.7	11.5	11.1	14.0	6.9	6.1	8.0
	3	18.3	16.7	21.9	10.2	9.2	11.0	6.0	5.5	6.5
III	1	18.7	16.5	21.8	9.5	8.9	11.5	5.7	4.3	5.8
	2	19.4	16.5	21.5	11.4	9.5	12.0	6.0	5.0	6.6
	3	16.8	14.4	19.3	9.8	8.0	10.9	5.0	4.6	5.9
IV	1	18.7	17.6	21.0	10.0	9.1	11.1	5.3	5.2	5.6
	2	20.0	16.7	21.3	11.5	9.7	11.5	5.7	5.2	6.3
	3	17.1	15.2	19.3	9.5	9.0	11.4	4.8	5.4	5.8

replication to form a split-plot. Then at each temperature–laboratory–replication combination, the three mixes are randomly applied forming what is called a *split-split-plot* indicating that two main effects (laboratory and temperature) are confounded with blocks.

Table 13.8 shows one possible order of the first nine experiments in one replication of (a) a completely randomized, (b) a split-plot design, and (c) a split-split-plot design. The resultant data will usually not reveal how the experiment was ordered, but the use of lines in the data given in Table 13.9 should help indicate the restrictive order of randomization.

The model and its EMS values are given in Table 13.10.

Table 13.10   EMS for Rubber Cure Rate Index Experiment

| | | | 4 | 3 | 3 | 3 | 1 | |
| | | | R | F | F | F | R | |
	Source	df	i	j	k	m	q	EMS
Whole plot	R_i	3	1	3	3	3	1	$\sigma_\varepsilon^2 + 27\sigma_R^2$
	L_j	2	4	0	3	3	1	$\sigma_\varepsilon^2 + 9\sigma_{RL}^2 + 36\phi_L$
	RL_{ij}	6	1	0	3	3	1	$\sigma_\varepsilon^2 + 9\sigma_{RL}^2$
Split-plot	T_k	2	4	3	0	3	1	$\sigma_\varepsilon^2 + 9\sigma_{RT}^2 + 36\phi_T$
	RT_{ik}	6	1	3	0	3	1	$\sigma_\varepsilon^2 + 9\sigma_{RT}^2$
	LT_{jk}	4	4	0	0	3	1	$\sigma_\varepsilon^2 + 3\sigma_{RLT}^2 + 12\phi_{LT}$
	RLT_{ijk}	12	1	0	0	3	1	$\sigma_\varepsilon^2 + 3\sigma_{RLT}^2$
Split-split-plot	M_m	2	4	3	3	0	1	$\sigma_\varepsilon^2 + 9\sigma_{RM}^2 + 36\phi_M$
	RM_{im}	6	1	3	3	0	1	$\sigma_\varepsilon^2 + 9\sigma_{RM}^2$
	LM_{jm}	4	4	0	3	0	1	$\sigma_\varepsilon^2 + 3\sigma_{RLM}^2 + 12\phi_{LM}$
	RLM_{ijm}	12	1	0	3	0	1	$\sigma_\varepsilon^2 + 3\sigma_{RLM}^2$
	TM_{km}	4	4	3	0	0	1	$\sigma_\varepsilon^2 + 3\sigma_{RTM}^2 + 12\phi_{TM}$
	RTM_{ikm}	12	1	3	0	0	1	$\sigma_\varepsilon^2 + 3\sigma_{RTM}^2$
	LTM_{jkm}	8	4	0	0	0	1	$\sigma_\varepsilon^2 + \sigma_{RLTM}^2 + 4\phi_{LTM}$
	$RLTM_{ijkm}$	24	1	0	0	0	1	$\sigma_\varepsilon^2 + \sigma_{RLTM}^2$
	$\varepsilon_{q(ijkm)}$	0	1	1	1	1	1	σ_ε^2 (not retrievable)
	Total	107						

The EMS column indicates that F tests can be made on all three fixed effects and their interactions without a separate error term. Since the effect of replication is often considered of no great interest and the interaction of replication with the other factors is often assumed to be nonexistent, this appears to be a feasible experiment. If examination of the EMS column should indicate that some F tests have too few degrees of freedom in their denominator (some statisticians say less than 6), the experimenter might consider increasing the number of replications to increase the precision of the test. In the above experiment, if five replications were used, the whole plot error RL would have 8 df instead of 6, the split-plot error RLT would have 16 instead of 12, and so on.

Another technique that is sometimes used to increase the degrees of freedom is to pool all interactions with replications into the error term for that section of the table. Here one might pool RT and RLT and RM, RLM and RTM with $RLTM$ if additional degrees of freedom are believed necessary. One way to handle experiments with several replications is to stop after two or three replications, compute the results, and see if significance has been achieved or that the F's are large even if not significant. Then add another replication, and so on, increasing the precision (and the cost) of the experiment in the hope of detecting significant effects if they are there.

The numerical results of the rubber cure rate index are given in Table 13.11.

Table 13.11   ANOVA for Rubber Cure Rate Index
                   Experiment

Source	df	SS	MS
R	3	9.41	3.14
L	2	40.66	20.33*
RL	6	16.11	2.68
T	2	3119.51	1559.76***
RT	6	2.07	0.34
LT	4	4.94	1.24
RLT	12	7.81	0.65
M	2	145.71	72.86***
RM	6	3.29	0.55
LM	4	0.35	0.09
RLM	12	2.93	0.24
TM	4	43.69	10.92***
RTM	12	2.22	0.18
LTM	8	1.07	0.14
$RLTM$	24	4.97	0.21
Totals	107	3404.74	

Here laboratory effect is significant at the 5 percent level and temperature, mixes, and temperature–mix interaction are all highly significant (<0.001).

13.3   SUMMARY

The summary of Chapter 12 may now be extended.

Experiment	Design	Analysis

II. Two or more factors
A. Factorial (crossed)
	1. Completely randomized
	2. Randomized block
	   a. Complete
	     $Y_{ijk} = \mu + R_k + A_i$
	        $+ B_j + AB_{ij} + \varepsilon_{ijk}$
	   b. Incomplete, confounding:
	     i. Main effect—split-plot
	       $Y_{ijk} = \mu + \underbrace{R_i + A_j + RA_{ij}}_{\text{whole plot}}$
	       $+ \underbrace{B_k + RB_{ik} + AB_{jk} + RAB_{ijk}}_{\text{split-plot}}$
		1.
		2.
		   a. Factorial
		     ANOVA
		
		   b.
		     i. Split-plot
		       ANOVA

PROBLEMS

13.1 As part of the experiment discussed in Example 13.2 a chamber was set at 50 percent relative humidity and two boxes from each of three strips were inserted and the pulloff force determined. This was repeated for 70 percent and 90 percent, the order of humidities being randomized. Two replications of the whole experiment were made and the results follow.

Replication	Strips	Humidity					
		50%		70%		90%	
	1	1.12	1.75	3.50	0.50	1.00	0.75
I	2	1.13	3.50	0.75	1.00	0.50	0.50
	3	2.25	3.25	1.75	1.88	1.50	0.00
	1	1.75	1.88	1.75	0.75	1.50	0.75
II	2	5.25	5.25	0.75	2.25	1.50	1.50
	3	1.50	3.50	1.62	2.50	0.75	0.50

Set up a split-plot model for this experiment and outline the ANOVA table, including EMS.

13.2   Analyze Problem 13.1.

13.3   The following data layout was proposed to handle a factorial experiment designed to determine the effects of orifice size and flow rate on environmental chamber pressure in millimeters of mercury.

Flow Rate	Orifice Size (Inches)		
	8	10	12
3.5			
4.0			
4.5			

a. If this experiment is run as a completely randomized design, how might data be collected to fill in the above table?
b. If this experiment were run as a split-plot design, how might data be collected?
c. Assuming three complete replications of the above run as a split-plot design, outline its ANOVA table. (You need *not* include the EMS column.)
d. Suggest how this experiment might be improved upon.

13.4   Interest centered on the yields of alfalfa in tons per acre from three varieties following four dates of final cutting. A large section of land was divided into three strips, and each strip was planted with one of the three varieties of alfalfa. Once a strip had been selected, one fourth of the plants were cut on the first cutting date, one fourth on the next, and so on, for the four dates studied. This whole experiment could be replicated several times as necessary by choosing sections at random.
a. Identify the type of design implied here.
b. Set up an appropriate ANOVA table for this situation with the minimum number of replications you feel might be adequate to detect differences, if they exist.

13.5   In a time study two types of stimuli are used—two- and three-dimensional film—and combinations of stimuli and job are repeated once to give eight strips of film. The order of these eight is completely randomized on one long film and presented to five analysts for rating at one sitting. Since this latter represents a restriction on the randomization, the whole experiment is repeated at three sittings (four in all) and the experiment is considered a split-plot design. Set up a model for this experiment and indicate the tests to be made. (Consider analysts as random.)

13.6   Determine an F' test for any effects in Problem 13.5 that cannot be run directly.

13.7   Three replications are made of an experiment to determine the effect of days of the week and operators on screen-color difference of a TV tube in °K. On a given day, each of four operators measured the screen-color difference on a given tube. The order of measuring by the four operators is randomized each day for five days of the week. The results are

Replication	Operator	Monday	Tuesday	Wednesday	Thursday	Friday
I	A	800	950	900	740	880
	B	760	900	820	740	960
	C	920	920	840	900	1020
	D	860	940	820	940	980
II	A	780	810	880	960	920
	B	820	940	900	780	820
	C	740	900	880	840	880
	D	800	900	800	820	900
III	A	800	1030	800	840	800
	B	900	920	880	920	1000
	C	800	1000	920	760	920
	D	900	980	900	780	880

Set up the model and outline the ANOVA table for this problem.

13.8 Considering days as fixed, operators as random, and replications as random, analyze Problem 13.7.

13.9 Explain how a split-plot design differs from a nested design, since both have a factor within levels of another factor.

13.10 From the data of Table 13.9 on the rubber cure rate index experiment verify the results of Table 13.11.

13.11 Since two qualitative factors in Problem 13.10 show significance, compare their means by a Newman–Keuls test.

13.12 Since temperature in Problem 13.10 is a highly significant quantitative effect, fit a proper polynomial (or polynomials) to the data.

13.13 Redo Table 13.10, assuming that the three laboratories are chosen at random from a large number of possible laboratories.

13.14 Analyze the results of Table 13.9, based on the EMS values of Problem 13.13.

13.15 In a study of the effect of baking temperature and recipe (or mix) on the quality of cake, three mixes were used: A, B, and C. One of these three was chosen at random and the mix made up and five cakes were poured, each placed in a different oven and baked at temperatures of 300°, 325°, 350°, 375°, and 400°. After these five cakes had been baked, removed, and their quality (Y) evaluated, another mix was chosen and again five cakes were poured and baked at the five temperatures. Then this was repeated on the third mix. This experiment was replicated r times.
a. Identify this design. Show its data layout.

b. Write the mathematical model for this situation and determine the ANOVA outline indicating tests to be made.

c. Recommend how many replications r you would run to make this an acceptable experiment and explain why.

13.16 If, in Problem 13.15, only one oven was available but cakes of all three recipes could be inserted in the oven when set at a given temperature, how would the problem differ from the setup in Problem 13.15? Explain which of the two designs you would prefer and why.

13.17 Running two random replications of a problem cited in Cochran and Cox [Experimental Designs, 2nd edition, Wiley, 1957] on the effect of temperature and mix on cake quality similar to Problem 13.15 resulted in the following data.

		Temperatures					
Replication	Mix	175°	185°	195°	205°	215°	225°
	1	47	29	35	47	57	45
I	2	35	46	47	39	52	61
	3	43	43	43	46	47	58
	1	26	28	32	25	37	33
II	2	21	21	28	26	27	20
	3	21	28	25	25	31	25

Outline the ANOVA table for this problem and show what tests can be run and discuss these tests.

13.18 Analyze the data given in Problem 13.17 and state your conclusions.

13.19 What next steps would you recommend after your conclusions in Problem 13.18?

14

Factorial Experiment—
Confounding in Blocks

14.1 INTRODUCTION

As seen in Chapter 13, there are many situations in which randomization is restricted. The split-plot design is an example of a restriction on randomization where a main effect is confounded with blocks. Sometimes these restrictions are necessary because the complete factorial cannot be run in one day, or in one environmental chamber. When such restrictions are imposed on the experiment, a decision must be made as to what information may be sacrificed and thus on what is to be confounded. A simple example may help to illustrate this point.

A chemist was studying the disintegration of a chemical in solution. He had a beaker of the mixture and wished to determine whether or not the depth in the beaker would affect the amount of disintegration and whether material from the center of the beaker would be different than material from near the edge. He decided on a 2^2 factorial experiment—one factor being depth, the other radius. A cross section of the beaker (Figure 14.1) shows the four treatment combinations to be considered.

In Figure 14.1 (1) represents radius zero, depth at lower level (near bottom of beaker); a represents the radius at near maximum (near the edge of the beaker) and a low depth level; b is radius zero, depth at a higher level; and ab is both radius and depth at higher levels. In order to get a reading on

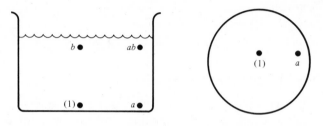

Figure 14.1

the amount of disintegration of material, samples of solution must be drawn from these four positions. This was accomplished by dipping into the beaker with an instrument designed to "capture" a small amount of solution. The only problem was that the experimenter had only two hands and could get only two samples at the same time. By the time he returned for the other two samples conditions might have changed and the amount of disintegration would probably be greater. Thus there is a physical restriction on this 2^2 factorial experiment; only two observations can be made at one time. The question is: Which two treatment combinations should be taken on the first dip? Considering the two dips as blocks, we are forced to use an incomplete block design, as the whole factorial cannot be performed in one dip.

Three possible block patterns are shown in Table 14.1.

Table 14.1   Three Possible Blockings of a 2^2 Factorial

Dip	Plan I		Plan II		Plan III	
1	(1)	b	(1)	a	(1)	ab
2	a	ab	b	ab	a	b

If plan I of Table 14.1 is used, the blocks (dips) are confounded with the radius effect, since all zero-radius readings are in dip 1 and all maximum radius readings are in dip 2. In plan II the depth effect is confounded with the dips, as low-level depth readings are both in dip 1 and high-level depth readings are in dip 2. In plan III neither main effect is confounded, but the interaction is confounded with the dips. To see that the interaction is confounded, recall the expression for the AB interaction in a 2^2 factorial found in Chapter 6.

$$AB = \tfrac{1}{2}[(1) - a - b + ab]$$

Note that the two treatment effects with the plus sign, (1) and ab, are both in dip 1, and the two with a minus sign, a and b, are in dip 2. Hence we cannot distinguish between a block effect (dips) and the interaction effect (AB). In most cases it is better to confound an interaction than to confound a main effect. The hope is, of course, that there is no interaction, and that information on the main effects can still be found from such an experiment.

14.2 CONFOUNDING SYSTEMS

To design an experiment in which the number of treatments that can be run in a block is less than the total number of treatment combinations, the experimenter must first decide on what effects he is willing to confound. If, as in the example above, he has only one interaction in the experiment and decides that he can confound this interaction with blocks, the problem is simply which treatment combinations to place in each block. One way to accomplish this is to place in one block those treatment combinations with a plus sign in the effect to be confounded, and those with a minus sign in the other block. A more general method is necessary when the number of blocks and the number of treatments increase.

First a *defining contrast* is set up. This is merely an expression stating which effects are to be confounded with blocks. In the simple example above, to confound AB, write AB as the defining contrast. Once the defining contrast has been set up, several methods are available for determining which treatment combinations will be placed in each block. One method was shown above—place in one block the treatment combinations that have a plus sign in AB and those with a minus sign in the other block. Another method is to consider each treatment combination. Those that have an even number of letters in common with the effect letters in the defining contrast go in one block. Those that have an odd number of letters in common with the defining contrast go into the other block. Here (1) has no letters in common with AB or an even number. Type a has one letter in common with AB (a and A) or an odd number. b also has one letter in common with AB. ab has two letters in common with AB. Thus the block contents are those in Figure 14.2.

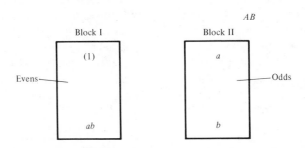

Figure 14.2

One disadvantage of the two methods given above is that they are only good on 2^f factorials. A more general method is attributed to Kempthorne [13]. Consider the linear expression

$$A_1 X_1 + A_2 X_2 + \cdots + A_n X_n = L$$

Here A_i is the exponent on the ith factor appearing in each independent defining contrast, and X_i is the level of the ith factor appearing in a given treatment combination. Every treatment combination with the same L value will be placed in the same block. For the simple example above where the defining contrast is AB, $A_1 = 1$, $A_2 = 1$, and all other A_i's $= 0$. Hence

$$L = 1 \cdot X_1 + 1 \cdot X_2$$

For each treatment combination the L values are

$$(1): L = 1 \cdot 0 + 1 \cdot 0 = 0$$
$$a: L = 1 \cdot 1 + 1 \cdot 0 = 1$$
$$b: L = 1 \cdot 0 + 1 \cdot 1 = 1$$
$$ab: L = 1 \cdot 1 + 1 \cdot 1 = 2$$

Hence 0 and 2 are both 0, since a 2^f factorial is of modulus 2 (see Section 9.2). The assignment of treatment combinations to blocks is then

$L = 0$ block 1	(1)	ab

$L = 1$ block 2	a	b

For a more complex defining contrast such as ABC^2, the expression is

$$L = X_1 + X_2 + 2X_3$$

and this can be used to decide which treatment combinations in a 3^3 experiment go into a common block.

The block containing the treatment combination (1) is called the *principal block*. The treatment combinations in this block are elements of a group where the operation on the group is multiplication, modulus 2. The elements of the other block or blocks may be generated by multiplying one element in the new block by each element in the principal block. Multiplying elements together within the principal block will generate more elements within the principal block. This is best seen with a slightly more complex factorial, say 2^3. If the highest order interaction is to be confounded and the eight treatment combinations are to be placed into two blocks of four treatments each, then confounding ABC gives

$$L = X_1 + X_2 + X_3$$

Testing each of the eight treatment combinations gives

$$(1): L = 0 + 0 + 0 = 0$$
$$a: L = 1 + 0 + 0 = 1$$
$$b: L = 0 + 1 + 0 = 1$$
$$ab: L = 1 + 1 + 0 = 2 = 0$$
$$c: L = 0 + 0 + 1 = 1$$
$$ac: L = 1 + 0 + 1 = 2 = 0$$
$$bc: L = 0 + 1 + 1 = 2 = 0$$
$$abc: L = 1 + 1 + 1 = 3 = 1$$

The blocks are then

block 1	block 2
(1)	a
ab	b
ac	c
bc	abc
$L = 0$	$L = 1$

Group theory can simplify this procedure. Determine two elements in the principal block, for example, ab and bc. Multiply these two to get $ab^2c = ac$, the fourth element. Generate block 2 by finding one element, say a. Multiply this element by each element in the principal block, thus generating all the elements of the second block as follows:

$$a \cdot (1) = a$$
$$a \cdot ab = a^2b = b$$
$$a \cdot ac = a^2c = c$$
$$a \cdot bc = abc$$

These concepts may be extended to more complex confounding examples, as shown in later sections.

14.3   BLOCK CONFOUNDING—NO REPLICATION

In many cases an experimenter cannot afford several replications of an experiment, and is further restricted in that the complete factorial cannot be run in one block or at one time. Again, the experiment may be blocked and information recovered on all but some high-order interactions, which may be confounded. Designs for such experiments will be considered for the special cases of 2^f and 3^f factorials because they appear often in practice.

Confounding in 2^f Factorials

The methods given earlier in this chapter can be used to determine the block composition for a specific confounding scheme. If only one replication is possible, one is run, and some of the higher order interaction terms must be used as experimental error unless some independent measure of error is available from previous data. This type of design is used mostly when there are several factors involved (say four or more), some high-order interactions may be confounded with blocks, and some others are yet available for an error estimate. For example, consider a 2^4 factorial where only eight treatment combinations can be run at one time. One possible confounding scheme would be

confound $ABCD$ and $L = X_1 + X_2 + X_3 + X_4$

block 1

(1)	ab	bc	ac	$abcd$	cd	ad	bd

$L = 0$

block 2

a	b	abc	c	bcd	acd	d	abd

$L = 1$

and the analysis would be as shown in Table 14.2. The three-way interactions may be pooled here to give 4 df for error, assuming that these interactions are actually nonexistent. All main effects and first-order interactions could then be tested with 1 and 4 df. After examining the results it might be possible to pool some of the two-way interactions that have small mean squares with the three-way interactions if more degrees of freedom are desired in the error term.

If only four treatment combinations can be run in a block, one may confound the 2^4 factorial in four blocks of four treatment combinations each. For four blocks 3 df must be confounded with blocks. If two interactions are confounded, the product (modulus 2) of these two is also confounded, since the product of the signs in two effects gives the signs for the product. Thus, if AB and CD are confounded, so also will $ABCD$ be confounded. This scheme would confound two first-order interactions and one third order. It might be better to confound two second-order interactions and one first order:

$$ABC \qquad BCD \qquad AD$$

Table 14.2 2^4 Factorial in Two Blocks

Source	df
A	1
B	1
C	1
D	1
AB	1
AC	1
AD	1
BC	1
BD	1
CD	1
ABC	1 ⎫
ABD	1 ⎬ 4 df for error
ACD	1 ⎥
BCD	1 ⎭
Blocks or $ABCD$	1
Total	15 df

Note here that $(ABC)(BCD) = AB^2C^2D = AD$ (modulus 2). Hence 3 df are confounded with blocks. Before proceeding to the block compositions, consider a problem where four factors (A, B, C, and D) are involved, each at two levels, and only four treatment combinations can be run in one day. The measured variable is yield. The confounding scheme given above gives two independent equations

$$L_1 = X_1 + X_2 + X_3$$
$$L_2 = X_2 + X_3 + X_4$$

Each treatment combination will give one of the following four pairs when substituted into both L_1 and L_2 = 00, 01, 10, 11. For example,

$$\begin{array}{ll} (1): L_1 = 0 & L_2 = 0 \\ a: L_1 = 1 & L_2 = 0 \\ b: L_1 = 1 & L_2 = 1 \\ ab: L_1 = 2 = 0 & L_2 = 1 \end{array}$$

and so on. All treatment combinations with the same pair of L values will be placed together in one block. Thus the 16 treatment combinations will be assigned to four blocks as in Table 14.3.

Here the order of experimentation within each block is randomized and the resulting responses are given for each treatment combination. The proper treatment combinations for each block can be generated from the principal block. If bc and acd are in the principal block, then their product

$$(bc)(acd) = abc^2d = abd$$

is also in the principal block. For another block, consider a and multiply a by all the combinations in the principal block, giving

$$a(1) = a, \qquad a(bc) = abc, \qquad a(acd) = cd, \qquad a(abd) = bd$$

the entries in the second block. In the same manner, the other two blocks may be generated. This means that only a few readings in the principal block need

Table 14.3   Blocking a 2^4 Factorial in Four Blocks of Four
Confounding ABC, BCD, and AD

block 1	block 2	block 3	block 4
(1) = 82	a = 76	b = 79	ab = 85
bc = 55	abc = 74	c = 71	ac = 84
acd = 81	cd = 72	$abcd$ = 89	bcd = 84
abd = 88	bd = 73	ad = 79	d = 80
$L_1 = 0$	$L_1 = 1$	$L_1 = 1$	$L_1 = 0$
$L_2 = 0$	$L_2 = 0$	$L_2 = 1$	$L_2 = 1$

to be determined by the L_1, L_2 values, and all the rest of the design can be generated from these few treatment combinations. Tables are also available for designs when certain effects are confounded [9].

This experiment is then a 2^4 factorial experiment. The design is a randomized incomplete block design where blocks are confounded with the interactions ABC, BCD, and AD. An example will illustrate the method.

Example 14.1   A systematic test was to be made on the effects of four variables, each at two levels, on coil breakdown voltage. The variables were A—firing furnace 1 or 3; B—firing temperature 1650°C or 1700°C; C—gas humidification, no or yes; and D—coil outside diameter small (<0.0300 inch) or large (>0.0305 inch). Since only four coils could be tested in a given short time interval, it was necessary to run four blocks of four coils for such testing. Only one complete test of 2^4 in four blocks of four was run and the results (after dividing the voltages through by 10) are shown in Table 14.3, which confounds ABC, BCD, and AD with the four blocks. For the analysis one might use Yates' method given in Chapter 6 to give Table 14.4.

In Table 14.4 the sums of squares in the last column correspond to the effect for the treatment combination on the left. That is,

$$SS_A = 225.00 \qquad SS_B = 0.25 \qquad \text{and so on}$$

Table 14.4   Yates' Method Analysis for 2^4 Example

Treatment Combination	Response	(1)	(2)	(3)	(4)	SS
(1)	82	158	322	606	1252	—
a	76	164	284	646	+60	225.00
b	79	155	320	+32	2	0.25
ab	85	129	326	28	30	56.25
c	71	159	0	−20	−32	64.00
ac	84	161	+32	22	+32	64.00
bc	55	153	14	+18	−14	12.25
abc	74	173	14	12	−26	42.25*
d	80	−6	6	−38	40	100.00
ad	79	+6	−26	6	−4	1.00*
bd	73	+13	2	+32	42	110.25
abd	88	+19	20	0	−6	2.25
cd	72	−1	12	−32	44	121.00
acd	81	15	+6	18	−32	64.00
bcd	84	9	16	−6	50	156.25*
abcd	89	5	−4	−20	−14	12.25
Total						1031.00

* Confounded with blocks.

Note that the effects ABC, BCD, and AD are confounded with blocks. Their total sum of squares is $42.25 + 156.25 + 1.00 = 199.50$. If the above block totals are computed, they yield

$$\text{block } 1 — 306 \qquad \text{block } 3 — 318$$
$$\text{block } 2 — 295 \qquad \text{block } 4 — 333$$

and the sum of squares between these blocks is

$$\frac{(306)^2 + (295)^2 + (318)^2 + (333)^2}{4} - \frac{(1252)^2}{16}$$

$$\text{SS}_{\text{block}} = 98{,}168.50 - 97{,}969.00 = 199.50$$

which is identical with the sum of the three interactions with which blocks are confounded. If all three-way and four-way interactions that are not confounded are pooled for an error term, the resulting analysis is that shown in Table 14.5.

Table 14.5   ANOVA for 2^4 Example in Four Blocks

Source	df	SS	MS
A	1	225.00	225.00
B	1	0.25	0.25
C	1	64.00	64.00
D	1	100.00	100.00
AB	1	56.25	56.25
AC	1	64.00	64.00
BC	1	12.25	12.25
BD	1	110.25	110.25
CD	1	121.00	121.00
Error or ABD, ACD, $ABCD$	3	78.50	26.17
Blocks or ABC, BCD, AD	3	199.50	66.50
Totals	15	1031.00	

With only 1 and 3 df, none of the four main effects or the five two-way interactions can be declared significant at the 5 percent significance level. Since the B effect and the BC interaction are not significant even at the 25 percent level, we might wish to pool these with the error to increase the error degrees of freedom. The resulting error sum of squares would then be

$$0.25 + 12.25 + 78.50 = 91.00$$

with 5 df. The error mean square then $= 91.00/5 = 18.20$. Using this error mean square, A and CD are now significant at the 5 percent level. We might conclude that the effect of factor A is important, that B is not present, and that there may be a CD interaction. As 2^f experiments are often run to get an overall picture of the important factors, each set at the extremes of its range, another experiment might now be planned without factor B, since its effect is negligible over the range considered here. The next experiment might include factors A, C, and D at, perhaps, three levels each, or a 3^3 experiment. This experiment might also require blocking.

14.4 BLOCK CONFOUNDING WITH REPLICATION

Whenever an experiment is restricted so that all treatments cannot appear in one block, some interaction is usually confounded with blocks. If several replications of the whole experiment (all blocks) are possible, as in the case of the split-plot, the same interaction may be confounded in all replications. In this case, the design is said to be *completely confounded*. If, on the other hand, one interaction is confounded in the first replication, a different interaction is confounded in the second replication, and so on, the design is said to be *partially confounded*.

Complete Confounding

Considering a 2^3 factorial experiment in which only four treatment combinations can be finished in one day yields a 2^3 factorial run in two incomplete blocks of four treatment combinations each. We might confound the highest order interaction ABC as follows:

$$ABC$$

As seen in Section 14.2, the blocks would be

block 1	block 2
(1)	a
ab	b
ac	c
bc	abc

If this whole experiment (2^3 in two blocks of four each) can be replicated, say three times, the layout might be

replication I		replication II		replication III	
block 1	block 2	block 1	block 2	block 1	block 2
ac	a	c	(1)	ab	c
(1)	c	abc	ac	(1)	b
ab	abc	b	bc	ac	abc
bc	b	a	ab	bc	a

The confounding scheme in the above (ABC) is the same for all three replications, but the order of experimentation has been randomized within each block. Also, the decision as to which block is to be run first in each replication is made at random. An analysis layout for this experiment appears in Table 14.6.

Table 14.6 Analysis Layout for Completely Confounded 2^3 Factorial

Source	df	df
Replications	2	
Blocks or ABC	1	
Replications × block interactions	2	5 between plots
A	1	
B	1	
AB	1	
C	1	
AC	1	
BC	1	
Replications × all others	12	18 within plots
Totals	23	23

Here the replication interaction with all three main effects and their interactions are usually taken as the error term for testing the important effects. The replication effect and the block (or ABC) effect could be tested against the replication × block interaction, but the degrees of freedom are low and the power of such a test is poor. Such a design is quite powerful, however, in testing the main effects A, B, and C and their first-order interactions. However, it will give no clean information on the ABC interaction.

Partial Confounding

In the example just considered, we might be concerned with some test on the ABC interaction. This could be found by confounding some interactions other than ABC in some of the replications. We might use four replications and confound ABC in the first one, AB in the second, AC in the third, and BC in the fourth. Thus the four replications will yield full information on A, B, and C but three-fourths information on AB, AC, BC, and ABC, since we can compute an unconfounded interaction such as AB in three out of four of the replications. The following layout shows the proper block entries for each replication.

replication I		replication II		replication III		replication IV	
confound ABC		AB		AC		BC	
(1)	a	(1)	a	(1)	a	(1)	b
ab	b	c	b	ac	c	bc	c
ac	c	ab	ac	abc	bc	a	ab
bc	abc	abc	bc	b	ab	abc	ac

$$L = X_1 + X_2 \qquad L = X_1 + X_2 \qquad L = X_1 + X_3 \qquad L = X_2 + X_3$$
$$+ \; X_3$$

The analysis layout would be that shown in Table 14.7.

The residual term with its 17 df can be explained as follows. In all four replications the main effects A, B, and C can be isolated. This gives 3×1 df for each replication by main-effect interaction, or a total of 9 df. The AB, AC, BC, and ABC interactions can be isolated in only three out of the four replications. Hence the replications by AC have $2 \times 1 = 2$ df, or a total of 8 df for all such interactions. This gives $9 + 8 = 17$ df for effects by replications, which is usually taken as the error estimate. The blocks and replications are separated out in the top of Table 14.7 only to reduce the experimental error. The numerical analysis of problems such as this is carried out in the same manner as given in earlier chapters.

Example 14.2   Coils were wound from each of two winding machines using either of two wire stocks and their outside diameters were measured at two

Table 14.7   ANOVA for Partially Confounded 2^3 Factorial

Source	df		df
Replications	3		
Blocks within replications	4		7 between plots
ABC (Replication I)		1	
AB (Replication II)		1	
AC (Replication III)		1	
BC (Replication IV)		1	
A	1		
B	1		
C	1		
AB	1	only from	
AC	1	replications	
BC	1	where not	
ABC	1	confounded	
Replications $\times$ all effects	17		24 within plots
Totals	31		31

positions on an optical comparator. Since only four readings could be made in a given time period, the $2^3 = 8$ treatment combinations were divided into two blocks of four. However, four replications were possible. At first the experimenter decided to confound ABC in all four replicates. Here A represents machines, B stocks, and C positions. The results are shown in Table 14.8.

Table 14.8 Layout of Example 14.2 in Four Blocks of Four Confounding ABC in All Replicates

replication I		replication II	
block 1	block 2	block 1	block 2
$ab = 2173$	$b = 2300$	$a = 2228$	$ac = 3495$
$(1) = 2249$	$c = 3538$	$abc = 3592$	$(1) = 2094$
$ac = 3532$	$abc = 3524$	$c = 3116$	$bc = 2249$
$bc = 2948$	$a = 2319$	$b = 2386$	$ab = 2373$

replication III		replication IV	
block 1	block 2	block 1	block 2
$(1) = 2382$	$c = 3528$	$abc = 2996$	$ab = 2393$
$ab = 2240$	$b = 2118$	$c = 1934$	$bc = 2814$
$bc = 3495$	$abc = 3350$	$b = 2234$	$ac = 3400$
$ac = 2995$	$a = 2272$	$a = 2215$	$(1) = 2297$

All measurements multiplied by 10^5 so 2173 is an outside diameter of 0.02173 inch.

Analyzing by computer or by direct methods gives Table 14.9.

Table 14.9 ANOVA for Example 14.2

Source	df	SS	MS	F	Prob.
Replications	3	409,719.59			
Blocks or ABC	1	8,482.53			
$R \times B$ interaction	3	515,407.85			
	7				
A	1	364,444.53	364,444.53	3.13	0.094
B	1	5,227.53	5,227.53	<1	
AB	1	18,963.78	18,963.78	<1	
C	1	6,330,571.53	6,330,571.53	54.44**	0.000
AC	1	302,058.78	302,058.78	2.60	0.124
BC	1	16,698.78	16,698.78	<1	
Replications $\times$ all others	18	2,093,039.32	116,279.96		
	24				
Totals	31	10,064,614.22			

Table 14.9 shows only positions on the coil as having a highly significant effect on coil outside diameter.

Example 14.3   When coils from two other winding machines were to be tested, it was decided to partially confound the interactions. Results gave Table 14.10.

Table 14.10   Layout of Example 14.3 in Four Blocks of Four with Partial Confounding

replication I (confounding ABC)		replication II (confounding AB)	
block 1	block 2	block 3	block 4
(1) = 2208	a = 2196	(1) = 2004	a = 2179
ab = 2133	b = 2086	ab = 2112	b = 2073
ac = 2459	c = 3356	c = 3073	ac = 3474
bc = 3096	abc = 2776	abc = 2631	bc = 3360

replication III (confounding AC)		replication IV (confounding BC)	
block 5	block 6	block 7	block 8
c = 2839	(1) = 1916	a = 2056	b = 1878
a = 2189	ac = 2979	(1) = 2010	ab = 2156
bc = 3522	b = 2151	bc = 3209	c = 3423
ab = 2095	abc = 2500	abc = 3066	ac = 2524

Analysis gives Table 14.11. In this table position on the coil is again highly significant, but an AC interaction is also significant at the 5 percent significance level.

In both problems the effect of A—machines—is approaching significance. It may be that if more replications are run, this machine effect will emerge as significant.

Confounding in 3^f Factorials

When a 3^f-factorial experiment cannot be completely randomized, it is usually blocked in blocks that are multiples of 3. The use of the I and J components of interaction introduced in Section 9.2 is helpful in confounding only part of an interaction with blocks. Kempthorne's rule also applies, using such interactions as AB, AB^2, ABC^2, . . . , and treatment combinations in the form 00, 10, 01, 11 instead of (1), a, b, ab as in 2^f. Here treatment combinations can be multiplied that will actually add these exponents (modulus 3).

If a 3^2 factorial is restricted so that only three of the nine treatment combinations can be run in one block, we usually confound AB or AB^2, as each carries 2 df. The defining contrast might be AB^2, which gives $L = X_1 + 2X_2$.

Table 14.11   ANOVA for Example 14.3

Source	df		SS	MS	F	Probability
Replications	3		38,717.59			
Blocks in replications	4		401,060.13			
		7				
A	1		224,282.53	224,282.53	3.97	0.063
B	1		52.53	52.53	<1	
AB	1		737.04	737.04	<1	
C	1		6,886,688.28	6,886,688.28	121**	0.000
AC	1		416,066.67	416,066.67	7.36*	0.015
BC	1		2,667.04	2,667.04	<1	
ABC	1		70,742.04	70,742.04	1.25	0.279
Replications × all	17		961,283.32	56,546.08		
		24				
Totals	31		9,002,296.97			

Each of the nine treatment combinations then yields

$$00, L = 0 \qquad 01, L = 2 \qquad 02, L = 4 = 1$$
$$10, L = 1 \qquad 11, L = 3 = 0 \qquad 12, L = 5 = 2$$
$$20, L = 2 \qquad 21, L = 4 = 1 \qquad 22, L = 6 = 0$$

Placing treatment combinations with the same L value in common blocks gives

block 1	block 2	block 3
$L = 0$	$L = 1$	$L = 2$
00	10	20
11	21	01
22	02	12

Block 1 with 00 in it is the principal block. Block 2 is generated by *adding* one treatment combination in block 2 to all those in block 1, giving

$$00 + 10 = 10 \qquad 11 + 10 = 21 \qquad 22 + 10 = 32 = 02$$

Similarly for block 3,

$$00 + 20 = 20 \qquad 11 + 20 = 31 = 01 \qquad 22 + 20 = 42 = 12$$

Since there are three blocks, 2 df must be confounded. Here AB^2 is confounded with blocks. If AB with its 2 df were confounded, $L = X_1 + X_2$, and the blocking would be

block 1	block 2	block 3
$L = 0$	$L = 1$	$L = 2$
00	10	20
12	22	02
21	01	11

If the data of Table 9.1 are used as an example, and AB^2 is confounded, the responses are

block 1	block 2	block 3
00 = 1	10 = -2	20 = 3
11 = 4	21 = 1	01 = 0
22 = 2	02 = 2	12 = -1

The block sums of squares are determined from the three block totals of 7, 1, and 2:

$$\text{SS}_{\text{block}} = \frac{7^2 + 1^2 + 2^2}{3} - \frac{(10)^2}{9} = 6.89$$

which is identical with the $I(AB)$ interaction component computed in Section 9.2 (Table 9.6) as it should be. The remainder of the analysis proceeds the same as in Section 9.2, and the resulting ANOVA table is that exhibited in Table 14.12.

If these were real data, the AB part of the interaction might be considered as error, but then neither main effect is significant. The purpose here is only to illustrate the blocking of a 3^2 experiment.

Consider now a 3^3 experiment in which the 27 treatment combinations (such as given in Table 9.7) cannot all be completely randomized. If nine can be randomized and run on one day, nine on another day, and so on, we might use a 3^3 factorial in three blocks of nine treatment combinations each. This requires 2 df to be confounded with blocks. Since the ABC interaction with 8 df can be partitioned into ABC, ABC^2, AB^2C, and AB^2C^2, each with 2 df,

Table 14.12   ANOVA for 3^2 Example

Source	df	SS	MS
A	2	4.22	2.11
B	2	1.56	0.78
AB	2	16.22	8.11
Blocks or AB^2	2	6.89	3.44
Totals	8	28.89	

one of these four could be confounded with blocks. If AB^2C is confounded,

$$L = X_1 + 2X_2 + X_3$$

and the three blocks are

$L = 0$	$L = 1$	$L = 2$
000	100	200
011	111	211
110	210	010
121	221	021
102	202	002
212	012	112
220	020	120
022	122	222
201	001	101

The analysis of data for this design would yield Table 14.13.

This is a useful design, since we can retrieve all main effects (A, B, C) and all two-way interactions, if we are willing to pool the 6 df in ABC as error. The determination of sums of squares, and so forth, is the same as for a complete 3^3 given in Example 9.1. Confounding other parts of the ABC interaction, of course, yield different blocking arrangements, although the outline of the analysis would be the same. In practice, different blocking arrangements might yield different responses.

More complex confounding schemes can be found in tables such as in Chapter 9 of Davies [9].

For practice, consider confounding a 3^3 in nine blocks of three treatment combinations each. Here 8 df must be confounded with blocks. One confounding scheme might be

$$AB^2C^2 \qquad AB \qquad BC^2 \qquad AC$$

Table 14.13 3^3 Factorial in Three Blocks

Source	df
A	2
B	2
AB	4
C	2
AC	4
BC	4
Error or ABC, ABC^2, AB^2C^2	6
Blocks or AB^2C	2
Total	26

Note that

$$(AB^2C^2)(AB) = A^2C^2 = AC$$

and

$$(AB)(BC^2) = AB^2C^2$$

These are not all independent. In fact, only two of the expressions are. For this reason we need only two expressions for the L values:

$$L_1 = X_1 + 2X_2 + 2X_3$$
$$L_2 = X_1 + X_2$$

as these two will yield nine pairs of numbers—one pair for each block. First determine the principal block, where both L_1 and L_2 are zero. One treatment combination is obviously 000. Another is 211, as

$$L_1 = 1(2) + 2(1) + 2(1) = 6 = 0$$
$$L_2 = 1(2) + 1(1) = 3 = 0$$

A third is 122, as

$$L_1 = 1(1) + 2(2) + 2(2) = 9 = 0$$
$$L_2 = 1(1) + 1(2) = 3 = 0$$

The other eight blocks can now be generated from this principal block, giving

block	1	2	3	4	5	6	7	8	9
	000	001	002	010	020	100	200	110	101
	211	212	210	221	201	011	111	021	012
	122	120	121	102	112	222	022	202	220

An analysis layout would be that shown in Table 14.14.

Table 14.14 3^3 Factorial in Nine Blocks

Source	df
A	2
B	2
C	2
AB^2	2
AC^2	2
BC	2
ABC	2 ⎫
AB^2C	2 ⎬ 6 df for error
ABC^2	2 ⎭
Blocks or AB, BC^2, AC, AB^2C^2	8
Total	26

This design might be reasonable if interest centered only in the main effects A, B, and C. Such a design might be necessary where three factors are involved, each at three levels, but only three treatment combinations can be run in one day. These three might involve an elaborate environmental test where at best only three sets of environmental conditions—temperature, pressure, and humidity—can be simulated in one day.

14.5 SUMMARY

The summary of Chapter 13 may now be extended.

Experiment	Design		Analysis
II. Two or more factors			
A. Factorial (crossed)	1. Completely randomized		
	2. Randomized block	2.	
	a. Complete		a. Factorial ANOVA
	$Y_{ijk} = \mu + R_k + A_i$ $+ B_j + AB_{ij} + \varepsilon_{ijk}$		
	b. Incomplete, confounding:		b.
	i. Main effect—split-plot		i. Split-plot ANOVA
	$Y_{ijk} = \mu + \underbrace{R_i + A_j + RA_{ij}}_{\text{whole plot}}$ $+ \underbrace{B_k + RB_{ik} + AB_{jk} + RAB_{ijk}}_{\text{split-plot}}$		
	ii. Interactions in 2^f and 3^f		ii.
	(1) Several replications		(1) Factorial ANOVA with replications R_i and blocks β_j or confounded interaction
	$Y_{ijkq} = \mu + R_i + \beta_j$ $+ R\beta_{ij} + A_k + B_q$ $+ AB_{kq} + \varepsilon_{ikq}$		
	(2) One replication only		(2) Factorial ANOVA with blocks β_j or confounded interaction
	$Y_{ijk} = \mu + \beta_i + A_j$ $+ B_k + AB_{jk}, \ldots$		

PROBLEMS

14.1 Assuming that a 2^3 factorial such as the problem of Chapters 1, 5, and 6 involving two tool types, two bevel angles, and two types of cut cannot all be run on one lathe in one time period but must be divided into four treatment combinations per lathe, set up a confounding scheme for confounding the AC interaction in these two blocks of four. Assuming the results of one replication give $(1) = 5$, $a = 0$, $b = 4$, $ab = 2$, $c = -3$, $ac = 0$, $bc = -1$, $abc = -2$, show that the scheme you have set up does indeed confound AC with blocks.

14.2 Assuming three replications of the 2^3 experiment in Problem 14.1 can be run but each one must be run in two blocks of four, set up a scheme for the complete confounding of ABC in all three replicates and show its ANOVA layout.

14.3 Repeat Problem 14.2 using partial confounding of AB, AC, and BC.

14.4 Data involving four replications of the 2^3 factorial described in the foregoing problems and in Section 14.3 gave the following numerical results.

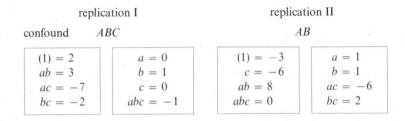

replication I			replication II		
confound	ABC			AB	
$(1) = 2$		$a = 0$	$(1) = -3$		$a = 1$
$ab = 3$		$b = 1$	$c = -6$		$b = 1$
$ac = -7$		$c = 0$	$ab = 8$		$ac = -6$
$bc = -2$		$abc = -1$	$abc = 0$		$bc = 2$

replication III			replication IV		
confound	AC			BC	
$(1) = 5$		$a = 0$	$(1) = -2$		$b = 9$
$ac = 0$		$c = -3$	$bc = -1$		$c = -3$
$abc = -2$		$bc = -1$	$a = -6$		$ab = 0$
$b = 4$		$ab = 2$	$abc = -4$		$ac = -4$

Analyze these data and state your conclusions.

14.5 A 2^3 factorial is to be used to study three factors and only four treatments can be run on each of two days. On the first day the following data are collected:

$$(1) = 2, \quad a = 1, \quad bc = 3, \quad abc = 4$$

and on the next day:

$$c = 1, \quad ac = 4, \quad b = 6, \quad ab = 0$$

By examining the variation between the two days and your knowledge of a 2^f experiment, find out which term in the model (main effect or interaction) is confounded with days.

14.6 For the data of Problem 7.6 consider running the 16 treatment combinations in four blocks of four (two observations per treatment). Confound ACD, BCD, and AB and then determine the block compositions.

14.7 From the results of Problem 7.6 show that the blocks are confounded with the interactions in Problem 14.6.

14.8 For Problem 9.5 consider this experiment as run in three blocks of nine with two replications for each treatment combination. Confound ABC (2 df) where A is surface thickness, B is base thickness, and C is subbase thickness, and determine the design. Also use the numerical results of Problem 9.7 to show that ABC is indeed confounded in your design.

14.9 Set up a scheme for confounding a 2^4 factorial in two blocks of eight each, confounding ABD with blocks.

14.10 Repeat Problem 14.9 and confound BCD.

14.11 Work out a confounding scheme for a 2^5 factorial, confounding in four blocks of eight each. Show the outline of an ANOVA table.

14.12 Repeat Problem 14.11 in four replications confounding different interactions in each replication. Show the ANOVA table outline.

14.13 Work out the confounding of a 3^4 factorial in three blocks of 27 each.

14.14 Repeat Problem 14.13 in nine blocks of nine each.

14.15 In Example 14.2, since ABC is completely confounded, the analysis could be easily run on a computer or by using Yates' method on the A, B, C sections and general methods for replications and blocks. Verify the results given in Table 14.9.

14.16 Since Example 14.3 has partial confounding, it may be difficult for the computer. However, a Yates method can be used if care is taken as to how it is applied. Try to verify Table 14.11 results using a modified Yates procedure.

14.17 A systematic test was made to determine the effects on coil breakdown voltage of the following six factors, each at two levels as indicated:

1. Firing furnace   Number 1 or 3
2. Firing temperature   1650°C or 1700°C
3. Gas humidification   No or yes
4. Coil outer diameter   Below 0.0300 inch or above 0.0305 inch
5. Artificial chipping   No or yes
6. Sleeve   Number 1 or 2

All 64 experiments could not be performed under the same conditions so the whole experiment was run in eight subgroups of eight experiments. Set up a reasonable confounding scheme for this problem and outline its ANOVA.

14.18 An experiment was run on the effects of several annealing variables on the magnetic characteristic of a metal. The following factors were to be considered.

1. Temperature   1375°, 1450°, 1525°, 1600°
2. Time in hours   2, 4

3. Exogas ratio   6.5 to 1, 8.5 to 1

4. Dew point $10°$, $30°$

5. Cooling rate in hours   4, 8

6. Core material A, B.

Since all factors are at two levels or multiples of two levels, outline a scheme for this experiment in two blocks of 64 observations.

14.19 Outline a Yates' method for handling the data of Problem 14.18.

14.20 If X is the surface rating of a steel sample and material is tested from eight heats, two chemical treatments, and three positions on the ingot, the resulting factorial is an $8 \times 2 \times 3$ or $2^3 \times 2 \times 3$ or $2^4 \times 3$, which is called a mixed factorial in the form $2^m \cdot 3^n$. Set up an ANOVA table for such a problem treating the eight heats as 2^3 pseudo factors.

14.21 Assuming Problem 14.20 results must be fired in furnaces that can handle only 24 steel samples at a time, devise a confounding scheme for the experiment.

14.22 Complete the analysis of Problem 14.21 for the following data.

	Position					
	1		2		3	
	Chemical Treatment		Chemical Treatment		Chemical Treatment	
Heat	1	2	1	2	1	2
1	1	0	0	0	0	2
2	−3	2	−1	−2	−2	−2
3	1	2	0	3	0	4
4	0	0	−2	1	0	1
5	−3	1	0	2	−2	0
6	−2	1	−3	2	0	0
7	3	3	3	3	1	−1
8	1	1	0	5	1	2

14.23 Consider a 2^5 factorial experiment with all fixed levels to be run in four blocks of eight treatment combinations each. The whole experiment can be replicated in a total of three replications. The following interactions are to be confounded in each replication.

Replication I — ABC, CDE, $ABDE$

Replication II — $BCDE$, ACD, ABE

Replication III — $ABCD$, BCE, ADE

Determine the block compositions for each block in each replication and outline the ANOVA for this problem.

15

Fractional Replication

15.1 INTRODUCTION

As the number of factors to be considered in a factorial experiment increases, the number of treatment combinations increases very rapidly. This can be seen with a 2^f factorial where $f = 5$ requires 32 experiments for one replication, $f = 6$ requires 64, $f = 7$ requires 128, and so on. Along with this increase in the amount of experimentation comes an increase in the number of high-order interactions. Some of these high-order interactions may be used as error, as those above second order (three way) would be difficult to explain if found significant. Table 15.1 gives some idea of the number of main effects, first-order, second-order, . . . , interactions that can be recovered if a complete 2^f factorial can be run.

Considering $f = 7$, there will be 7 df for the seven main effects, 21 df for the 21 first-order interactions, 35 df for the 35 second-order interactions, leaving

$$35 + 21 + 7 + 1 = 64 \text{ df}$$

for an error estimate, assuming no blocking. Even if this experiment were confounded in blocks, there is still a large number of degrees of freedom for the error estimate. In such cases, it may not be economical to run a whole replicate of 128 observations. Nearly as much information can be gleaned

Table 15.1 Buildup of 2^f-Factorial Effects

| f | 2^f | Main Effect | Order of Interaction | | | | | | |
			1st	2nd	3rd	4th	5th	6th	7th
5	32	5	10	10	5	1			
6	64	6	15	20	15	6	1		
7	128	7	21	35	35	21	7	1	
8	256	8	28	56	70	56	28	8	1

303

from half as many observations. When only a fraction of a replicate is run, the design is called a *fractional replication* or a *fractional factorial*.

For running a fractional replication, the methods of Chapter 14 are used to determine a confounding scheme such that the number of treatment combinations in a block is within the economic range of the experimenter. If, for example, the experimenter can run 60 or 70 experiments and is interested in seven factors, each at two levels, a one-half replicate of a 2^7 factorial can be used. The complete $2^7 = 128$ experiment is laid out in two blocks of 64 by confounding some high-order interaction. Then only one of these two blocks is run; the decision is made as to which one by the toss of a coin.

15.2 ALIASES

To see how a fractional replication is run, consider a simple case. Three factors are of interest, each at two levels. However, the experimenter cannot afford $2^3 = 8$ experiments, but will, however, settle for four. This suggests a one-half replicate of a 2^3 factorial. Suppose ABC is confounded with blocks. The two blocks are then

$$I = ABC$$

block 1	(1)	ab	bc	ac

block 2	a	b	c	abc

A coin is flipped and the decision is made to run only block 2. What information can be gleaned from block 2 and what information is lost when only half of the experiment is run?

Referring to Table 6.7, which shows the proper coefficients ($+1$ or -1) for the treatment combinations in a 2^3 factorial to give the effects desired, box only those treatment combinations in block 2 that are to be run (Table 15.2).

Note, in the boxed area, that ABC has all plus signs in block 2, which is a result of confounding blocks with ABC. Note also that the effect of A is

$$A = +a - b - c + abc$$

and

$$BC = +a - b - c + abc$$

so that we cannot distinguish between A and BC in block 2. Two or more effects that have the same numerical value are called *aliases*. We cannot tell them apart. B and AC likewise are aliases, as are C and AB. Because of this confounding when only a fraction of the experiment is run, we must check the aliases and be reasonably sure they are not both present if such a design is to be of value. An analysis of this one-half replication of a 2^3 would be as shown in Table 15.3.

Table 15.2 One-Half Replication of a 2^3 Factorial

Treatment Combination	Effect						
	A	B	AB	C	AC	BC	ABC
(1)	−	−	+	−	+	+	−
a	+	−	−	−	−	+	+
b	−	+	−	−	+	−	+
ab	+	+	+	−	−	−	−
c	−	−	+	+	−	−	+
ac	+	−	−	+	+	−	−
bc	−	+	−	+	−	+	−
abc	+	+	+	+	+	+	+

Table 15.3 One-Half Replication of a 2^3 Factorial

Source	df
A (or BC)	1
B (or AC)	1
C (or AB)	1
Total	3

This would hardly be a practical experiment unless the experimenter is sure that no first-order interactions exist and has some external source of error to use in testing A, B, and C. The real advantage of such fractionating will be seen on designs with larger f values.

If block 1 were run for the experiment instead of block 2, A would have the value $-(1) + ab + ac - bc$, and BC the value $+(1) - ab - ac + bc$, which gives the same total except for sign. The sum of squares due to A and BC are therefore the same; again they are said to be aliases. Here,

$$A = -BC \qquad B = -AC \qquad C = -AB$$

The definition given above still holds where one effect is the alias of another if they have the same *numerical* value, or value regardless of sign.

A quick way to find the aliases of an effect in a fractional replication of a 2^f factorial experiment is to multiply the effect by the terms in the defining contrast, modulus 2. The results will be aliases of the original effect. In the example above,

$$I = ABC$$

The alias of A is

$$A(ABC) = A^2BC = BC$$

The alias of B is

$$B(ABC) = AB^2C = AC$$

The alias of C is

$$C(ABC) = ABC^2 = AB$$

This simple rule works for any fractional replication for a 2^f factorial. It works also with a slight modification for a 3^f factorial run as a fractional replication.

15.3   FRACTIONAL REPLICATIONS

2^f Factorials

As an example, consider the problem suggested in the introduction. An experimenter wishes to study the effect of seven factors, each at two levels, but cannot afford to run all 128 experiments and so will settle for 64, or a one-half replicate of a 2^7. Deciding to confound the highest order interaction with blocks, the experimenter has

$$I = ABCDEFG$$

The two blocks are found by placing (1) and all pairs, quadruples, and sextuples of the seven letters in one block and the single letters, triples, quintuples, and one septuple of the seven letters in the other block. One of the two blocks is chosen at random and run. Before carrying out this experiment, the experimenter should check on the aliases. From the defining contrast

$$I = ABCDEFG$$

the alias of A is $A(ABCDEFG) = BCDEFG$, and all main effects are likewise aliased with fifth-order interactions.

AB is aliased with $AB(ABCDEFG) = CDEFG$, a fourth-order interaction. So all first-order interactions are aliased with fourth-order interactions. A second-order interaction such as ABC is aliased with a third-order interaction as

$$ABC(ABCDEFG) = DEFG$$

If the second- and third-order interactions are taken as error, the analysis will be as in Table 15.4.

This is a very practical design, as there are good tests on all main effects and first-order interactions, assuming all higher order interactions are zero. The degrees of freedom for each test would be 1 and 35. If, for some reason

Table 15.4   One-Half Replication of a 2^7 Factorial

Source	df	
Main effects $A, B, \ldots, G$ (or fifth order)	1 each for 7	
First-order interaction AB, AC (or fourth order)	1 each for 21	
Second-order interaction ABC, ABD (or third order)	1 each for 35 $\Big\}$	use as error
Total	63	

known to the experimenter, some second-order interaction were suspected, the experimenter could leave it out of the error estimate and still have sufficient degrees of freedom for the error estimate. The analysis of such an experiment follows the methods given in Chapter 6.

If this same experimenter is further restricted and can only afford to run 32 experiments, a one-fourth replication of a 2^7 might be tried. Here 3 df must be confounded with blocks. If two fourth-order interactions are confounded with blocks, one third-order is automatically confounded also, as

$$I = ABCDE = CDEFG = ABFG$$

which confounds 3 df with the four blocks. In this design, if only one of the four blocks of 32 observations is run, each effect has three aliases. These are

$$A = BCDE = ACDEFG = BFG \qquad CD = ABE = EFG = ABCDFG$$
$$B = ACDE = BCDEFG = AFG \qquad CE = ABD = DFG = ABCEFG$$
$$C = ABDE = DEFG = ABCFG \qquad CF = ABDEF = DEG = ABCG$$
$$D = ABCE = CEFG = ABDFG \qquad CG = ABDEG = DEF = ABCF$$
$$E = ABCD = CDFG = ABEFG \qquad DE = ABC = CFG = ABDEFG$$
$$F = ABCDEF = CDEG = ABG \qquad DF = ABCEF = CEG = ABDG$$
$$G = ABCDEG = CDEF = ABF \qquad DG = ABCEG = CEF = ABDF$$
$$AB = CDE = ABCDEFG = FG \qquad EF = ABCDF = CDG = ABEG$$
$$AC = BDE = ADEFG = BCFG \qquad EG = ABCDG = CDF = ABEF$$
$$AD = BCE = ACEFG = BDFG \qquad ACF = BDEF = ADEG = BCG$$
$$AE = BCD = ACDFG = BEFG \qquad ACG = BDEG = ADEF = BCF$$
$$AF = BCDEF = ACDEG = BG \qquad ADF = BCEF = ACEG = BDG$$
$$AG = BCDEG = ACDEF = BF \qquad ADG = BCEG = ACEF = BDF$$
$$BC = ADE = BDEFG = ACFG \qquad AEG = BCDG = ACDF = BEF$$
$$BD = ACE = BCEFG = ADFG \qquad BEG = ACDG = BCDF = AEF$$
$$BE = ACD = BCDFG = AEFG$$

Table 15.5   One-Fourth Replication of a 2^7 Factorial

Source	df
Main effects $A, B, \ldots, G$	1 each for 7
First-order interaction $AC, AD, \ldots$	1 each for 15
AB (or FG), AF (or BG), AG (or BF), $\ldots$	1 each for 3
Second-order interaction or higher ($ACF, \ldots$)	1 each for 6
Total	31

This is quite a formidable list of aliases; but when only one block of 32 is run, there are 31 df within the block. The above list accounts for these 31 df. If, in this design, all second-order (three-way) and higher interactions can be considered negligible, the main effects are all clear of first-order interactions. Three first-order interactions (AB, AF, and AG) are each aliased with another first-order interaction (FG, BG, and BF). If the choice of factors A, B, F, and G can be made with the assurance that the above interactions are either negligible or not of sufficient interest to be tested, these 3 df can be either pooled with error or left out of the analysis. The remaining 15 first-order interactions are all clear of other interactions except second order or higher. There are also 6 df left over for error involving only second-order interactions or higher. An analysis might be that shown in Table 15.5.

Here tests can be made with 1 and 6 df or 1 and 9 df if the last two lines can be pooled for error. This is a fairly good design when it is necessary to run only a one-fourth replication of a 2^7 factorial.

Example 15.1   In an experiment on slicing yield from pork bellies, four factors were considered as possible sources of significant variation. These were A—press effect; B—molding temperature; C—slicing temperature; D—side of the hog. If each of these four factors were set at two levels, it would require $2^4 = 16$ experiments to run one complete replication. If only eight bellies could be handled at one time, it was decided to run a one-half replicate of this 2^4 factorial and confound $ABCD$ with blocks. If $I = ABCD$, the blocks are

block I	(1)	ab	cd	abcd	ac	bc	ad	bd

block II	a	b	acd	bcd	c	abc	d	abd

Now if only block I is run, Table 15.6 shows the data where entries are for the treatment combinations from block I only.

The analysis could be handled by a Yates' method using only the one-half replication data available as shown in Table 15.7. Note that the numerical alias of A is $A(ABCD) = BCD$, of $B = ACD$, etc. (shown by the arrows). The analysis follows in Table 15.8.

Table 15.6 Slicing Yields in Percent from Pork Belly Experiment (Example 15.1)

		Press (A)			
		Normal		High	
		Molding Temperature (B)		Molding Temperature	
Slicing Temperature	Side	23°	40°	23°	40°
23°	L	(1) = 95.29	b	a	ab = 86.58
	R	d	bd = 89.38	ad = 96.45	abd
30°	L	c	bc = 86.79	ac = 88.70	abc
	R	cd = 90.35	bcd	acd	abcd = 89.57

Table 15.7 Yates' Solution of Example 15.1

Treatment Combination	Y	(1)	(2)	(3)	(4)	SS
(1)	95.29	95.29	181.87	357.36	723.11	
a		86.58	175.49	365.75	−0.69	0.06
b		88.70	185.83	−6.80	−18.47	42.64
ab	86.58	86.79	179.92	6.29	0.47	0.03
c		96.45	−8.71	−10.62	−12.29	18.88
ac	88.70	89.38	1.91	−7.85	2.77	0.96
bc	86.79	90.35	7.07	6.38	13.09	21.42
abc		89.57	−0.78	−5.91	8.39	8.78
d		−95.29	−8.71	−6.38	8.39	
ad	96.45	86.58	−1.91	−5.91	13.09	
bd	89.38	88.70	−7.07	10.62	2.77	
abd		−86.79	−0.78	−7.85	−12.29	
cd	90.35	96.45	181.87	6.80	0.47	
acd		−89.38	−175.49	6.29	−18.47	
bcd		−90.35	−185.83	−357.36	−0.69	
abcd	89.57	89.57	179.92	365.75	723.11	

Table 15.8 ANOVA for Example 15.1

Source	df	SS	MS	F	Prob.
A (or BCD)	1	0.06	0.06	<1	
B (or ACD)	1	42.64	42.64	5.71	0.097
C (or ABD)	1	18.88	18.88	2.53	0.210
D (or ABC)	1	8.78	8.78	1.18	0.357
AB or CD	1 ⎫	0.03 ⎫			
AC or BD	1 ⎬ 3	0.96 ⎬ 22.41	7.47		
AD or BC	1 ⎭	21.42 ⎭			
Total	7				

This experiment is not too good because the two way interactions are hopelessly confounded and the degrees of freedom on error are so small. Nothing shows significance but the analysis concept is seen by this example.

3^f Factorials

The simplest 3^f factorial is a 3^2 requiring nine experiments for a complete factorial. If only a fraction of these can be run, we might consider a one-third replication of a 3^2 factorial. First we confound either AB or AB^2 with the three blocks. Confounding AB^2 gives

$$I = AB^2$$
$$L = X_1 + 2X_2$$

and the three blocks are

$L = 0$	$L = 1$	$L = 2$
00	10	20
11	21	01
22	02	12

If one of these three is run, the aliases are

$$A(AB^2) = A^2B^2 = A^4B^4 = AB$$

since the exponent on the first element is never left greater than one. A^2B^2 is squared to give $A^4B^4 = AB$ (modulus 3). However, the alias of B is

$$B(AB^2) = AB^3 = A$$

Hence all three effects A, B, and AB are aliases and mutually confounded. To get both aliases from one multiplication using the defining contrast, one multiplies the effect by the square of I as well as I. This gives

$$A[(AB^2)]^2 = A(A^2B^4) = A^3B = B$$

Hence $A = AB = B$; this is a poor experiment, since, within the block that is run, both degrees of freedom confound A, B, and AB. In addition, no degrees of freedom are left for error. This example is cited only to show the slight modification necessary when determining the aliases in a 3^f factorial run as a fractional replication. The value of these fractional replications is seen when more factors are involved.

Consider a 3^3 factorial in three blocks of nine treatment combinations each. The effects can be broken down into 13 2-df effects: A, B, C, AB, AB^2, AC, AC^2, BC, BC^2, ABC, AB^2C, ABC^2, and AB^2C^2. Confounding ABC^2 with the three blocks gives

$$I = ABC^2 \qquad L = X_1 + X_2 + 2X_3$$

and the blocks are

$L = 0$	000	011	022	101	112	120	210	221	202

$L = 1$	100	111	122	201	212	220	010	021	002

$L = 2$	200	211	222	001	012	020	110	121	102

If only one of these blocks is now run, the aliases are

$$A = A(ABC^2) = A^2BC^2 = AB^2C$$

and

$$A = A(ABC^2)^2 = A^3B^2C^4 = B^2C = B^4C^2 = BC^2$$
$$B = B(ABC^2) = AB^2C^2$$

and

$$B = B(ABC^2)^2 = A^2B^3C^4 = A^4C^8 = AC^2$$
$$C = C(ABC^2) = ABC^3 = AB$$

and

$$C = C(ABC^2)^2 = A^2B^2C^5 = A^4B^4C^{10} = ABC$$
$$AB^2 = AB^2(ABC^2) = A^2B^3C^2 = A^4B^6C^4 = AC$$

and

$$AB^2 = AB^2(ABC^2)^2 = A^3B^4C^4 = BC$$

The analysis would be that in Table 15.9.

Table 15.9 One-Third Replication of a 3^3 Factorial

Source	df
A (or BC^2 or AB^2C)	2
B (or AB^2C^2 or AC^2)	2
C (or AB or ABC)	2
AC (or AB^2 or BC)	2
Total	8

This design would be somewhat practical if we could consider all interactions negligible and be content with 2 df for error.

These methods are easily extended to higher order 3^f experiments. Analysis proceeds as in Chapter 9.

It might be instructive to examine somewhat further this design of a one-third replication of a 3^3. A layout for 3^3 as a complete factorial was given in Table 9.7. If only one third of this were run, say block $L = 1$, suppose the other entries in Table 9.7 were crossed out, giving the results shown in Table 15.10.

Table 15.10   One-Third Replication of a 3^3 Factorial

		Factor A		
Factor B	Factor C	0	1	2
0	0	000	100	200
	1	001	101	201
	2	002	102	202
1	0	010	110	210
	1	011	111	211
	2	012	112	212
2	0	020	120	220
	1	021	121	221
	2	022	122	222

The remaining nine entries in Table 15.10 can now be summarized by levels of factor C only, to give the result shown in Table 15.11.

A second glance at Table 15.11 reveals that this design is none other than a Latin square! In a sense our designs have now come full circle, from a Latin square as two restrictions on a single-factor experiment (Chapter 4) to a Latin square as a one-third replication of a 3^3 factorial. The designs come out the same, but they result from entirely different objectives. In a single-factor experiment the general model

$$Y_{ij} = \mu + \tau_j + \varepsilon_{ij}$$

is partitioned by refining the error term due to restrictions on the randomization to give $Y_{ijk} = \mu + \tau_j + \beta_i + \gamma_k + \varepsilon'_{ij}$ where the block and position effects are taken from the original error term ε_{ij}.

In the one-third replication of a 3^3, there are three factors of interest that comprise the treatments, giving

$$Y_{ijk} = \mu + A_i + B_j + C_k + \varepsilon_{ijk}$$

Table 15.11   One-Third Replication of a 3^3 in
Terms of Factor C

	Factor A		
Factor B	0	1	2
0	2	0	1
1	0	1	2
2	1	2	0

where A, B, and C are taken from the treatment effect. Both models look alike but derive from different types of experiments. In both cases the assumption is made that there is no interaction among the factors. This is more reasonable when the factors represent only randomization restrictions such as blocks and positions, rather than when each of the three factors is of vital concern to the experimenter.

By choosing each of the other two blocks in the confounding of ABC^2 ($L = 0$ or $L = 2$), two other Latin squares are obtained. If another confounding scheme is chosen, such as AB^2C, ABC, or AB^2C^2, more and different Latin squares can be generated. In fact, with four parts of the three-way interaction and three blocks per confounding scheme, 12 distinct Latin squares may be generated. This shows how it is possible to select a Latin square at random for a particular design.

It cannot be overstressed that we must be quite sure that no interaction exists before using these designs on a 3^3 factorial. It is not enough to say that there is no interest in the interactions, because, unfortunately, the interactions are badly confounded with main effects as aliases.

Example 15.2   Torque after preheat in pounds was to be measured on some rubber material with factors A—temperature at 145°C, 155°C, and 165°C; B—mix I, II, and III; and C—laboratory 1, 2, and 3. The $3^3 = 27$ experiments seemed too costly, so a one-third replicate was run. Confounding AB^2C gives $L = X_1 + 2X_2 + X_3$ and block composition as follows:

$L = 0$	000 110 011 121 212 022 220 201 102
$L = 1$	100 210 111 221 012 122 020 001 202
$L = 2$	200 010 211 021 112 222 120 101 002

If, now, block $L = 1$ only is run, the numerical results are as given in Table 15.12 and the analysis as in Table 15.13.

Table 15.12   Rubber Torque Data of Example 15.2

| | Temperature (A) | | | | | | | | |
| | 145° Mix | | | 155° Mix | | | 165° Mix | | |
Laboratory	I	II	III	I	II	III	I	II	III
1			16.8	11.2				9.9	
2	15.8				14.4				17.8
3		17.1				20.5	15.7		

Table 15.13 ANOVA for Example 15.2

Source	df	SS	MS
A (or ABC^2 or BC^2)	2	6.66	3.33
B (or AC or ABC)	2	38.13	19.06
C (or AB^2 or AB^2C^2)	2	40.81	20.40
AB (or AC^2 or BC)	2	0.12	0.06
Totals	8	85.72	

Because of the tiny error term all main effects are highly significant even with only 2 and 2 degrees of freedom.

15.4 SUMMARY

Continuing the summary at the end of Chapter 14 under incomplete randomized blocks gives

Experiment	Design	Analysis

II. Two or more factors
   A. Factorial
      (crossed)		

	1. Completely randomized	
	2. Randomized block	
	a. Incomplete, confounding	
	i. Main effect—split-plot	
	ii. Interactions in 2^f and 3^f	
	1. Several replications $$Y_{ijkq} = \mu + R_i + \beta_j \\ + R\beta_{ij} + A_k + B_q \\ + AB_{kq} + \varepsilon_{ikq}$$	1. Factorial ANOVA with replications and blocks
	2. One replication only $$Y_{ijk} = \mu + \beta_i + A_j \\ + B_k + AB_{jk}, \ldots$$	2. Factorial ANOVA with blocks or confounded interaction

	3. Fractional replication—aliases
	         a. in 2^f
	         b. in 3^f; $f = 3$:
	           Latin square
	$$Y_{ijk} = \mu + A_i + B_j \\ + AB_{ij}, \ldots$$
		3. Factorial ANOVA with aliases or aliased interactions

PROBLEMS

15.1   Assuming that not all of Problem 14.1 can be run, determine the aliases if a one-half replication is run. Analyze the data for the principal block only.

15.2   Assuming only a one-fourth replication can be run in Problem 14.6, determine the aliases.

15.3   For Problem 14.8 use a one-third replication and determine the aliases.

15.4   Run an analysis on the block in Problem 15.3 that contains treatment combination 022.

15.5   What would be the aliases in Problem 14.9 if a one-half replication were run?

15.6   Determine the aliases in a one-fourth replication of Problem 14.11.

15.7   Determine the aliases in Problem 14.13 where a one-third replication is run.

15.8   If a one-eighth replication of Problem 14.17 is run, what are the aliases? Comment on its ANOVA table.

15.9   What are the aliases and the ANOVA outline if only one block of Problem 14.18 is run?

15.10 If data are taken from one furnace only in Problem 14.21, what are the aliases and the ANOVA outline?

15.11 What are the numerical results and conclusions from just one furnace of Problem 14.22?

15.12 The experiment of Example 15.1 was later continued and the second block was now run with results as follows:

Slicing Temperature	Side	Normal Molding Temperature (B) 23°	Normal Molding Temperature (B) 40°	High Molding Temperature 23°	High Molding Temperature 40°
			Press (A)		
23°	L		$b = 88.08$	$a = 89.26$	
	R	$d = 90.87$			$abd = 93.16$
30°	L	$c = 85.95$			$abc = 85.68$
	R		$bcd = 83.65$	$acd = 95.25$	

Analyze these results.

15.13 Take the results of Example 15.1 and Problem 15.12 and combine into the full factorial in two blocks of eight and analyze the results and comment.

15.14 a. A four-factor experiment is to be run with each factor at two levels only. However, only eight treatment combinations can be run in each block. If one wishes to confound the ABD interaction, what are the block compositions?

b. Outline the ANOVA for just one run of this experiment. Consider all factors fixed and note what you might use as an error term.

c. If only one block of this experiment can be run, explain what complications this makes in the experiment and how the ANOVA table must be altered.

15.15 Consider a 2^5 factorial experiment with all fixed levels to be run in four blocks of eight treatment combinations each. This whole experiment can be replicated in a total of three replications. The following interactions are to be confounded in each replication:

Replication I — *ABC, CDE, ABDE*
Replication II — *BCDE, ACD, ABE*
Replication III — *ABCD, BCE, ADE*

a. Determine the block composition for the principal block in replication I.

b. If only a one-fourth replication were run of replication II, what would be the aliases of *BC* in this block?

c. Determine the number of degrees of freedom for the replication by three-way interactions in this problem.

15.16 Consider an experiment designed to study the effect of six factors—*A, B, C, D, E,* and *F*—each at two levels.

a. Determine at least six entries in the principal block if this experiment is to be run in two blocks of 32 and the highest order interaction is to be confounded.

b. If only one of the two blocks can be run, what are the aliases of the effects *A, BC,* and *ABC*?

c. Outline the ANOVA table for part (b) and indicate what effects might be tested and how they would be tested.

15.17 In determining which of four factors *A, B, C,* or *D* might affect the underpuffing of cereal in a plant, the following one-fourth replicate was run.

		A_1		A_2	
		B_1	B_2	B_1	B_2
C_1	D_1			96.6	125.7
	D_2	43.5	22.4		
C_2	D_1	14.1	9.5		
	D_2			28.8	52.5

a. Determine which interaction was confounded in this design.

b. Analyze these data and comment on the results and on the design.

15.18 A company manufacturing engines was concerned about the emission characteristics of these engines. To try to minimize various undesirable emission variables, the company proposed to build and operate experimental engines varying five factors: throat diameter to be set at three levels, ignition system at three levels, temperature at three levels, velocity of the jet stream at three levels, and timing system at three levels. As this would require 243 experiments, it was decided that a one-third replicate would be run. Set up a confounding scheme for this problem and indicate a few terms in the principal block. Also show an ANOVA outline when running one of the three blocks, indicating some aliases. Comment on the design.

15.19 In studying the surface finish of steel five factors were varied each at two levels; B—slab width; C—percent sulfur; D—tundish temperature; E—casting speed; and F—mold level. The following data were recorded on the number of longitudinal face cracks observed.

D Tundish temperature, °F	E Casting speed, 1 pm	F Mold level, inches		B Slab width, inches			
				<60		>62	
				C Sulfur, percent			
				<0.015	>0.017	<0.015	>0.017
>2845	>59	>10			2	3	
		5–8		0			0
	<54	>10		5			0
		5–8			6	0	
<2840	>59	>10		6			3
		5–8			2	0	
	<54	>10			7	0	
		5–8		1			0

Determine what interaction is confounded here in order to run this one-half replicate. Outline its ANOVA and comment on the design.

15.20 Using the data of Problem 15.19, do the analysis and state your conclusions.

Miscellaneous Topics

16.1 INTRODUCTION

Several techniques have been developed, based on the general design principles presented in the first 15 chapters of this book and on other well-known
statistical methods. No attempt will be made to discuss these in detail, as
they are presented very well in the references. It is hoped that a background
in experimental design as given in the first 15 chapters will make it possible
for the experimenter to read and understand the references.

The methods to be discussed are covariance analysis, response surface
experimentation, evolutionary operation, analysis of attribute data, incomplete block design, and Youden squares.

16.2 COVARIANCE ANALYSIS

Philosophy

Occasionally, when a study is being made of the effect of one or more factors
on some response variable, say Y, there is another variable, or variables,
that vary along with Y. It is often not possible to control this other variable
(or variables) at some constant level throughout the experiment, but the
variable can be measured along with the response variable. This variable is
referred to as a *concomitant variable* X as it "runs along with" the response
variable Y. In order, then, to assess the effect of the treatments on Y, one
should first attempt to remove the effects of this concomitant variable X.
This technique of removing the effect of X (or several X's) on Y and then
analyzing the residuals for the effect of the treatments on Y is called *covariance analysis*.

Several examples of the use of this technique might be cited. If one
wishes to study the effect on student achievement of several teaching methods,
it is customary to use the difference between a pretest and a posttest score
as the response variable. It may be, however, that the gain Y is also affected

by the student's pretest score X as gain, and pretest scores may be correlated. Covariance analysis will provide for an adjustment in the gains due to differing pretest scores assuming some type of regression of gain on pretest scores. The advantage of using this technique is that one is not limited to running an experiment on only those pupils who have approximately the same pretest scores or matching pupils with the same pretest scores and randomly assigning them to control and experimental groups. Covariance analysis has the effect of providing "handicaps" as if each student had the same pretest scores while actually letting the pretest scores vary along with the gains.

This technique has often been used to adjust weight gains in animals by their original weights in order to assess the effect of certain feed treatments on these gains. Many industrial examples might be cited such as the one given below where original weight of a bracket may affect the weight of plating applied to the bracket by different methods.

Snedecor [22] gives several examples of the use of covariance analysis. Ostle [19] is an excellent reference on the extension of covariance analysis to many different experimental designs. The *Biometrics* article of 1957 [5] devotes most of its pages to a discussion of many of the fine points of covariance analysis. The problem discussed below will illustrate the technique for the case of a single-factor experiment run in a completely randomized design. Since the covariance technique represents a marriage between regression and the analysis of variance, the methods of Chapter 3 and Chapter 7 will be used.

Example 16.1   Several steel brackets were sent to three different vendors to be zinc plated. The chief concern in this process is the thickness of the zinc plating and whether or not there was any difference in this thickness between the three vendors. Data on this thickness in hundred thousandths of an inch (10^{-5}) for four brackets plated by the three vendors are given in Table 16.1.

Assuming a single-factor experiment and a completely randomized design, an analysis of variance model for this experiment would be

$$Y_{ij} = \mu + \tau_j + \varepsilon_{ij} \tag{16.1}$$

Table 16.1   Plating Thickness in 10^{-5} Inches from Three Vendors

	Vendor	
A	B	C
40	25	27
38	32	24
30	13	20
47	35	13

Table 16.2 ANOVA on Plating Thickness

Source	df	SS	MS
Between vendors	2	665.2	332.6
Within vendors	9	543.5	60.4
Totals	11	1208.7	

where Y_{ij} is the observed thickness of the ith bracket from the jth vendor, μ is a common effect, τ_j represents the vendor effect, and ε_{ij} represents the random error. Considering vendors as a fixed factor and random errors as normally and independently distributed with mean zero and common variance σ_ε^2, an ANOVA table was compiled (Table 16.2).

The F statistic equals 5.51 and is significant at the 5 percent significance level with 2 and 9 df. One might conclude, therefore, that there is a real difference in average plating thickness among these three vendors, and steps should be taken to select the most desirable vendor.

During a discussion of these results, it was pointed out that some of this difference among vendors might be due to unequal thickness in the brackets before plating. In fact, there might be a correlation between the thickness of the brackets before plating and the thickness of the plating. To see whether or not such an idea is at all reasonable, a scatterplot was made on the thickness of the bracket before plating X and the thickness of the zinc plating Y. Results are shown in Figure 16.1.

A glance at this scattergram would lead one to suspect that there is a positive correlation between the thickness of the bracket and the plating

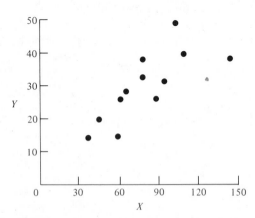

Figure 16.1 Scattergram of bracket thickness X versus plating thickness Y in 10^{-5} inches.

thickness. Since another variable, X, can be measured on each piece and may be related to the variable of interest Y, a covariance analysis may be run in order to remove the linear effect of X on Y when comparing vendors. A covariance analysis is used to remove the linear or higher order relationship of one or more independent variables from a dependent variable in order to assess the effect of a factor on the dependent variable.

Covariance

Since a positive correlation is suspected between X and Y above, a regression model might be written as

$$Y_{ij} = \mu + \beta(X_{ij} - \bar{X}) + \varepsilon_{ij} \tag{16.2}$$

where β is the true linear regression coefficient (or slope) between Y and X over all the data; $\bar{X}$ is the mean of the X values. In such a model it is assumed that a measure of variable X can be made on each unit along with a corresponding measure of Y. For a covariance analysis models (16.1) and (16.2) can be combined to give the covariance model

$$Y_{ij} = \mu + \beta(X_{ij} - \bar{X}) + \tau_j + \varepsilon_{ij} \tag{16.3}$$

In this model the error term ε_{ij} should be smaller than in model (16.1) because of the removal of the effect of the covariate X_{ij}.

To determine how to adjust the analysis of variance to provide for the removal of this covariate, consider the regression model (16.2) in deviation form

$$y = bx + e \tag{16.4}$$

where x and y are now deviations from their respective means $(x = X_{ij} - \bar{X},$ $y = Y_{ij} - \bar{Y})$ and b is the sample slope. By the method of least squares it will be recalled that

$$b = \frac{\sum xy}{\sum x^2}$$

where summations are over all points. The sum of the squares of the errors of estimate, which have been minimized, is now, from Equation (16.4),

$$e = y - bx$$
$$\sum e^2 = \sum (y - bx)^2$$
$$= \sum y^2 - 2b \sum xy + b^2 \sum x^2$$

Substituting for b

$$b = \frac{\sum xy}{\sum x^2}$$

and

$$\sum e^2 = \sum y^2 - 2\frac{\sum xy}{\sum x^2} \cdot \sum xy + \left(\frac{\sum xy}{\sum x^2}\right)^2 \sum x^2$$

$$= \sum y^2 - \frac{(\sum xy)^2}{\sum x^2} \tag{16.5}$$

In this expression the term $(\sum xy)^2/\sum x^2$ is the amount of reduction in the sum of squares of the y variable due to its linear regression on x. This term is then the sum of squares due to linear regression. It involves the sum of the cross products of the two variables $(\sum xy)$ and the sum of squares of the x variable alone $(\sum x^2)$. If a term of this type is subtracted from the sum of squares for the dependent variable y, the result will be a corrected or adjusted sum of squares for y.

The sums of squares and cross products can be computed on all the data (total), within each vendor or between vendors. In each case it will be helpful to recall that the sums of squares and cross products in terms of the original data are

$$\sum x^2 = \sum X^2 - \frac{(\sum X)^2}{N}$$

$$\sum y^2 = \sum Y^2 - \frac{(\sum Y)^2}{N}$$

$$\sum xy = \sum XY - \frac{(\sum X)(\sum Y)}{N}$$

Data on both variables appear with the appropriate totals in Table 16.3.

From Table 16.3 the sums of squares and cross products for the total data are[1]

$$T_{xx} = (110)^2 + (75)^2 + \cdots + (59)^2 - \frac{(944)^2}{12} = 9240.7$$

$$T_{yy} = (40)^2 + (38)^2 + \cdots + (13)^2 - \frac{(344)^2}{12} = 1208.7$$

$$T_{xy} = (110)(40) + (75)(38) + \cdots + (59)(13) - \frac{(944)(344)}{12} = 2332.7$$

The between-vendors sums of squares and cross products are computed by

---

[1] Here T, V, and E are used to denote sum of squares and cross products for totals, vendors, and error respectively.

Table 16.3 Bracket Thickness X and Plating Thickness Y in 10^{-5} Inches from Three Vendors

	Vendor							
	A		B		C		Total	
	X	Y	X	Y	X	Y	X	Y
	110	40	60	25	62	27		
	75	38	75	32	90	24		
	93	30	38	13	45	20		
	97	47	140	35	59	13		
Totals	375	155	313	105	256	84	944	344

vendor totals as in an ANOVA:

$$V_{xx} = \frac{(375)^2 + (313)^2 + (256)^2}{4} - \frac{(944)^2}{12} = 1771.2$$

$$V_{yy} = \frac{(155)^2 + (105)^2 + (84)^2}{4} - \frac{(344)^2}{14} = 665.2$$

$$V_{xy} = \frac{(375)(155) + (313)(105) + (256)(84)}{4} - \frac{(944)(344)}{12} = 1062.2$$

By subtraction the error sums of squares are

$$E_{xx} = T_{xx} - V_{xx} = 9240.7 - 1771.2 = 7469.5$$
$$E_{yy} = T_{yy} - V_{yy} = 1208.7 - 665.2 = 543.5$$
$$E_{xy} = T_{xy} - V_{xy} = 2332.7 - 1062.2 = 1270.5$$

Using this information, the sums of squares of the dependent variable Y may be adjusted for regression on X. On the totals, the adjusted sum of squares is then

$$\text{adjusted } \sum y^2 = T_{yy} - \frac{T_{xy}^2}{T_{xx}} \qquad \text{as in Equation (16.5)}$$

$$= 1208.7 - \frac{(2332.7)^2}{9240.7} = 619.8$$

and the adjusted sum of squares within vendors is

$$\text{adjusted } \sum y^2 = E_{yy} - \frac{(E_{xy})^2}{E_{xx}}$$

$$= 543.5 - \frac{(1270.5)^2}{7469.5} = 327.4$$

Then, by subtraction, the adjusted sum of squares between vendors is

$$619.8 - 327.4 = 292.4$$

These results are usually displayed as in Table 16.4.

Table 16.4   Analysis of Covariance for Bracket Data

Source	df	SS and Products $\sum y^2$	$\sum xy$	$\sum x^2$	Adjusted $\sum y^2$	df	MS
Between vendors	2	665.2	1062.2	1771.2	—	—	—
Within vendors	9	543.5	1270.5	7469.5	327.4	8	40.9
Totals	11	1208.7	2332.7	9240.7	619.8	10	
Between vendors					292.4	2	146.2

With the adjusted sums of squares, the new F statistic is

$$F = \frac{146.2}{40.9} = 3.57$$

with 2 and 8 df. This is now not significant at the 5 percent significance level. This means that after removing the effect of bracket thickness, the vendors no longer differ in the average plating thickness on the brackets.

Table 16.4 should have a word or two of explanation. The degrees of freedom on the adjusted sums of squares are reduced by 2 instead of 1 as estimates of both the mean and the slope are necessary in their computation. Adjustments are made here on the totals and the within sums of squares rather than on the between sum of squares as we are interested in making the adjustment based on an overall slope of Y on X and a within-group average slope of Y on X as explained in more detail later.

In this analysis three different types of regression can be identified: the "overall" regression of all the Y's on all the X's, the within-vendors regression, and the regression of the three Y means on the three X means. This last regression could be quite different from the other two and is of little interest since we are attempting to adjust the Y's and the X's within the vendors. An estimate of this "average within vendor" slope is

$$b = \frac{\sum xy}{\sum x^2} = \frac{E_{xy}}{E_{xx}} = \frac{1270.5}{7469.5} = 0.17$$

If this slope or regression coefficient is used to adjust the observed Y means for the effect of X on Y, one finds

$$\text{adjusted } \bar{Y}_{.j} = \bar{Y}_{.j} - b(\bar{X}_{.j} - \bar{X}_{..})$$

For vendor A,

$$\text{adjusted } \bar{Y}_{.1} = \frac{155}{4} - (0.17)\left(\frac{375}{4} - \frac{944}{12}\right)$$

$$= 38.75 - (0.17)(93.75 - 78.67) = 36.19$$

For vendor B,

$$\text{adjusted } \bar{Y}_{.2} = 26.25 - (0.17)(78.25 - 78.67) = 26.31$$

For vendor C,

$$\text{adjusted } \bar{Y}_{.3} = 21.00 - (0.17)(64.00 - 78.67) = 23.49$$

A comparison of the last column above with the column of unadjusted means just to the right of the equal signs shows that these adjusted means are closer together than the unadjusted means which is confirmed in the preceding analysis. This can be seen graphically by plotting each vendor's data in a different symbol and "sliding" the means along lines parallel to this regression slope (Figure 16.2).

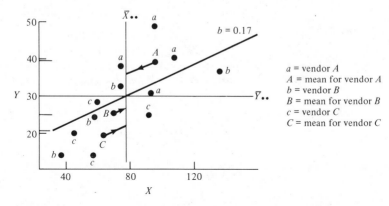

Figure 16.2 Plating thickness Y versus bracket thickness X plotted by vendors.

At the left-hand ordinate in Figure 16.2, the three unadjusted Y means are marked off. By moving these three means along lines parallel to the general slope ($b = 0.17$), a comparison is seen between the three means (on the $\bar{X}_{..}$ ordinate) after adjustment. This shows graphically the results of the covariance analysis. By examining each group of observations for a given vendor, it is seen that all three slopes within vendors are in the same direction and of similar magnitude. By considering all 12 points, the overall slope can be visualized. On the other hand, if the three means (indicated by the capital letters) are considered, the slope of these $\bar{Y}$'s on their corresponding $\bar{X}$'s is much steeper. Hence adjustments are made on the basis of the overall slope and the "average" or "pooled" within-groups slope.

One of the problems in covariance analysis is that there are many assumptions made in its application. These should be examined in some detail.

First, there are the usual assumptions of analysis of variance on the dependent variable Y: an additive model, normally and independently distributed error, and homogeneity of variance within the groups. This last assumption is often tested with Barlett's or a similar test. In this problem, since the ranges of the three vendor readings are 17, 22, and 14, the assumption seems tenable.

In covariance analysis it is further assumed that the regression is linear, that the slope is not zero (covariance was necessary), that the regression coefficients within each group are homogeneous so that the "average" or "pooled" within-groups regression can be used on all groups, and finally that the independent variable X is not affected by the treatments given to the groups.

The linearity assumption may be tested if the experiment is planned in such a way that there is more than one Y observation for each X. Since this was not done here, a look at the scattergram will have to suffice for this linearity assumption.

To test the hypothesis that the true slope β in Equation (16.3) is zero, consider whether or not the reduction in the error sum of squares is significant when compared with the error sum of squares. By an F test the ratio would be

$$F_{1, N-k-1} = \frac{E_{xy}^2/E_{xx}}{\text{adjusted } E_{yy}/(N-k-1)}$$

$$F_{1,8} = \frac{216.1}{327.4/8} = \frac{216.1}{40.9} = 5.28$$

As the 5 percent F is 5.32, this result is nearly significant and certainly casts considerable doubt on the hypothesis that $\beta = 0$. Since the results after covariance adjustment were not significant as compared with the results before adjustment, it seems reasonable to conclude that covariance was necessary.

To test the hypothesis that all three regression coefficients are equal, let us compute each sample coefficient and adjust the sum of squares within each group by its own regression coefficient.

For vendor A,

$$b_A = \frac{\sum xy}{\sum x^2} = \frac{\sum XY - (\sum X)(\sum Y)/n}{\sum X^2 - (\sum X)^2/n}$$

Within group A,

$$b_A = \frac{14{,}599 - 58{,}125/4}{35{,}783 - 140{,}625/4} = 0.108$$

$$\text{adjusted } \sum y^2 = \sum y^2 - \frac{(\sum xy)^2}{\sum x^2} = 146.75 - 7.32 = 139.43$$

For vendor B,

$$b_B = \frac{9294 - 32{,}865/4}{30{,}269 - 24{,}492.25/4} = 0.187$$

$$\text{adjusted } \sum y^2 = 286.75 - 201.72 = 85.03$$

For vendor C,

$$b_C = \frac{5501 - 21{,}504/4}{17{,}450 - 65{,}536/4} = 0.117$$

$$\text{adjusted } \sum y^2 = 110.00 - 14.66 = 95.34$$

Summarizing, we obtain Table 16.5.

Table 16.5   Adjusted Sums of Squares within Each Vendor

Within Vendor	df	$\sum y^2$	b	Adjusted $\sum y^2$	df
A	3	146.75	0.108	139.43	2
B	3	286.75	0.187	85.03	2
C	3	110.00	0.117	95.34	2
Totals	9	543.50		319.80	6

Adjusting each within-vendor sum of squares by its own regression coefficient reduces the degrees of freedom to 2 per vendor. Adding the new adjusted sums of squares, Table 16.5 shows a within-vendor sum of squares of 319.80 based on 6 df. When the within-vendor sum of squares was adjusted by a "pooled" within-vendor regression, Table 16.4 showed an adjusted sum of squares within vendors of 327.4 based on 8 df. If there were significant differences in regression within the three vendors, this would show as a difference in these two figures: $327.4 - 319.8 = 7.6$. An F test for this hypothesis would then be

$$F_{k-1,\,N-2k} =$$

$$\frac{\left[\begin{array}{l}\text{adjusted } \sum y^2 \text{ (based on pooled within groups regression)}\\ \quad - \text{ adjusted } \sum y^2 \text{ (based on regressions within each group)}\end{array}\right]\Big/(k-1)}{[\text{adjusted } \sum y^2 \text{ (within each group)}]/(N-2k)}$$

and

$$F_{2,6} = \frac{(327.4 - 319.8)/2}{319.8/6} = \frac{3.8}{53.3} = \ <1$$

hence nonsignificant, and we conclude that the three regression coefficients are homogeneous.

The final assumption that the vendors do not affect the covariate X is tenable here on practical grounds as the vendor has nothing to do with the bracket thickness before plating. However, in some applications this assumption needs to be checked by an F test on the X variable. For our data from Table 16.4 such an F test would give

$$F_{2,9} = \frac{1771.2/2}{7469.5/9} = \frac{885.6}{829.9} = 1.07$$

which is obviously nonsignificant.

From this discussion of the assumptions, one notes that it is often a problem in the use of covariance analysis to satisfy all of the assumptions. This may be the reason many people avoid covariance although it is applicable whenever one suspects that another measured variable or variables affect the dependent variable.

The above techniques may be extended to handle several covariates, nonlinear regression, and several factors and interactions of interest. In handling randomized blocks, Latin squares, or experiments with several factors, Ostle [19] points out that the proper technique is to add the sum of squares and sum of cross products of the term of interest to the sum of squares and sum of cross products of the error, adjust this total, then adjust the error and determine the adjusted treatment effect by subtraction.

When several covariates are involved, many questions arise concerning how many covariates can be handled, how many really affect the dependent variable, how a significant subset can be found, and so on. All of these questions are of concern in any multiple regression problem and some discussion of these and other problems can be found in [5].

16.3   RESPONSE-SURFACE EXPERIMENTATION

Philosophy

The concept of a response surface involves a dependent variable Y, called the response variable, and several independent or controlled variables, $X_1, X_2, \ldots, X_k$. If all of these variables are assumed to be measurable, the response surface can be expressed as

$$Y = f(X_1, X_2, \ldots, X_k)$$

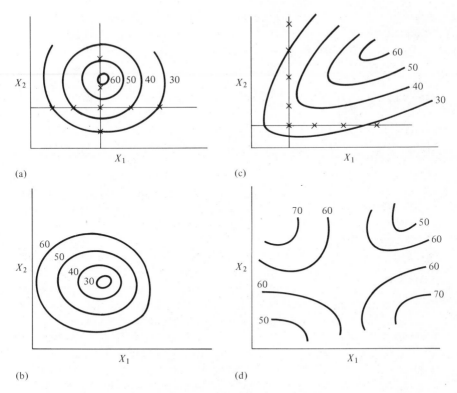

Figure 16.3   Some typical response surfaces in two dimensions: (a) mound;
(b) depression; (c) rising ridge; (d) saddle.

For the case of two independent variables such as temperature X_1 and time X_2, the yield Y of a chemical process can be expressed as

$$Y = f(X_1, X_2)$$

This surface can be plotted in three dimensions, with X_1 on the abscissa, X_2 on the ordinate, and Y perpendicular to the $X_1 X_2$ plane. If the values of X_1 and X_2 that yield the same Y are connected, we can picture the surface with a series of equal-yield lines, or contours. These are similar to the contours of equal height on topographic maps and the isobars on weather maps. Some response surfaces might be of the types in Figure 16.3. Two excellent references for understanding response-surface experimentation are [9] and [12].

The Twofold Problem

The problem involved in the use of response-surface experimentation is two-fold: (1) to determine, on the basis of one experiment, where to move in the next experiment toward the optimal point on the underlying response surface; (2) having located the optimum or near optimum of the surface, to determine the equation of the response surface in an area near this optimum point.

One method of experimentation that seeks the optimal point of the response surface might be the traditional one-factor-at-a-time method. As shown in Figure 16.3(a), if X_2 is fixed and X_1 is varied, we find the X_1 optimal (or near optimal) value of response Y at the fixed value of X_2. Having found this X_1 value, experiments can now be run at this fixed X_1, and the X_2 for optimal response can be found. In the case of the mound in Figure 16.3(a), this method would lead eventually to the peak of the mound or near it. However, this same method, when applied to a surface such as the rising ridge in Figure 16.3(c), fails to lead to the maximum point on the response surface. In experimental work the type of response surface is usually unknown, so that a better method is necessary if the optimum set of conditions is to be found for any surface.

The method developed by those who have worked in this area is called the *path of steepest ascent* method. The idea here is to run a simple experiment over a small area of the response surface where, for all practical purposes, the surface may be regarded as a plane. We then determine the equation of this plane and from it the direction we should take from this experiment in order to move toward the optimum of the surface. Since the next experiment should be in a direction in which we hope to scale the height the fastest, this is referred to as the *path of steepest ascent*. This technique does not determine how far away from the original experiment succeeding sequential experiments should be run, but it does indicate to the experimenter the direction along which the next experiment should be performed. A simple example will illustrate this method.

To determine the equation of the response surface, several special experimental designs have been developed that attempt to approximate this equation using the smallest number of experiments possible. In two dimensions the simplest surface is a plane given by

$$Y = B_0 X_0 + B_1 X_1 + B_2 X_2 + \varepsilon \qquad (16.6)$$

where Y is the observed response, X_0 is taken as unity, and estimates of the B's are to be determined by the method of least squares, which minimizes the sum of the squares of the errors ε. Such an equation is referred to as a first-order equation, since the power on each independent variable is unity.

If there is some evidence that the surface is not planar, a second-order equation in two dimensions may be a more suitable model

$$Y = B_0 X_0 + B_1 X_1 + B_2 X_2 + B_{11} X_1^2 + B_{12} X_1 X_2 + B_{22} X_2^2 + \varepsilon \quad (16.7)$$

Here the $X_1 X_2$ term represents an interaction between the two variables X_1 and X_2.

If there are three independent or controlled variables, the first-order equation is again a plane or hyperplane

$$Y = B_0X_0 + B_1X_1 + B_2X_2 + B_3X_3 + \varepsilon \qquad (16.8)$$

and the second-order equation is

$$\begin{aligned} Y = {} & B_0X_0 + B_1X_1 + B_2X_2 + B_3X_3 + B_{11}X_1^2 + B_{22}X_2^2 \\ & + B_{33}X_3^2 + B_{12}X_1X_2 + B_{13}X_1X_3 + B_{23}X_2X_3 + \varepsilon \quad (16.9) \end{aligned}$$

As the complexity of the surface increases, more coefficients must be estimated, and the number of experimental points must necessarily increase. Several very clever designs have been developed that minimize the amount of work necessary to estimate these response-surface equations.

To determine the coefficients for these more complex surfaces and to interpret their geometric nature, both multiple regression techniques and the methods of solid analytical geometry are used. In the example that follows, only the simplest type of surface will be explored. The references quoted will give many more complex examples.

Example 16.2   Consider an example in which an experimenter is seeking the proper values for both concentration of filler to epoxy resin X_1 and position in the mold X_2 to minimize the abrasion on a plastic die. This abrasion or wear is measured as a decrease in thickness of the material after 10,000 cycles of abrasion. Since the maximum thickness is being sought, the first experiment should attempt to discover the direction in which succeeding experiments should be run in order to approach this maximum by the steepest path. Assuming the surface to be a plane in a small area, the first experiment will be used to determine the equation of this plane. The response surface is then

$$Y = B_0X_0 + B_1X_1 + B_2X_2 + \varepsilon$$

As there are three parameters to be estimated, B_0, B_1, and B_2, at least three experimental points must be taken to estimate these coefficients. Such a design might be an equilateral triangle, but since there are two factors, X_1 and X_2, each can be set at two levels and a 2^2 factorial may be used. Two concentrations were chosen, $\frac{1}{2}$:1 and 1:1 (ratio of filler to resin), and two positions, 1 inch and 2 inches from a reference point, with responses Y, the thickness of the material in 10^{-4} inches, as shown in Figure 16.4.

To determine the equation of the best fitting plane for these four points, consider the error in the prediction equation

$$\varepsilon = Y - B_0X_0 - B_1X_1 - B_2X_2$$

The sum of the squares for this error is

$$\sum \varepsilon^2 = \sum (Y - B_0X_0 - B_1X_1 - B_2X_2)^2$$

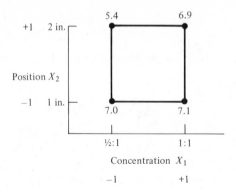

Figure 16.4 2^2 factorial example on response surface.

where the summation is over all points given in the design. To find the B's, we differentiate this expression with respect to each parameter and set these three expressions equal to zero. This provides three least squares normal equations, which can be solved for the best estimates of the B's. These estimates are designated as b's.

Differentiating the expression above gives

$$\frac{\partial(\sum \varepsilon^2)}{\partial B_0} = -2 \sum (Y - B_0 X_0 - B_1 X_1 - B_2 X_2) X_0 = 0$$

$$\frac{\partial(\sum \varepsilon^2)}{\partial B_1} = -2 \sum (Y - B_0 X_0 - B_1 X_1 - B_2 X_2) X_1 = 0$$

$$\frac{\partial(\sum \varepsilon^2)}{\partial B_2} = -2 \sum (Y - B_0 X_0 - B_1 X_1 - B_2 X_2) X_2 = 0$$

and from these results we get

$$\sum X_0 Y = b_0 \sum X_0^2 + b_1 \sum X_0 X_1 + b_2 \sum X_0 X_2$$
$$\sum X_1 Y = b_0 \sum X_0 X_1 + b_1 \sum X_1^2 + b_2 \sum X_1 X_2 \qquad (16.10)$$
$$\sum X_2 Y = b_0 \sum X_0 X_2 + b_1 \sum X_1 X_2 + b_2 \sum X_2^2$$

as the least squares normal equations.

By a proper choice of experimental variables, it is possible to reduce these equations considerably for simple solution. More complex models can be solved best by the use of matrix algebra. For the 2^2 factorial, the following coding scheme simplifies the solution of Equation (16.10). Set

$$X_1 = 4C - 3 \quad \text{where } C \text{ is the concentration}$$
$$X_2 = 2P - 3 \quad \text{where } P \text{ is the position}$$

Table 16.6   Orthogonal Layout for
Figure 16.4

Y	X_0	X_1	X_2
7.0	1	-1	-1
5.4	1	-1	1
7.1	1	1	-1
6.9	1	1	1

X_0 is always taken as unity. For the experimental variables X_1 and X_2, the responses can be recorded as in Table 16.6. These experimental variables are also indicated on Figure 16.4.

An examination of the data as presented in Table 16.6 shows that X_0, X_1, and X_2 are all orthogonal to each other as

$$\sum X_1 = \sum X_2 = 0 \quad \text{and} \quad \sum X_1 X_2 = 0$$

Hence the least squares normal equations become

$$\sum X_0 Y = b_0 n + b_1 \cdot 0 + b_2 \cdot 0$$
$$\sum X_1 Y = b_0 \cdot 0 + b_1 \cdot \sum X_1^2 + b_2 \cdot 0 \tag{16.11}$$
$$\sum X_2 Y = b_0 \cdot 0 + b_1 \cdot 0 + b_2 \sum X_2^2$$

Solving gives

$$b_0 = \sum X_0 Y / n$$
$$b_1 = \sum X_1 Y / \sum X_1^2 \tag{16.12}$$
$$b_2 = \sum X_2 Y / \sum X_2^2$$

For this problem

$$b_0 = \frac{26.4}{4} = 6.60$$

$$b_1 = \frac{1.6}{4} = 0.40$$

$$b_2 = \frac{-1.8}{4} = -0.45$$

and the response surface can be approximated as

$$\hat{Y} = 6.60 + 0.40 X_1 - 0.45 X_2$$

To determine the sum of squares due to each of these terms in the model, we can use the sum of squares due to b_i's

$$\mathrm{SS}_{b_i} = b_i \cdot \sum X_i Y$$

Here

$$SS_{b_0} = (6.60)(26.4) = 174.24$$
$$SS_{b_1} = (0.40)(1.6) = 0.64$$
$$SS_{b_2} = (-0.45)(-1.8) = 0.81$$

each carrying 1 df. The ANOVA table is shown as Table 16.7.

Table 16.7 ANOVA for 2^2 Factorial Response-Surface Example

Source	df	SS
b_0	1	174.24
b_1	1	0.64
b_2	1	0.81
Residual	1	0.49
Totals	4	176.18

Here, all 4 df are shown, since the b_0 term represents the degree of freedom usually associated with the mean, that is, the correction term in many examples. The total sum of squares is $\sum Y^2$ of the responses and the residual is what is left over. With this 1-df residual, no good test is available on the significance of each term in the model, nor is there any way to assess the adequacy of the planar model to describe the surface.

To decide on the direction for the next experiment, plot contours of equal response using the equation of the plane determined above

$$\hat{Y} = 6.60 + 0.40X_1 - 0.45X_2$$

Solve for X_2,

$$X_2 = \frac{6.60 - \hat{Y} + 0.40X_1}{0.45}$$

If $\hat{Y} = 5.5$,

$$X_2 = \frac{1.10 + 0.40X_1}{0.45}$$

when

$$X_1 = -1 \qquad X_2 = 1.56$$
$$X_1 = 1 \qquad X_2 = 3.33$$

If $\hat{Y} = 6.0$,

$$X_2 = \frac{0.60 + 0.40X_1}{0.45}$$

when

$$X_1 = -1 \qquad X_2 = 0.44$$
$$X_1 = 1 \qquad X_2 = 2.22$$

If $\hat{Y} = 6.5$,

$$X_2 = \frac{0.10 + 0.40X_1}{0.45}$$

when

$$X_1 = -1 \qquad X_2 = -0.67$$
$$X_1 = 1 \qquad X_2 = 1.11$$

If $\hat{Y} = 7.0$,

$$X_2 = \frac{-0.40 + 0.40X_1}{0.45}$$

when

$$X_1 = -1 \qquad X_2 = -1.67$$
$$X_1 = 1 \qquad X_2 = 0$$

If $\hat{Y} = 7.5$,

$$X_2 = \frac{-0.90 + 0.40X_1}{0.45}$$

when

$$X_1 = -1 \qquad X_2 = -2.89$$
$$X_1 = 1 \qquad X_2 = -1.11$$

Plotting these five contours on the original diagram gives the pattern shown in Figure 16.5.

By moving in a direction normal to these contours and "up" the surface, we can anticipate larger values of response until the peak is reached. To

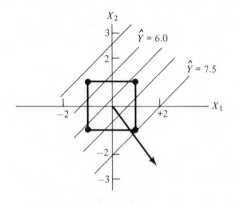

Figure 16.5 Contours on 2^2 factorial response-surface example.

decide on a possible set of conditions for the next experiment, consider the equation of a normal to these contours through the point (0, 0). The contours are

$$X_2 = \frac{6.60 - \hat{Y} + 0.40X_1}{0.45}$$

and their slope is $0.40/0.45 = 8/9$. The normal will have a slope $= -9/8$, and its equation is

$$X_2 - 0 = \frac{-9}{8}(X_1 - 0)$$

$$X_2 = \frac{-9}{8} X_1 \qquad \text{(see arrow in Figure 16.5)}$$

This technique will not tell how far to go in this direction, but we might run the next experiment with the center of the factorial at $(+1, -9/8)$. The four points would be

X_1	X_2
0	$-\frac{1}{8}$
0	$-2\frac{1}{8}$
2	$-\frac{1}{8}$
2	$-2\frac{1}{8}$

These points are expressed in terms of the experimental variables, but must be decoded to see where to set the concentration and position. Using the coding,

$$X_1 = 4C - 3 \quad \text{or} \quad C = \frac{X_1 + 3}{4}$$

$$X_2 = 2P - 3 \quad \text{or} \quad P = \frac{X_2 + 3}{2}$$

the four new points would be

X_1	X_2	C	P
0	$-\frac{1}{8}$	$\frac{3}{4}{:}1$	$1\frac{7}{16}$
0	$-2\frac{1}{8}$	$\frac{3}{4}{:}1$	$\frac{7}{16}$
2	$-\frac{1}{8}$	$1\frac{1}{4}{:}1$	$1\frac{7}{16}$
2	$-2\frac{1}{8}$	$1\frac{1}{4}{:}1$	$\frac{7}{16}$

if these settings are possible. Responses may now be taken at these four points on a new 2^2 factorial, and after analysis we can again decide the direction of steepest ascent. This prodecure continues until the optimum is obtained.

The above design is sufficient to indicate the direction for subsequent experiments, but it does not provide a good measure of experimental error to test the significance of b_0, b_1, and b_2, nor is there any test of how well the plane approximates the surface. One way to improve on this design is to take two or more points at the center of the square. By replication at the same point, an estimate of experimental error is obtained and the average of the center-point responses will provide an estimate of "goodness of fit" of the plane. If the experiment is near the maximum response, this center point might be somewhat above the four surrounding points, which would indicate the need for a more complex model.

To see how this design would help, consider two observations of response at the center of the example above. Using the same responses at the vertices of the square, the results might be those shown in Figure 16.6.

Y	X_0	X_1	X_2
7.0	1	-1	-1
5.4	1	-1	1
7.1	1	1	-1
6.9	1	1	1
6.6	1	0	0
6.8	1	0	0

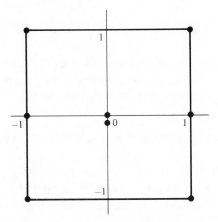

Figure 16.6 2^2 factorial with two points in center.

The coefficients in the model are still estimated by Equation (16.12):

$$b_0 = \frac{\sum X_0 Y}{n} = \frac{39.8}{6} = 6.63$$

$$b_1 = \frac{\sum X_1 Y}{\sum X_1^2} = \frac{1.6}{4} = 0.40$$

$$b_2 = \frac{\sum X_2 Y}{\sum X_2^2} = \frac{-1.8}{4} = -0.45$$

and

$$SS_{b_0} = 6.63(39.8) = 263.87$$
$$SS_{b_1} = 0.40(1.6) = 0.64$$
$$SS_{b_2} = -0.45(-1.8) = 0.81$$

For the sum of squares of the error at (0, 0),

$$SS_e = (6.6)^2 + (6.8)^2 - \frac{(13.4)^2}{2} = 89.80 - 89.78 = 0.02$$

The analysis is shown in Table 16.8.

Testing these effects against error gives

$$b_0: F_{1,1} = \frac{236.87}{0.02} = 11{,}843.5$$

$$b_1: F_{1,1} = \frac{0.64}{0.02} = 32$$

$$b_2: F_{1,1} = \frac{0.81}{0.02} = 40.5$$

$$\text{lack of fit}: F_{2,1} = \frac{0.32}{0.02} = 16$$

With such a small number of degrees of freedom, only b_0 shows significance at the 5 percent level. However, the tests on the b terms are larger than the test on lack of fit, which may indicate that the plane

$$\hat{Y} = 6.63 + 0.40X_1 - 0.45X_2$$

is a fair approximation to the surface where this first experiment was run.

More Complex Surfaces

For more factors—controlled variables—the model for the response variable is more complex, but several designs have been found useful in estimating the coefficients of these surfaces.

Table 16.8 ANOVA for 2^2 + Two Center Points

Source	df	SS	MS
Total	6	265.98	
b_0	1	263.87	236.87
b_1	1	0.64	0.64
b_2	1	0.81	0.81
Residual	3	0.66	—
Error	1	0.02	0.02
Lack of fit	2	0.64	0.32

When three variables are involved, a first approximation is again a plane or hyperplane of the form

$$Y = B_0 X_0 + B_1 X_1 + B_2 X_2 + B_3 X_3 + \varepsilon$$

For this first-order surface, at least four points must be taken to estimate the four B's. Since three dimensions are involved, we might consider a 2^3 factorial. As this design has eight experimental conditions at its vertices, the design often used is a one-half replication of a 2^3 factorial with two or more points at the center of the cube. These six points (two at the center) are sufficient to estimate all four B's and to test for lack of fit of this plane to the surface in three dimensions. After this initial one-half replication of the 2^3 the other half might be run, giving more information for a better fit.

If a plane is not a good fit in two dimensions, a second-order model might be tried. Such a response surface has the form

$$Y = B_0 X_0 + B_1 X_1 + B_2 X_2 + B_{11} X_1^2 + B_{12} X_1 X_2 + B_{22} X_2^2 + \varepsilon$$

Here six B's are to be estimated. The simplest design for this model would be a pentagon (five points) plus center points. This would yield six or more responses, and all six B's could be estimated.

In developing these designs Box and others found that the calculations can be simplified if the design can be rotated. A *rotatable design* is one that has equal predictability in all directions from the center, and the points are at a constant distance from the center. All first-order designs are rotatable, as the square and one-half replication of the cube in the cases above. The simplest second-order design that is rotatable is the pentagon with a point at the center, as given above.

In three dimensions a second-order surface is given by

$$Y = B_0 X_0 + B_1 X_1 + B_2 X_2 + B_3 X_3 + B_{11} X_1^2 + B_{22} X_2^2$$
$$+ B_{33} X_3^2 + B_{12} X_1 X_2 + B_{13} X_1 X_3 + B_{23} X_2 X_3 + \varepsilon \quad (16.13)$$

which has ten unknown coefficients. The cube for a three-dimensional model has only eight points, so a special design has been developed, called a *central composite* design. It is a 2^3 factorial with points along each axis at a distance from the center equal to the distance to each vertex. This gives $8 + 6 = 14$ points plus a point at the center makes 15, which is adequate for estimating the B's in Equation (16.13). This design is pictured in Figure 16.7.

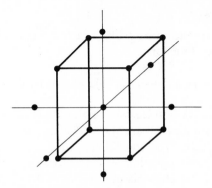

Figure 16.7   Central composite design.

In attempting to determine the equation of the response surface, some concepts in solid analytical geometry are often helpful. If a second-order model in two dimensions is

$$Y = B_0 X_0 + B_1 X_1 + B_2 X_2 + B_{11} X_1^2 + B_{12} X_1 X_2 + B_{22} X_2^2 + \varepsilon$$

the shape of this surface can be determined by reducing this equation to what is called *canonical form*. This would be

$$Y = B'_{11} X'^2_1 + B'_{22} X'^2_2$$

This is accomplished by a translation of axes to remove terms in X_1 and X_2, then a rotation of axes to remove the $X_1 X_2$ term. From B'_{11} and B'_{22} we can determine the shape of the surface, whether it is a sphere, ellipsoid, paraboloid, hyperboloid, or other. In higher dimensions this becomes more complicated, but it can still be useful in describing the response surface.

16.4   EVOLUTIONARY OPERATION (EVOP)

Philosophy

Evolutionary operation is a method of process operation that has a built-in procedure to increase productivity. The technique was developed by Box [6]. This book and Barnett's article [2] should be read to understand how the method works. Many chemical companies have reported considerable success using EVOP.

The procedure consists of running a simple experiment, usually a factorial, within the range of operability of a process as it is currently running. It is assumed that the variables to be controlled are measurable and can be set within a short distance of the current settings without disturbing production quality. The idea is to gather data on a response variable, usually yield, at the various points of an experimental design. When one set of data has been taken at all the points, one *cycle* is said to have been completed. One cycle is usually not sufficient to detect any shift in the response, so a second cycle is taken. This continues until the effect of one or more control variables, their interactions, or a change in the mean shows up as significant when compared with a measure of experimental error. This estimate of error is obtained from the cycle data, thus making the experiment self-contained. After a significant increase in yield has been detected, one *phase* is said to have been completed, and at this point a decision is usually made to change the basic operating conditions in a direction that should improve the yield. Several cycles may be necessary before a shift can be detected. The objective here, as with response surfaces, is to move in the direction of an optimum response. Response surface experimentation is primarily a laboratory or research technique; evolutionary operation is a production-line method.

To facilitate the EVOP procedure, a simple form has been developed to be used on the production line for each cycle of a 2^2 factorial with a point at the center. In the sections that follow an example will be run using these forms, and later the details of the form will be developed.

Example 16.3   To illustrate the EVOP procedure, consider a chemical process in which temperature and pressure are varied over short ranges, and the resulting chemical yield is recorded. Since two controlled variables affect the yield, a 2^2 factorial should indicate the effect of each factor as well as a possible interaction between them. By taking a point at the center of a 2^2 factorial, we can also check on a change in the mean (CIM) by comparing this point at the center with the four points around the vertices of the square. If the process should be straddling a maximum, the center point should eventually (after several cycles) be significantly above the peripheral points at the vertices. The standard form for EVOP locates the five points in this design as indicated in Figure 16.8.

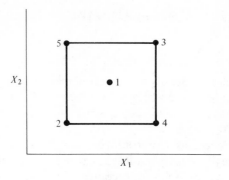

Figure 16.8

By comparing the responses (or average responses) at points 3 and 4 with those at 2 and 5, an effect of variable X_1 may be detected. Likewise, by comparing responses at 3 and 5 with those at 2 and 4, we may assess the X_2 effect. Comparing the responses at 2 and 3 with those at 4 and 5 will indicate an interaction effect, and comparing the responses at 2, 3, 4, and 5 with those at 1 will indicate a change in the mean, if present. The forms shown in Table 16.9 are filled in with data to show their use, which is self-explanatory.

Not much can be learned from the first cycle unless some separate estimate of standard deviation is available. The second cycle (Table 16.10) will begin to show the method for testing the effects.

Since, in the second cycle, none of the effects are numerically larger than their error limits, the true effect could easily be zero. In this case another cycle must be run. The only items that might need explaining are under the calculation of standard deviation. The range referred to here is the range of the differences (iv), and $f_{k,n}$ is found in a table where $k = 5$ for these five-point designs, and n is the cycle number. Part of such a table shows

$$n = \quad 2 \quad\quad 3 \quad\quad 4 \quad\quad 5 \quad\quad 6 \quad\quad 7 \quad\quad 8$$
$$f_{5,n} = 0.30 \quad 0.35 \quad 0.37 \quad 0.38 \quad 0.39 \quad 0.40 \quad 0.40$$

The third cycle is shown in Table 16.11. Here the temperature effect is seen to be significant, with an increase in temperature giving a higher yield. Once a significant effect has been found, the first *phase* of the EVOP procedure has been completed. The average results at this point are usually displayed on an EVOP bulletin board as in Figure 16.9.

An EVOP committee usually reviews these data and decides whether or not to reset the operating conditions. If they change the operating conditions (point 1), then EVOP is reinstated around this new point and the

Table 16.9 EVOP Work Sheet—First Cycle

Project: 424
Phase: 1
Date: 10/12

Cycle: $n = 1$
Response: Yield

Calculation of Averages

Operating Conditions	(1)	(2)	(3)	(4)	(5)
(i) Previous cycle sum					
(ii) Previous cycle average					
(iii) New observations	94.0	94.5	96.5	94.5	94.5
(iv) Differences [(ii) less (iii)]					
(v) New sums	94.0	94.5	96.5	94.5	94.5
(vi) New averages $\bar{Y}_i$	94.0	94.5	96.5	94.5	94.5

Calculation of Standard Deviation

Previous sum s =
Previous average s =
Range =
New s = range $\times f_{k,n}$ =
New sum s =
New average $s = \dfrac{\text{new sum } s}{n-1}$

Calculation of Effects

Temperature effect $= \frac{1}{2}(\bar{Y}_3 + \bar{Y}_4 - \bar{Y}_2 - \bar{Y}_5) = 1.00$

Pressure effect $= \frac{1}{2}(\bar{Y}_3 + \bar{Y}_5 - \bar{Y}_2 - \bar{Y}_4) = 1.00$

$T \times P$ interaction effect $= \frac{1}{2}(\bar{Y}_2 + \bar{Y}_3 - \bar{Y}_4 - \bar{Y}_5) = 1.00$

Change in mean effect $= \frac{1}{5}(\bar{Y}_2 + \bar{Y}_3 + \bar{Y}_4 + \bar{Y}_5 - 4\bar{Y}_1) = 0.80$

Calculations of Error Limits

For new average $= \dfrac{2}{\sqrt{n}}\, s$ =

For new effects $= \dfrac{2}{\sqrt{n}}\, s$ =

For change in mean $= \dfrac{1.78}{\sqrt{n}}\, s$ =

Table 16.10 EVOP Work Sheet—Second Cycle

Cycle: $n = 2$
Response: Yield

Project: 424
Phase: 1
Date: 10/12

Calculations of Averages

Operating Conditions	(1)	(2)	(3)	(4)	(5)
(i) Previous cycle sum	94.0	94.5	96.5	94.5	94.5
(ii) Previous cycle average	94.0	94.5	96.5	94.5	94.5
(iii) New observations	96.0	95.0	95.0	96.5	94.0
(iv) Differences [(ii) less (iii)]	−2.0	−0.5	1.5	−2.0	−0.5
(v) New sums	190.0	189.5	191.5	191.0	188.5
(vi) New averages $\bar{Y}_i$	95.0	94.7	95.7	95.5	94.2

Calculation of Standard Deviation

Previous sum $s =$
Previous average $s =$
Range $= 3.5$
New $s = \text{range} \times f_{k,n} = 1.05$
New sum $s = 1.05$
New average $s = \dfrac{\text{new sum } s}{n-1} = 1.05$

Calculation of Effects

Temperature effect $= \frac{1}{2}(\bar{Y}_3 + \bar{Y}_4 - \bar{Y}_2 - \bar{Y}_5) = 1.15$

Pressure effect $= \frac{1}{2}(\bar{Y}_3 + \bar{Y}_5 - \bar{Y}_2 - \bar{Y}_4) = -0.15$

$T \times P$ interaction effect $= \frac{1}{2}(\bar{Y}_2 + \bar{Y}_3 - \bar{Y}_4 - \bar{Y}_5) = 0.35$

Change in mean effect $= \frac{1}{5}(\bar{Y}_2 + \bar{Y}_3 + \bar{Y}_4 + \bar{Y}_5 - 4\bar{Y}_1) = 0.02$

Calculation of Error Limits

For new average $= \dfrac{2}{\sqrt{n}}\, s = 1.48$

For new effects $= \dfrac{2}{\sqrt{n}}\, s = 1.48$

For change in mean $= \dfrac{1.78}{\sqrt{n}}\, s = 1.32$

Table 16.11 EVOP Work Sheet—Third Cycle

Cycle: $n = 3$
Response: Yield

Project: 424
Phase: 1
Date: 10/12

Calculations of Averages

Operating Conditions	(1)	(2)	(3)	(4)	(5)
(i) Previous cycle sum	190.0	189.5	191.5	191.0	188.5
(ii) Previous cycle average	95.0	94.7	95.7	95.5	94.2
(iii) New observations	94.5	93.5	96.0	97.0	94.0
(iv) Differences [(ii) less (iii)]	0.5	1.2	−0.3	−1.5	0.2
(v) New sums	284.5	283.0	287.2	286.5	282.7
(vi) New averages $\bar{Y}_i$	94.8	94.3	95.7	95.5	94.2

Calculation of Standard Deviation

Previous sum $s = 1.05$

Previous average $s = 1.05$

Range $= 2.7$

New $s = $ range $\times f_{k,n} = 0.95$

New sum $s = 2.00$

New average $s = \dfrac{\text{new sum } s}{n-1} = 1.00$

Calculation of Effects

Temperature effect $= \frac{1}{2}(\bar{Y}_3 + \bar{Y}_4 - \bar{Y}_2 - \bar{Y}_5) = 1.35*$

Pressure effect $= \frac{1}{2}(\bar{Y}_3 + \bar{Y}_5 - \bar{Y}_2 - \bar{Y}_4) = 0.05$

$T \times P$ interaction effect $= \frac{1}{2}(\bar{Y}_2 + \bar{Y}_3 - \bar{Y}_4 - \bar{Y}_5) = -0.15$

Change in mean effect $= \frac{1}{5}(\bar{Y}_2 + \bar{Y}_3 + \bar{Y}_4 + \bar{Y}_5 - 4\bar{Y}_1) = 0.10$

Calculation of Error Limits

For new average $\dfrac{2}{\sqrt{n}} s = 1.16$

For new effects $= \dfrac{2}{\sqrt{n}} s = 1.16$

For change in mean $= \dfrac{1.78}{\sqrt{n}} s = 1.02$

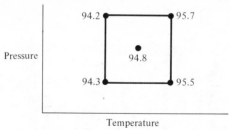

Figure 16.9

second phase is begun. EVOP is continued again until significant changes are detected. In fact, it goes on continually, seeking to optimize a process.

This is a very simple example with just two independent variables. The references should be consulted for more complex situations.

EVOP Form Rationale

Most of the steps in the EVOP form above are quite clear. It may be helpful to see where some of the constants come from.

In estimating the standard deviation from the range of differences in step (iv), consider a general expression for these differences

$$D_p = \frac{X_{p1} + X_{p2} + \cdots + X_{p,\,n-1}}{n-1} - X_{p,n}$$

where D_p represents the difference at any point p of the design. X_{pi} represents the observation at point p in the ith cycle. n is, of course, the number of cycles. Since the variance of a sum equals the sum of the variances for independent variables, and since the variance of a constant times a random variable equals the square of the constant times the variance of the random variable,

$$\sigma^2_{D_p} = \frac{1}{(n-1)^2}\left[\sigma^2_{xp1} + \sigma^2_{xp2} + \cdots + \sigma^2_{xp,\,n-1}\right] + \sigma^2_{xp,\,n}$$

Since all these x's represent the same population, their variances are all alike, and then

$$\sigma^2_D = \frac{1}{(n-1)^2}\left[(n-1)\sigma^2_x\right] + \sigma^2_x = \frac{n}{n-1}\sigma^2_x$$

and

$$\sigma_D = \sqrt{\frac{n}{n-1}}\,\sigma_x$$

The standard deviation of the population can then be written in terms of the standard deviation of these differences

$$\sigma_x = \sqrt{\frac{n-1}{n}}\,\sigma_D$$

σ_D can now be estimated from the range of these differences R_d. From the quality control field

$$\sigma_D = \frac{R_d}{d_2}$$

where d_2 depends on the number of differences in the range, which is 5 on this form. Here $d_2 = 2.326$ for samples of 5,

$$\sigma_D = \frac{R_d}{2.326}$$

and the standard deviation of the population is estimated by

$$\sigma_x = \sqrt{\frac{n-1}{n}}\,\frac{R_d}{2.326}$$

The quantity

$$\sqrt{\frac{n-1}{n}}\,\frac{1}{2.326}$$

is called $f_{k,n}$ in the EVOP form where $k = 5$.

Note that

$$f_{5,2} = \sqrt{\frac{1}{2}}\,\frac{R_d}{2.326} = 0.30 R_d$$

$$f_{5,6} = \sqrt{\frac{5}{6}}\,\frac{R_d}{2.326} = 0.39 R_d$$

which tallies with the table values given.

To determine the error limits for the effects, two standard deviation limits are used, as they represent the approximate 95 percent confidence limits on the parameter being estimated.

For any effect such as

$$E = \tfrac{1}{2}(\bar{Y}_3 + \bar{Y}_4 - \bar{Y}_2 - \bar{Y}_5)$$

its variance would be

$$V_E = \tfrac{1}{4}(\sigma_{\bar{Y}_3}^2 + \sigma_{\bar{Y}_4}^2 + \sigma_{\bar{Y}_2}^2 + \sigma_{\bar{Y}_5}^2) = \tfrac{1}{4}(4\sigma_{\bar{Y}}^2) = \sigma_{\bar{Y}}^2 = \frac{\sigma_{\bar{Y}}^2}{n}$$

and two standard deviation limits on an effect would be

$$\pm 2 \frac{\sigma_Y}{\sqrt{n}} \quad \text{estimated by} \quad \pm 2 \frac{s}{\sqrt{n}}$$

For the change in mean effect

$$\text{CIM} = \tfrac{1}{5}(\bar{Y}_2 + \bar{Y}_3 + \bar{Y}_4 + \bar{Y}_5 - 4\bar{Y}_1)$$

$$V_{CIM} = \tfrac{1}{25}(\sigma_{\bar{Y}_2}^2 + \sigma_{\bar{Y}_3}^2 + \sigma_{\bar{Y}_4}^2 + \sigma_{\bar{Y}_5}^2 + 16\sigma_{\bar{Y}_1}^2)$$

$$= \frac{20}{25}(\sigma_{\bar{Y}}^2) = \frac{20}{25} \frac{\sigma_Y^2}{n}$$

$$\sigma_{CIM} = \sqrt{\frac{4}{5}} \cdot \frac{\sigma_Y}{\sqrt{n}}$$

and two standard deviation limits would be

$$\pm 2 \cdot \sqrt{\frac{4}{5}} \frac{s}{\sqrt{n}} = \pm 1.78 \frac{s}{\sqrt{n}}$$

as given on the EVOP form.

Example 16.4   This EVOP procedure often has three factors, each at two levels. The $2^3 = 8$ experimental points are run as well as two points in the center of the cube, giving ten experimental points per cycle. Usually this design is run in two blocks where the *ABC* interaction is confounded with blocks. The standard form for the location of the ten points is shown in Figure 16.10.

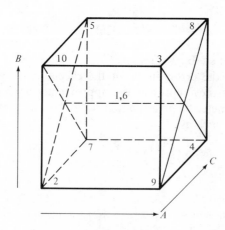

Figure 16.10

Table 16.12 EVOP Work Sheet—Block 1 in 2^3

Cycle: $n = 2$
Response: Mean surface factor
    for grinder slivers
Factors: A = Drop-out temperature 2145° and 2165°
      B = Back-zone temperature 2140° and 2160°
      C = Atmosphere: Reducing and oxidizing

Project: 478
Phase: 1
Date: 11/6

Calculation of Averages—Block 1

Operating Conditions (Block 1)	(1)	(2)	(3)	(4)	(5)
(i) Previous cycle sum	3.9	0.0	1.6	9.8	4.3
(ii) Previous cycle sum	3.9	0.0	1.6	9.8	4.3
(iii) New observations	5.6	4.0	6.7	15.0	3.4
(iv) Differences [(ii) less (iii)]	−1.7	−4.0	−5.1	−5.2	−0.9
(v) New sums	9.5	4.0	8.3	24.8	7.7
(vi) New averages $\bar{Y}_i$	4.8	2.0	4.2	12.4	3.8

Calculation of Standard Deviation

Previous sum s = (all blocks)
Previous average s =
Range = **4.3**
New sum s = range $\times f_{k,n}$ = **1.29**
New sum s = 1.29 (all blocks)
New average $s = \dfrac{\text{new sum } s}{2n - 3}$ = **1.29**

Calculation of Effects—Block 1

$A - BC = \frac{1}{2}(\bar{Y}_3 + \bar{Y}_4 - \bar{Y}_2 - \bar{Y}_5) = 5.40$

$B - AC = \frac{1}{2}(\bar{Y}_3 + \bar{Y}_5 - \bar{Y}_2 - \bar{Y}_4) = -3.20$

$C - AB = \frac{1}{2}(\bar{Y}_4 + \bar{Y}_5 - \bar{Y}_2 - \bar{Y}_3) = 5.00$

Change in mean effect $= \frac{1}{5}(\bar{Y}_2 + \bar{Y}_3 + \bar{Y}_4 + \bar{Y}_5 - 4\bar{Y}_1) = 0.64$

Table 16.13 EVOP Work Sheet—Block 2 in 2^3

Cycle: $n = 2$

		Calculation of Averages—Block 2				Calculation of Standard Deviation
Operating Conditions (Block 2)	(6)	(7)	(8)	(9)	(10)	
(i) Previous cycle sum	4.2	4.4	2.8	10.1	0.0	Previous sum $s = 1.29$ (all blocks)
(ii) Previous cycle average	4.2	4.4	2.8	10.1	0.0	Previous average $s = 1.29$
(iii) New observations	3.2	3.4	6.6	15.0	1.9	Range = 5.9
(iv) Differences [(ii) less (iii)]	1.0	0.9	−3.8	−4.9	−1.9	New sum s = range $\times f_{k,n} = 1.77$
(v) New sums	7.4	7.7	9.4	25.1	1.9	New sum $s = 3.06$ (all blocks)
(vi) New averages $\bar{Y}_i$	3.7	3.8	4.7	12.6	1.0	New average $s = \dfrac{\text{new sum } s}{2n - 2} = 1.53$

Calculation of Effects—Block 2

$A + BC = \frac{1}{2}(\bar{Y}_8 + \bar{Y}_9 - \bar{Y}_7 - \bar{Y}_{10}) = 6.25$

$B + AC = \frac{1}{2}(\bar{Y}_8 + \bar{Y}_{10} - \bar{Y}_7 - \bar{Y}_9) = -5.35$

$C + AB = \frac{1}{2}(\bar{Y}_7 + \bar{Y}_8 - \bar{Y}_9 - \bar{Y}_{10}) = -2.55$

Change in mean effect $= \frac{1}{5}(\bar{Y}_7 + \bar{Y}_8 + \bar{Y}_9 + \bar{Y}_{10} - 4\bar{Y}_6) = 1.46$

Calculation of Error Limits

For new averages $= \dfrac{2s}{\sqrt{n}} = \pm 2.16$

For new effects $= \dfrac{1.42s}{\sqrt{n}} = \pm 1.53$

For change in mean $= \dfrac{1.26s}{\sqrt{n}} = \pm 1.36$

Calculation of Effects—Both Blocks

$A = \frac{1}{2}[(A + BC) + (A - BC)] = 5.82*$
$B = \frac{1}{2}[(B + AC) + (B - AC)] = -4.28*$
$C = \frac{1}{2}[(C + AB) + (C - AB)] = 1.23$
Change in mean effect $= \frac{1}{2}(\text{CIM}_1 + \text{CIM}_2) = 1.05$

$AB = \frac{1}{2}[(C + AB) - (C - AB)] = -3.78*$
$AC = \frac{1}{2}[(B + AC) - (B - AC)] = -1.08$
$BC = \frac{1}{2}[(A + BC) - (A - BC)] = -0.42$

The first five points (1 through 5) are run as the first block and the last five points (6 through 10) are run as the second block. Tables 16.12 and 16.13 illustrate this EVOP technique for the second cycle of an example on the mean surface factor for grinder slivers where the factors are drop-out temperature A, back-zone temperature B, and atmosphere C. The work sheets should be self-explanatory as they are based on the design principles of confounding a 2^3 factorial in two blocks of four observations. One could terminate this process with the second cycle as it shows drop-out temperature A, back-zone temperature B, and their AB interaction to be significant in producing changes in the mean surface factor for grinder slivers.

16.5   ANALYSIS OF ATTRIBUTE DATA

Philosophy

Two assumptions used in the application of the analysis of variance technique are that (1) the response variable is normally distributed and (2) the variances of the experimental errors are equal throughout the experiment. In practice it is often necessary to deal with attribute data where the response variable is either 0 or 1. In such cases one often records the number of occurrences of a particular phenomenon or the percentage of such occurrences.

It is well known that the number of occurrences per unit such as defects per piece, errors per page, customers per unit time, often follow a Poisson distribution where such response variables are not only nonnormal but variances and means are equal. When proportions or percentages are used as the response variable, the data are binomial and again variances are related to means and the basic assumptions of ANOVA do not hold. Some studies have shown that lack of normality is not too serious in applying the ANOVA technique, but most statisticians recommend that a transformation be made on the original data if it is known to be nonnormal. Recommendations for proper transformations are given in Davies [9].

Another technique, called *factorial chi square*, has been found very useful in treating attribute data in industrial problems. This technique is described by Batson [3], who gives several examples of its application to actual problems. Although this factorial chi-square technique may not be as precise as a regular ANOVA on transformed data, its simplicity makes it well worth consideration for many applied problems.

To illustrate the techniques mentioned above, consider the following example from a consulting-firm study.

Example 16.5   Interest centered on undesirable marks on steel samples from a grinding operation. The response variable was simply whether or not these marks occurred. Four factors were believed to affect these marks: blade size A, centering B, leveling C, and speed D. Each of these four factors was set at two levels giving a 2^4 factorial experiment. Each of the 16 treatment combinations was run in a completely randomized order, and 20 steel samples were produced under each of the 16 experimental conditions. The number of damaged samples in each group of 20 is the recorded variable in Table 16.14.

Table 16.14 Damaged Steel Samples in 2^4 Experiment

		Factor A			
		0		1	
		Factor B		Factor B	
Factor C	Factor D	0	1	0	1
0	0	0	0	16	20
	1	0	0	10	20
1	0	0	0	10	14
	1	1	0	12	20

Table 16.15 Yates' Method on Transformed Data of Steel Sample Experiment

Treatment	Proportion p	Transformed Variable $X \equiv \arcsin \sqrt{p}$	(1)	(2)	(3)	(4)	SS
(1)	0.00	0.00	1.11	2.68	4.46	9.51	—
a	0.80	1.11	1.57	1.78	5.05	9.05	5.12
b	0.00	0.00	0.79	2.36	4.46	1.89	0.22
ab	1.00	1.57	0.99	2.69	4.59	2.35	0.35
c	0.00	0.00	0.79	2.68	0.66	−1.23	0.09
ac	0.50	0.79	1.57	1.78	1.23	−1.03	0.06
bc	0.00	0.00	1.12	2.36	0.66	−0.59	0.03
abc	0.70	0.99	1.57	2.23	1.69	−0.13	0.01
d	0.00	0.00	1.11	0.46	−0.90	0.59	0.03
ad	0.50	0.79	1.57	0.20	−0.33	0.13	0.01
bd	0.00	0.00	0.79	0.78	−0.90	0.57	0.02
abd	1.00	1.57	0.99	0.45	−0.13	1.03	0.06
cd	0.05	0.23	0.79	0.46	−0.26	0.57	0.02
acd	0.60	0.89	1.57	0.20	−0.33	0.77	0.04
bcd	0.00	0.00	0.66	0.78	−0.26	−0.07	0.00
$abcd$	1.00	1.57	1.57	0.91	0.13	0.39	0.01

If the above responses are each divided by 20, the proportions of damaged samples can be determined. Since proportions follow a binomial distribution, an arc sine transformation is appropriate (see [9]). Taking the arc sin $\sqrt{p}$ as the response variable, Yates' method can be used to determine the sums of squares as shown in Table 16.15.

If the three- and four-way interactions are assumed to be nonexistent, the resulting ANOVA table is shown as Table 16.16.

Table 16.16 ANOVA for Steel Sample Experiment

Source	df	SS	MS
A	1	5.12	5.12***
B	1	0.22	0.22*
AB	1	0.35	0.35**
C	1	0.09	0.09
AC	1	0.06	0.06
BC	1	0.03	0.03
D	1	0.03	0.03
AD	1	0.01	0.01
BD	1	0.02	0.02
CD	1	0.02	0.02
Error, pooled	5	0.12	0.02

This analysis shows the highly significant blade size effect, which is obvious from a glance at Table 16.14. It also shows a significant centering effect and a blade size–centering interaction that is not as obvious from Table 16.14.

To apply the factorial chi-square technique to this same example, Table 16.17 is filled in. The plus and minus signs simply give the proper weights for the main effects and two-way interactions and T is the value of the corresponding contrast. D is the total number of samples in the total experiment; here 20 per cell or $20 \times 2^4 = 320$ samples. The chi-square (χ^2) value is determined by the formula

$$\chi^2_{[1]} = \frac{N^2}{(S)(F)} \times \frac{T^2}{D}$$ (16.14)

Here N is the total number of items in the whole experiment and S is the number of occurrences (successes) in the N. F is the number of nonoccurrences (or failures). In the example $N = 320$ samples, $S = 123$ with un-

Table 16.17 Factorial Chi-Square Work Sheet for Table 16.14

																	Σ+	Σ−	T	T²	D	T²/D	χ²
Blade A	0	0	0	0	0	0	0	0	1	1	1	1	1	1	1	1							
Center B	0	0	0	0	1	1	1	1	0	0	0	0	1	1	1	1							
Level C	0	0	1	1	0	0	1	1	0	0	1	1	0	0	1	1							
Speed D	0	1	0	1	0	1	0	1	0	1	0	1	0	1	0	1							
S/20	0	0	0	1	0	0	0	0	16	10	10	12	20	20	14	20							
Main effects																							
Blade A	−	−	−	−	−	−	−	−	+	+	+	+	+	+	+	+	122	−1	121	14,641	320	45.75	193.52*
Center B	−	−	−	−	+	+	+	+	−	−	−	−	+	+	+	+	74	−49	25	625	320	1.95	8.25*
Level C	−	−	+	+	−	−	+	+	−	−	+	+	−	−	+	+	57	−66	−9	81	320	0.25	1.06
Speed D	−	+	−	+	−	+	−	+	−	+	−	+	−	+	−	+	63	−60	3	9	320	0.03	0.12
Interactions																							
AB	+	+	+	+	−	−	−	−	−	−	−	−	+	+	+	+	75	−48	27	729	320	2.28	9.65*
AC	+	+	−	−	+	+	−	−	−	−	+	+	−	−	+	+	56	−67	−9	81	320	0.25	1.06
AD	+	−	+	−	+	−	+	−	−	+	−	+	−	+	−	+	62	−61	1	1	320	0.01	0.04
BC	+	+	−	−	−	−	+	+	+	+	−	−	−	−	+	+	60	−63	−3	9	320	0.03	0.12
BD	+	−	+	−	−	+	−	+	+	−	+	−	−	+	−	+	66	−57	9	81	320	0.25	1.06
CD	+	−	−	+	+	−	−	+	+	−	−	+	+	−	−	+	69	−54	+15	225	320	0.70	2.96

desirable marks. For factor A, then

$$\chi_{[1]}^2 = \frac{(320)^2}{(123)(197)} (45.75)$$

$$= (4.23)(45.75) = 193.52$$

The 4.23 is constant for these data. The resulting chi-square values are then compared with χ^2 with 1 df at the significance level desired. For the 5 percent level, $\chi_{[1]}^2 = 3.84$ and again factors A, B, and AB are significant.

As seen in Table 16.17, the calculations for this technique are quite simple and the results are consistent with those using a more precise method. Extensions to problems other than 2^f factorials can be found in Batson [3].

16.6   RANDOMIZED INCOMPLETE BLOCKS— RESTRICTION ON EXPERIMENTATION

Method for Balanced Blocks

In some randomized block designs it may not be possible to apply all treatments in every block. If there were, for example, six brands of tires to test, only four could be tried on a given car (not using the spare), and such a block would be incomplete, having only four out of the six treatments in it.

Example 16.6   Take the problem of determining the effect on current flow of four treatments applied to the coils of TV tube filaments. As each treatment application requires some time, it is not possible to run several observations of these treatments in one day. If days are taken as blocks, all four treatments must be run in random order on each of several days in order to have a randomized block design. After checking it is found that even four treatments cannot be completed in a day; three are the most that can be run. The question then is: Which treatments are to be run on the first day, which on the second, and so forth, if information is desired on all four treatments?

The solution to this problem is to use a balanced incomplete block design. An *incomplete block design* is simply one in which there are more treatments than can be put in a single block. A *balanced incomplete block design* is an incomplete block design in which every pair of treatments occurs the same number of times in the experiment. Tables of such designs may be found in Fisher and Yates [11]. The number of blocks necessary for balancing will depend upon the number of treatments that can be run in a single block.

For the example mentioned there are four treatments and only three treatments can be run in a block. The balanced design for this problem requires four blocks (days) as shown in Table 16.18.

Table 16.18 Balanced Incomplete Block Design
for TV Filament Example

Block	Treatment				
(days)	A	B	C	D	$T_{i.}$
1	2	—	20	7	29
2	—	32	14	3	49
3	4	13	31	—	48
4	0	23	—	11	34
$T_{.j}$	6	68	65	21	$160 = T_{..}$

In this design only treatments A, C, and D are run on the first day; B, C, and D on the second day, and so forth. Note that each pair of treatments, such as AB, occurs together twice in the experiment. A and B occur together on days 3 and 4; C and D occur together on days 1 and 2; and so on. As in randomized complete block designs, the order in which the three treatments are run on a given day is completely randomized.

The analysis of such a design is easier if some new notation is introduced. Let

b = number of blocks in the experiment ($b = 4$)

t = number of treatments in the experiment ($t = 4$)

k = number of treatments per block ($k = 3$)

r = number of replications of a given treatment throughout the experiment ($r = 3$)

N = total number of observations

   = $bk = tr (N = 12)$

λ = number of times each pair of treatments appears together throughout the experiment

   = $r(k - 1)/(t - 1)$ ($\lambda = 2$)

Table 16.18 has the current readings after they have been coded by subtracting 513 milliamperes and the block and treatment totals. The analysis for a balanced incomplete block design proceeds as follows:

1. Calculate the total sum of squares as usual:

$$SS_{total} = \sum_i \sum_j Y_{ij}^2 - \frac{T_{..}^2}{N}$$

$$= 3478 - \frac{(160)^2}{12} = 1344.67$$

2. Calculate the block sum of squares, ignoring treatments:

$$SS_{block} = \sum_{i=1}^{b} \frac{T_{i.}^2}{k} - \frac{T_{..}^2}{N}$$

$$= \frac{(29)^2 + (49)^2 + (48)^2 + (34)^2}{3} - \frac{(160)^2}{12} = 100.67$$

3. Calculate treatment effects, adjusting for blocks:

$$SS_{treatment} = \frac{\sum_{j=1}^{t} Q_j^2}{k\lambda t} \tag{4.8}$$

where

$$Q_j = kT_{.j} - \sum_{i} n_{ij}T_{i.} \tag{4.9}$$

where $n_{ij} = 1$ if treatment j appears in block i, and $n_{ij} = 0$ if treatment j does *not* appear in block i. Note that $\sum_i n_{ij}T_{i.}$ is merely the sum of all block totals which contain treatment j.

For the data given,

$$Q_1 = 3(6) - (29 + 48 + 34) = 18 - 111 = -93$$
$$Q_2 = 3(68) - 131 = 73$$
$$Q_3 = 3(65) - 126 = 69$$
$$Q_4 = 3(21) - 112 = -49$$
$$\overline{0}$$

Note that

$$\sum_{j=1}^{t} Q_j = 0$$

which is always true. Then

$$SS_{treatment} = \frac{(-93)^2 + (73)^2 + (69)^2 + (-49)^2}{3(2)4} = 880.83$$

4. Calculate the error sum of squares by subtraction

$$SS_{error} = SS_{total} - SS_{block} - SS_{treatment}$$
$$= 1344.67 - 100.67 - 880.83 = 363.17$$

Table 16.19 summarizes in an ANOVA table.

An F test gives $F_{3,5} = 293.61/72.63 = 4.04$, which is not significant at the 5 percent level (Appendix, Table D).

In Table 16.19 the error degrees of freedom are determined by subtraction rather than as the product of the block and treatment degrees of freedom. However, this error degrees of freedom is seen to be the product of treatment

Table 16.19 ANOVA for Incomplete Block Design Example

Source	df	SS	MS
Blocks (days)	3	100.67	—
Treatments (adjusted)	3	880.83	293.61
Error	5	363.17	72.63
Totals	11	1344.67	

and block degrees of freedom (9) if 4 is subtracted for the four missing values in the design.

In some incomplete block designs it may be desirable to test for a block effect. The mean square for blocks was not computed in Table 16.19 since it had not been adjusted for treatments. In the case of a *symmetrical balanced incomplete randomized block design*, where $b = t$, the block sum of squares may be adjusted in the same manner as the treatment sums of squares

$$Q_i' = rT_{i.} - \sum_j n_{ij}T_{.j}$$

$$Q_1' = 3(29) - 92 = -5$$
$$Q_2' = 3(49) - 154 = -7$$
$$Q_3' = 3(48) - 139 = +5$$
$$Q_4' = 3(34) - 95 = +7$$

$$\sum_i Q_i' = 0$$

$$SS_{block} = \sum_{i=1}^{b} (Q_i')^2/r\lambda b = \frac{(-5)^2 + (-7)^2 + (5)^2 + (7)^2}{3(2)4} = 6.17$$

and

$$SS_{treatment (unadjusted)} = \frac{6^2 + 68^2 + 65^2 + 21^2}{3} - \frac{(160)^2}{12} = 975.34$$

The results of this adjustment and the one in treatments may now be summarized for this symmetrical case as shown in Table 16.20

The terms in parentheses are inserted only to show how the error term was computed for one adjusted effect and one unadjusted effect

$$SS_{error} = SS_{total} - SS_{treatment (adjusted)} - SS_{block}$$
$$= 1344.67 - 880.83 - 100.67 = 363.17$$

or

$$SS_{error} = SS_{total} - SS_{treatment} - SS_{block (adjusted)}$$
$$= 1344.67 - 975.34 - 6.17 = 363.17$$

Table 16.20   ANOVA for Incomplete Block Design Example for Both Treatments and Blocks

Source	df	SS	MS
Blocks (adjusted)	3	6.17	2.06
Blocks	(3)	(100.67)	—
Treatments (adjusted)	3	880.83	293.61
Treatments	(3)	(975.34)	—
Error	5	363.17	72.63
Totals	11	1344.67	

It should be noted also that the final sum of squares values used in Table 16.20 to get the mean square values do not add up to the total sum of squares. This is characteristic of a nonorthogonal design. The F test for blocks was not run because its value is obviously extremely small, which indicates no day-to-day effect on current flow.

For nonsymmetrical or unbalanced designs, the general regression method may be a useful alternative. If contrasts are to be computed for an incomplete block design, it can be shown that the sum of squares for a contrast is given by

$$SS_{C_m} = \frac{(C_m)^2}{(\sum_{j=1}^{t} c_{jm}^2)k\lambda t} \tag{4.10}$$

where contrasts C_m are made on the Q_j's rather than $T_{.j}$'s.

As an example consider the following orthogonal contrasts on the data of Table 16.18:

$$\begin{aligned}
C_1 &= Q_1 - Q_2 & &= -166 \\
C_2 &= Q_1 + Q_2 - 2Q_3 & &= -158 \\
C_3 &= Q_1 + Q_2 + Q_3 - 3Q_4 &= &\ 196
\end{aligned}$$

The corresponding sums of squares are

$$SS_{C_1} = \frac{(-166)^2}{(2)(3)(2)(4)} = 574.08$$

$$SS_{C_2} = \frac{(-158)^2}{(6)(3)(2)(4)} = 173.36$$

$$SS_{C_3} = \frac{(196)^2}{(12)(24)} = 133.39$$

$$\overline{880.83}$$

which checks with the adjusted sums of squares for treatments. Comparing each of the above sums of squares with their 1 df against the error mean square in Table 16.19 indicates that contrast C_1 is the only one of these three that is significant at the 5 percent level of significance. This indicates a real difference in current flow between treatments A and B, even though treatments in general showed no significant difference in current flow.

16.7   YOUDEN SQUARES

When the conditions for a Latin square are met except for the fact that only three treatments are possible (for example, because in one block only three positions are available) and where there are four blocks altogether, the design is an incomplete Latin square. This design is called a *Youden square.* One such Youden square is illustrated in Table 16.21.

Table 16.21   Youden Square Design

Block	Position 1	2	3
I	A	B	C
II	D	A	B
III	B	C	D
IV	C	D	A

Note that the addition of a column (D, C, A, B) would make this a Latin square if another position were available. A situation calling for a Youden square might occur if four materials were to be tested on four machines but there were only three heads on each machine whose orientation might affect the results.

Example 16.7   The analysis of a Youden square proceeds like the incomplete block analysis. Assuming hypothetical values for some measured variable Y_{ijk} where

$$Y_{ijk} = \mu + \beta_i + \tau_j + \gamma_k + \varepsilon_{ijk}$$

with

$$i = 1, \ldots, 4 \qquad j = 1, 2, \ldots, 4 \qquad k = 1, 2, 3$$

we might have the data of Table 16.22.

In Table 16.22

$$t = b = 4$$
$$r = k = 3$$
$$\lambda = 2$$

Treatment totals of A, B, C, D are

$$T_{.j.}: 6, \quad 2, \quad -1, \quad -9$$

Table 16.22 Youden Square Design Data

Block	Position 1	Position 2	Position 3	$T_{i..}$
I	$A = 2$	$B = 1$	$C = 0$	3
II	$D = -2$	$A = 2$	$B = 2$	2
III	$B = -1$	$C = -1$	$D = -3$	-5
IV	$C = 0$	$D = -4$	$A = 2$	-2
$T_{..k}$	-1	-2	$+1$	$-2 = T_{...}$

From this,

$$SS_{total} = \sum_i \sum_j \sum_k Y_{ijk}^2 - \frac{T_{...}^2}{N} = 48 - \frac{(-2)^2}{12} = 47.67$$

Position effect may first be ignored, since every position occurs once and only once in each block and once with each treatment, so that positions are orthogonal to blocks and treatments:

$$SS_{block \text{ (ignoring treatments)}} = \frac{(3)^2 + (2)^2 + (-5)^2 + (-2)^2}{3} - \frac{(-2)^2}{12}$$

$$= \frac{42}{3} - \frac{1}{3} = \frac{41}{3} = 13.67$$

For treatment sum of squares adjusted for blocks we get

$$Q_1 = 3(+6) - 3 = 15$$
$$Q_2 = 3(\ 2) - 0 = 6$$
$$Q_3 = 3(-1) - (-4) = 1$$
$$Q_4 = 3(-9) - (-5) = -22$$

and

$$\sum_i Q_i = 0$$

$$SS_{treatment} = \frac{(15)^2 + (6)^2 + (1)^2 + (-22)^2}{(4)(6)} = 31.08$$

$$SS_{position} = \frac{(-1)^2 + (-2)^2 + (1)^2}{4} - \frac{(-2)^2}{12} = 1.17$$

$$SS_{error} = SS_{total} - SS_{block} - SS_{treatment \text{ (adjusted)}} - SS_{position}$$
$$= 47.67 - 13.67 - 31.08 - 1.17$$
$$= 1.75$$

The analysis appears in Table 16.23.

Table 16.23   Youden Square ANOVA

Source	df	SS	MS
Treatments τ_j (adjusted)	3	31.08	10.36
Blocks β_i	3	13.67	—
Positions γ_k	2	1.17	0.58
Error ε_{ijk}	3	1.75	0.58
Totals	11	47.67	

The position effect here is not significant and it might be desirable to pool it with the error, getting

$$s_\varepsilon^2 = \frac{1.17 + 1.75}{2 + 3} = \frac{2.92}{5} = 0.58$$

as an estimate of error variance, with 5 df. Then the treatment effect is highly significant. The block mean square is not given, as blocks must be adjusted by treatments if block effects are to be assessed. The procedure is the same as shown for incomplete block designs.

PROBLEMS

16.1 One study reports on the breaking strength Y of samples of seven different starch films where the thickness of each film X was also recorded as a covariate. Results rounded off considerably are as follows:

Source	df	$\sum y^2$	$\sum xy$	$\sum x^2$	Adjusted $\sum y^2$	df	MS
Between starches		6,000,000	50,000	500			
Within starches	100						
Totals		8,000,000	60,000	600			

1. Complete this table and set up a test for the effect of starch film or breaking strength both before and after adjustment for thickness.

2. Write the mathematical model for this experiment, briefly describing each term.

3. State the assumptions implied in the tests of hypotheses made above and describe briefly how you would test their validity.

16.2 In studying the effect of two different teaching methods the gain score of each pupil in a public school test was used as the response variable. The pretest scores are also available. Run a covariance analysis on the data below to determine whether or not the new teaching method did show an improvement in average achievement as measured by this public school test.

	Method I			Method II	
	Pretest			Pretest	
Pupil	Score	Gain	Pupil	Score	Gain
1	28	5	1	24	7
2	43	−8	2	40	9
3	36	4	3	41	4
4	35	1	4	33	4
5	31	4	5	45	5
6	34	1	6	41	−2
7	33	4	7	41	6
8	38	8	8	33	11
9	39	1	9	41	−1
10	44	−1	10	30	15
11	36	−7	11	45	4
12	21	2	12	50	2
13	34	−1	13	42	9
14	33	1	14	47	3
15	25	7	15	43	−2
16	43	1	16	34	2
17	34	3	17	44	7
18	40	5	18	43	4
19	36	4	19	28	3
20	30	9	20	46	−1
21	35	−6	21	43	2
22	38	0	22	46	−4
23	31	−14	23	21	4
24	41	4			
25	35	10			
26	27	10			
27	45	1			

16.3 Data below are for five individuals who have been subjected to four conditions or treatments represented by the groups or lots 1–4. Y represents some measure of an individual supposedly affected by the variations in the treatments of the four lots. X represents another measure of an individual that may affect the value of Y even in the absence of the four treatments. The problem is to determine whether or not the Y means of the four groups differ significantly from each other after the effects of the X variable have been removed.

Groups or Lots							
1		2		3		4	
X	Y	X	Y	X	Y	X	Y
29	22	15	30	16	12	5	23
20	22	9	32	31	8	25	25
14	20	1	26	26	13	16	28
21	24	6	25	35	25	10	26
6	12	19	37	12	7	24	23

16.4 Consider a two-way classified design with more than one observation per cell.

1. Show the F test for one of the main effects in a covariance analysis if both factors are fixed.

2. Show the F test for the interaction in a covariance analysis if one factor is fixed and the other one is random.

16.5 For Example 16.4 in the text, another response variable was measured—the mean surface factor for heat slivers. Results of the first three cycles at the ten experimental points gave

Cycle	(1)	(2)	(3)	(4)	(5)	(6)	(7)	(8)	(9)	(10)
1	4.0	6.8	5.9	4.5	4.3	4.2	7.8	8.3	11.0	5.9
2	5.2	7.2	6.1	3.7	4.0	4.5	7.0	8.1	12.5	6.3
3	4.8	8.0	7.0	3.6	3.8	4.3	6.5	7.8	12.7	7.0

Set up EVOP work sheets for these data and comment on your results after three cycles.

16.6 Data on yield in pounds for varying percent concentration at levels 29, 30, and 31 percent, and varying power at levels 400, 450, and 500 watts gave the following results in four cycles at the five points in a 2^2 design.

Cycle	(1)	(2)	(3)	(4)	(5)
1	477	472	411	476	372
2	469	452	430	468	453
3	465	396	375	468	292
4	451	469	363	432	460

Analyze by EVOP methods.

16.7 The response variable in the data on page 365 is the number of defective knife handles in samples of 72 knives where the factors are machines, lumber grades, and replications. Analyze after making a suitable transformation of the data.

Lumber Grade	Replication	Machine 1	Machine 2
A	I	8	4
	II	4	6
B	I	9	3
	II	10	4

16.8 Analyze the data of Problem 16.7 using factorial chi square. Compare with the results of Problem 16.7.

16.9 Data on screen color difference on a television tube measured in degrees Kelvin are to be compared for four operators. On a given day only three operators can be used in the experiment. A balanced incomplete block design gave results as follows:

Day	Operator A	B	C	D
Monday	780	820	800	—
Tuesday	950	—	920	940
Wednesday	—	880	880	820
Thursday	840	780	—	820

Do a complete analysis of these data and discuss your findings with regard to differences between operators.

16.10 Run orthogonal contrasts on the operators in Problem 16.9.

16.11 At the Bureau of Standards an experiment was to be run on the wear resistance of a new experimental shoe leather as compared with the standard leather in use by the Army. It was decided to equip several men with one shoe of the experimental-type leather and the other shoe of standard leather and after many weeks in the field compare wear on the two types. Considering each man as a block, suggest a reasonable number of men to be used in this experiment and outline its possible analysis.

16.12 If three experimental types of leather were to be tested along with the standard in Problem 16.11, it is obvious that only two of the four types can be tested on one man. Set up a balanced incomplete block design for this situation. Insert some arbitrary numerical values and complete an ANOVA table for your data.

17
Summary and Special Problems

Throughout this book the three phases of an experiment have been emphasized: the experiment, the design, and the analysis. At the end of most chapters an outline of the designs considered up to that point has been presented. It may prove useful to the reader to have a complete outline as a summary of the designs presented in Chapters 1–15.

This is not the only way such an outline could be constructed, but it represents an attempt to see each design as part of the overall picture of designed experiments. A look at the outline will reveal that the same design is often used for an entirely different experiment, which simply illustrates the fact that much care must be taken to spell out the experiment and its design or method of randomization before an experiment is performed.

For each experiment, the chapter reference in this book is given.

After the summary several special problems are presented. Many do not fit any one pattern or design but may be of interest as they represent the kinds of problems actually encountered in practice. These problems have no "pat" solutions and the way they are set up may depend a great deal upon assumptions the reader may make. They should serve to evoke some good discussion about designing an experiment to solve a given practical problem.

Experiment	Design	Analysis	Chapter Reference
I. Single factor	1. Completely randomized $Y_{ij} = \mu + \tau_j + \varepsilon_{ij}$	1. One-way ANOVA	Chapter 3
	2. Randomized block $Y_{ij} = \mu + \beta_i + \tau_j + \varepsilon_{ij}$	2.	
	a. Complete	a. Two-way ANOVA	Chapter 4
	b. Incomplete, balanced	b. Special ANOVA	Chapter 16
	c. Incomplete, general	c. General regression method	
	3. Latin square $Y_{ijk} = \mu + \beta_i + \tau_j + \gamma_k + \varepsilon_{ijk}$	3.	
	a. Complete	a. Three-way ANOVA	Chapter 4
	b. Incomplete, Youden square	b. Special ANOVA (like 2b)	Chapter 16
	4. Graeco-Latin square $Y_{ijkm} = \mu + \beta_i + \tau_j + \gamma_k + \omega_m + \varepsilon_{ijkm}$	4. Four-way ANOVA	Chapter 4
II. Two or more factors A. Factorial (crossed)	1. Completely randomized $Y_{ijk} = \mu + A_i + B_j + AB_{ij} + \varepsilon_{k(ij)}, \cdots$ for more factors	1.	
	a. General case	a. ANOVA with interactions	Chapter 5
	b. 2^f case	b. Yates' method or general ANOVA; use: $(1), a, b, ab, \dots$	Chapter 6

	c. 3^f case	c. General ANOVA;
		    use 00, 10, 20, 01, 011, $\dots$
		    and
		    $A \times B = AB + AB^2, \dots$
		    for interaction
			Chapter 9

Experiment	Design	Analysis	Chapter Reference
	2. Randomized block	2.	
	a. Complete $Y_{ijk} = \mu + R_k + A_i + B_j + AB_{ij} + \varepsilon_{ijk}$	a. Factorial ANOVA with replications R_k	Chapter 12
	b. Incomplete, confounding: i. Main effect—split-plot $Y_{ijk} = \mu + \underbrace{R_i + A_j + RA_{ij}}_{\text{whole plot}} + \underbrace{B_k + RB_{ik} + AB_{jk} + RAB_{ijk}}_{\text{split-plot}}$	b. i. Split-plot ANOVA	Chapter 13
	ii. Interactions in 2^f and 3^f (1) Several replications $Y_{ijkq} = \mu + R_i + \beta_j + R\beta_{ij} + A_k + B_q + AB_{kq} + \varepsilon_{ijkq}$	ii. (1) Factorial ANOVA with replications R_i and blocks β_j or confounded interaction	Chapter 14
	(2) One replication only $Y_{ijk} = \mu + \beta_i + A_j + B_k + AB_{jk}, \ldots$	(2) Factorial ANOVA with blocks β_i or confounded interaction	Chapter 14
	(3) Fractional replication—aliases $Y_{ijk} = \mu + A_i + B_j + AB_{ij}, \ldots$ (a) in 2^f (b) in 3^f; $f = 3$: Latin square	(3) Factorial ANOVA with aliases or aliased interactions	Chapter 15

3. Latin square
   a. Complete
   $$Y_{ijkm} = \mu + R_k + \gamma_m + A_i + B_j + AB_{ij} + \varepsilon_{ijkm}$$

3.
   a. Factorial ANOVA with replications and positions   Chapter 12

B. Nested (hierarchical)

1. Completely randomized
   $$Y_{ijk} = \mu + A_i + B_{j(i)} + \varepsilon_{k(ij)}$$

   1. Nested ANOVA   Chapter 11

2. Randomized block
   a. Complete
   $$Y_{ijk} = \mu + R_k + A_i + B_{j(i)} + \varepsilon_{ijk}$$

   2.
   a. Nested ANOVA with blocks R_k Chapters 4 and 11

3. Latin square
   a. Complete
   $$X_{ijkm} = \mu + R_k + \gamma_m + A_i + B_{j(i)} + \varepsilon_{ijkm}$$

   3.
   a. Nested ANOVA with blocks and positions   Chapters 4 and 11

C. Nested factorial

1. Completely randomized
   $$Y_{ijkm} = \mu + A_i + B_{j(i)} + C_k + AC_{ik} + BC_{kj(i)} + \varepsilon_{m(ijk)}$$

   1. Nested-factorial ANOVA   Chapter 11

2. Randomized block
   a. Complete
   $$Y_{ijkm} = \mu + R_k + A_i + B_{j(i)} + C_m + AC_{im} + BC_{mj(i)} + \varepsilon_{ijkm}$$

   2.
   a. Nested-factorial ANOVA with blocks R_k Chapters 4 and 11

3. Latin square
   a. Complete
   $$Y_{ijkmq} = \mu + R_k + \gamma_m + A_i + B_{j(i)} + C_q + AC_{iq} + BC_{qj(i)} + \varepsilon_{ijkmq}$$

   3.
   a. Nested-factorial ANOVA with blocks and positions   Chapters 4 and 11

SPECIAL PROBLEMS

17.1 Three different time periods are chosen in which to plant a certain vegetable: early May, late May, and early June. Plots are selected in early May and five different fertilizers are randomly tried in each plot. This is repeated in late May and again in early June. The criterion is the number of bushels per acre at harvest time. Assuming r plots for each fertilizer can be used at each planting:
a. Discuss the nature of this experiment.
b. Recommend a value for r to make this an acceptable study. (Time periods and fertilizers are fixed.)

17.2 In the study of the factors affecting the surface finish of steel in Problem 15.19 there was a sixth factor—factor A (the flux type)—also set at two levels. It was possible to run all 64 (2^6) treatment combinations except for the fact that the experimenter wanted a complete replication of the experiment that was to be run. Thus, it was decided to do a one-half replicate of the 2^6 and then replicate the same block at another time. It was also learned that first a flux type had to be chosen, and then a slab width, and then the other four factors randomized. After this the experimenter would run the other slab width and then repeat the experiment with the other flux type.

  Outline a data format for this problem and show what its ANOVA table might look like.

17.3 In Problem 15.18 on the 81 engines it took about a year to gather all the data. When the data were ready they were found to be incomplete at the low-temperature level so the experiment was really a one-third replication of a $3^4 \times 2$ factorial instead of the 3^5 factorial. How would you analyze the data if one whole level in your answer to this problem were missing?

17.4 In a food industry there was a strong feeling that viscosity readings on a slurry were not consistent. Some claimed that different laboratories gave different readings. Others said it was the fault of the instruments in these laboratories. And still others claimed it was the fault of the operators who read the instruments. To determine where the main sources of variability were, slurries from two products with vastly different viscosities were taken. Each sample of slurry was mixed well and then divided into four jars, one of which was sent to each of four laboratories for analysis. In each laboratory the jar was divided into eight samples in order for each of two randomly chosen operators within that laboratory to test two samples on each of two randomly chosen test instruments. Considering the laboratories as fixed, the two products as fixed, operators as random, and instruments as random, show a data layout for this problem. Show its model and its ANOVA outline and discuss any concerns about the tests to be made.

17.5 A total of 256 seventh grade students with equal numbers of boys and girls and representing both high and low socioeconomic status (SES) was subjected to a mathematics lesson that was believed to be either interesting or uninteresting under one of four reinforcement conditions: no knowledge of results, immediate verbal knowledge of results, delayed verbal knowledge of results, and delayed written knowledge of results. Task interest and type of reinforcement emerged as significant at the 1 percent significance level. The delayed verbal condition was significantly superior to the other three, which did not differ from each other. Also, low SES children learned the high-interest task as quickly as high SES children, and more quickly than they learned the low-interest task. On the basis of this information answer the following.

a. What is the implied criterion Y?

b. List each independent variable and state the levels of each.

c. Set up an ANOVA table for this problem and indicate which results are significant.

d. Show a data layout indicating the number of children in each treatment.

e. For the significant results set some hypothetical means to show possible interpretation of such results.

17.6   In testing for a significant effect of treatments on some measured variable Y the resulting F test was almost, but not quite, significant at the 5 percent level (say, $F = 4.00$ with 3 and 8 df). Now the experimenter repeats the whole experiment on another set of independent observations under the same treatments and again the F fails to reach the 5 percent level (say, $F = 4.01$ with 3 and 8 df). What conclusion—if any—would you reach with regard to the hypothesis of equal treatment effects? Briefly explain your reasoning.

17.7   Six different formulas or mixes of concrete are to be purchased from five competing suppliers. Tests of crushing strength are to be made on two blocks constructed according to each formula–supplier combination. One block will be poured, hardened, and tested for each combination before the second block is formed, thus making two replications of the whole experiment. Assuming mixes and suppliers fixed and replications random, set up a mathematical model for this situation, outline its ANOVA table, and show what F tests are appropriate.

17.8   Four chemical treatments A, B, C, and D are prepared in three concentrations (3 to 1, 2 to 1, and 1 to 1) and two samples of each combination are poured; then three determinations of percent of pH are made from each sample. Considering treatments and concentrations fixed and samples and determinations random, set up the appropriate ANOVA table and comment on statistical tests to be made. Also explain in some detail how you would proceed to set 95 percent confidence limits on the average percent of pH for treatment B, concentration 2 to 1.

17.9   A graduate student carried out an experiment involving six experimental conditions. From an available group of subjects she assigned five men and five women at random to each of the six conditions, using 60 subjects in all. Upon completion of the experiment, she carried out the following analysis to test the differences among the condition means:

Source	df	
Between conditions	5	$F = \dfrac{MS_b}{MS_w}$ with 5 and 54 df
Within conditions	54	
Total	59	

At this point a statistical consultant was called in.

a. The consultant blew his top. Why?

b. You are now the consultant. Set up the analysis the student should use, with sources of variation, df, and F tests, and explain to her why she must use your analysis and why hers is incorrect.

c. As consultant you must point out the possibility that one of the significance tests in the appropriate analysis may affect the interpretation of the other. How? What should be done in such a case?

17.10 For the problem (Problem 6.6) on the effect of four factors on chemical yield, the original data had an unequal number of observations at each treatment combination. The numbers in each treatment combination below are the n's for each particular treatment. Explain how you might handle the analysis of this problem.

		A_1		A_2	
		B_1	B_2	B_1	B_2
C_1	D_1	2	2	7	5
	D_2	2	1	2	3
C_2	D_1	4	2	5	1
	D_2	1	1	2	2

Glossary of Terms

Alias. An effect in a fractionally replicated design that "looks like" or cannot be distinguished from another effect.

Alpha (α). Size of the type I error or probability of rejecting a hypothesis when true.

Beta (β). Size of the type II error or probability of accepting a hypothesis when some alternative hypothesis is true.

Canonical form. Form of a second-degree response-surface equation that allows determination of the type of surface.

Completely randomized design. A design in which all treatments are assigned to the experimental units in a completely random manner.

Confidence limits. Two values between which a parameter is said to lie with a specified degree of confidence.

Confounding. An experimental arrangement in which certain effects cannot be distinguished from others. One such effect is usually blocks.

Consistent estimator. An estimator of a parameter whose value comes closer to the parameter as the sample size is increased.

Contrast. A linear combination of treatment totals or averages where the sum of the coefficients is zero.

Correlation coefficient (Pearson r). The square root of the proportion of total variation accounted for by linear regression.

Correlation index R. The square root of the proportion of total variation accounted for by the regression equation of the degree being fitted to the data.

Covariance analysis. Technique for removing the effect of one or more variables on the response variable before assessing the effect of treatments on the response variable.

Critical region. A set of values of a test statistic where the hypothesis under test is rejected.

Defining contrast. An expression that indicates which effects are confounded with blocks in a factorial design that is confounded.

Effect of a factor. The change in response produced by a change in level of the factor.

Errors. Type I: Rejecting a hypothesis when true. Type II: Accepting a hypothesis when false.

Eta squared (η^2). Proportion of total variation accounted for by a regression line through all average responses for each value of the independent variable.

Evolutionary operation. An experimental procedure for collecting information to improve a process without disturbing production.

Expected value of a statistic. The average value of a statistic if it were calculated from an infinite number of equal-sized samples from a given population.

Ex-post-facto research. Research in which the variables have already acted and one is seeking for some possible relationships between the variables.

Factorial chi square. Technique for handling attribute data in an analysis of variance.

Factorial experiment. An experiment in which all levels of each factor in the experiment are combined with all levels of every other factor.

Fractional replication. An experimental design in which only a fraction of a complete factorial is run.

Graeco-Latin square. An experimental design in which four factors are so arranged that each level of each factor is combined only once with each level of the other three factors.

Incomplete block design. A randomized block design in which not all treatment combinations can be included in one block. Such a design is called a *balanced* design if each pair of treatments occurs together the same number of times.

Interaction. An interaction between two factors means that a change in response between levels of one factor is not the same for all levels of the other factor.

Latin square. An experimental design in which each level of each factor is combined only once with each level of two other factors.

Least squares method. Method of assigning estimates of parameters in a model such that the sum of the squares of the errors is minimized.

Max-min-con principle. A design principle in which the experimenter attempts to maximize the variance due to the manipulated variables, minimize the error variance, and control other possible sources of variance.

Mean.   Of a sample:

$$\bar{Y} = \sum_{i=1}^{n} Y_i/n$$

Of a population:

$$\mu = E(Y).$$

Mean square.   An unbiased estimate of a population variance. Determined by dividing a sum of squares by its degree of freedom.

Minimum variance estimate.   If two or more estimates $u_1, u_2, \ldots, u_k$ are made of the same parameter θ, the estimate with the smallest variance is called the minimum variance estimate.

Mixed model.   The model of a factorial experiment in which one or more factors are at fixed levels and at least one factor is at random levels.

Nested experiment.   An experiment in which the levels of one factor are chosen within the levels of another factor.

Nested-factorial experiment.   An experiment in which some factors are factorial or crossed with others and some factors are nested within others.

Operating characteristic curve.   A plot of the probability of acceptance of a hypothesis under test versus the true parameter of the population.

Orthogonal contrasts.   Two contrasts are said to be orthogonal if the products of their corresponding coefficients add to zero.

Orthogonal polynomials.   Sets of polynomials used in regression analysis such that each polynomial in the set is orthogonal to all other polynomials in the set.

Parameter.   A characteristic of a population, such as the population mean or variance.

Point estimate.   A single statistic used to estimate a parameter.

Power of a test.   A plot of the probability of rejection of a hypothesis under test versus the true parameter. It is the complement of the operating characteristic curve.

Principal block.   The block in a confounded design that contains the treatment combination in which all factors are at their lowest level.

Random model.   A model of a factorial experiment in which the levels of the factor are chosen at random.

Random sample.   A sample in which each member of the population sampled has an equal chance of being selected in this sample.

Randomized block design.   An experimental design in which the treatment combinations are randomized within a block and several blocks are run.

Repeated-measures design.   A design in which measurements are taken on the experimental units more than once.

Research.   A systematic quest for undiscovered truth.

Rotatable design.   An experimental design that has equal predictive power in all directions from a center point, and in which all experimental points are equidistant from this center point.

Split-plot design.   An experimental design in which a main effect is confounded with blocks due to the practical necessities of the order of experimentation.

Standard error of estimate.   The standard deviation of errors of estimate around a least squares fitted regression model.

Statistic u.   A measure computed from a sample.

Statistical hypothesis H_0.   An assumption about a population being sampled.

Statistical inference.   Inferring something about a population of measures from a sample of that population.

Statistics.   A tool for decision making in the light of uncertainty.

Steepest ascent method.   A method of sequential experiments that will direct the experimenter toward the optimum of a response surface.

Sum of squares (SS).   The sum of the squares of deviations of a random variable from its mean,

$$SS = \sum_{i=1}^{n} (Y_i - \bar{Y})^2$$

Test of a hypothesis.   A rule by which a hypothesis is accepted or rejected.

Test statistic.   A statistic used to test a hypothesis.

Treatment combination.   A given combination showing the levels of all factors to be run for that set of experimental conditions.

True experiment.   A study in which certain independent variables are manipulated, their effect on one or more dependent variables is observed, and the levels of the independent variables are assigned at random to the experimental units of the study.

Unbiased statistic.   A statistic whose expected value equals the parameter it is estimating.

Variance.   Of a sample:

$$s^2 = \frac{\sum_{i=1}^{n} (Y_i - \bar{Y})^2}{n - 1} = \frac{SS}{df}$$

Of a population:

$$\sigma^2 = E(Y - \mu)^2.$$

Youden square.   An incomplete Latin square.

References

1. Anderson, V. L., and R. A. McLean, "Restriction Errors: Another Dimension in Teaching Experimental Statistics," *The American Statistician* (Nov. 1974).

2. Barnett, E. H., "Introduction to Evolutionary Operation," *Industrial and Engineering Chemistry*, vol. 52 (June 1960), p. 500.

3. Batson, H. C., "Applications of Factorial Analysis to Experiments in Chemistry," *National Convention Transactions of the American Society for Quality Control* (1956), pp. 9–23.

4. Bennett, C. A., and N. L. Franklin, *Statistical Analysis in Chemistry and the Chemical Industry*. New York: John Wiley & Sons, Inc. (1954).

5. *Biometrics*, vol. 13 (Sept. 1957), pp. 261–405.

6. Box, G. E. P., and N. R. Draper, *Evolutionary Operation: A Statistical Method for Process Improvement*. New York: John Wiley & Sons, Inc. (1969).

7. Burr, I. W., *Statistical Quality Control Methods*, New York: Marcel Dekker (1976).

8. Burr, I. W., *Applied Statistical Methods*, New York: Academic Press (1974).

9. Davies, O. L., *Design and Analysis of Industrial Experiments*. New York: Hafner Publishing Company (1954).

10. Dixon, W. J., and F. J. Massey, *An Introduction to Statistical Analysis* (2nd Ed.). New York: McGraw-Hill Book Company (1957).

11. Fisher, R. A., and F. Yates, *Statistical Tables for Biological, Agricultural and Medical Research* (4th Ed.). Edinburgh and London: Oliver & Boyd, Ltd. (1953).

12. Hunter, J. S., "Determination of Optimum Operating Conditions by Experimental Methods," *Industrial Quality Control* (Dec.–Feb. 1958–1959).

13. Kempthorne, O., *The Design and Analysis of Experiments*. New York: John Wiley & Sons, Inc. (1952).

14. Kerlinger, F. N., *Foundations of Behavioral Research* (2nd Ed.) New York: Holt, Rinehart and Winston (1973).

15. Keuls, M., "The Use of the Studentized Range in Connection with an Analysis of Variance," *Euphytica*, vol. 1 (1952), pp. 112–122.

16. Leedy, P. D., *Practical Research: Planning and Design*. New York: Macmillan (1974).

17. McCall, C. H., Jr., "Linear Contrasts, Parts I, II, and III," *Industrial Quality Control* (July–Sept., 1960).

18. Miller, L. D., "An Investigation of the Machinability of Malleable Iron Using Ceramic Tools," unpublished MSIE thesis, Purdue University (1959).

19. Ostle, B., *Statistics in Research* (2nd Ed.). Ames, Iowa: Iowa State University Press (1963).

20. Owen, D. B., *Handbook of Statistical Tables*. Boston: Addison-Wesley Publishing Company, Inc. (1962).

21. Scheffe, H., "A Method for Judging All Contrasts in the Analysis of Variance," *Biometrics*, vol. XL (June, 1953).

22. Snedecor, G. W., and W. C. Cochran, *Statistical Methods* (6th Ed.). Ames, Iowa: Iowa State University Press (1967).

23. Tanur, Judith M., editor, and others, *Statistics: A Guide to the Unknown* (2nd Ed.) San Francisco: Holden-Day (1978).

24. Winer, B. J., *Statistical Principles in Experimental Design* (2nd Ed.). New York: McGraw-Hill Book Company (1971).

25. Wortham, A. W., and T. E. Smith, *Practical Statistics in Experimental Design*. Dallas: Dallas Publishing House (1960).

26. Yates, F., *Design and Analysis of Factorial Experiments*. London: Imperial Bureau of Soil Sciences (1937).

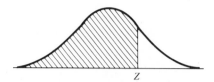

Z

Table A Areas under the Normal Curve* (Proportion of Total Area under the Curve from $-\infty$ to Designated Z Value)

Z	0.09	0.08	0.07	0.06	0.05	0.04	0.03	0.02	0.01	0.00
−3.5	0.00017	0.00017	0.00018	0.00019	0.00019	0.00020	0.00021	0.00022	0.00022	0.00023
−3.4	0.00024	0.00025	0.00026	0.00027	0.00028	0.00029	0.00030	0.00031	0.00033	0.00034
−3.3	0.00035	0.00036	0.00038	0.00039	0.00040	0.00042	0.00043	0.00045	0.00047	0.00048
−3.2	0.00050	0.00052	0.00054	0.00056	0.00058	0.00060	0.00062	0.00064	0.00066	0.00069
−3.1	0.00071	0.00074	0.00076	0.00079	0.00082	0.00085	0.00087	0.00090	0.00094	0.00097
−3.0	0.00100	0.00104	0.00107	0.00111	0.00114	0.00118	0.00122	0.00126	0.00131	0.00135
−2.9	0.0014	0.0014	0.0015	0.0015	0.0016	0.0016	0.0017	0.0017	0.0018	0.0019
−2.8	0.0019	0.0020	0.0021	0.0021	0.0022	0.0023	0.0023	0.0024	0.0025	0.0026
−2.7	0.0026	0.0027	0.0028	0.0029	0.0030	0.0031	0.0032	0.0033	0.0034	0.0035
−2.6	0.0036	0.0037	0.0038	0.0039	0.0040	0.0041	0.0043	0.0044	0.0045	0.0047
−2.5	0.0048	0.0049	0.0051	0.0052	0.0054	0.0055	0.0057	0.0059	0.0060	0.0062
−2.4	0.0064	0.0066	0.0068	0.0069	0.0071	0.0073	0.0075	0.0078	0.0080	0.0082
−2.3	0.0084	0.0087	0.0089	0.0091	0.0094	0.0096	0.0099	0.0102	0.0104	0.0107
−2.2	0.0110	0.0113	0.0116	0.0119	0.0122	0.0125	0.0129	0.0132	0.0136	0.0139
−2.1	0.0143	0.0146	0.0150	0.0154	0.0158	0.0162	0.0166	0.0170	0.0174	0.0179
−2.0	0.0183	0.0188	0.0192	0.0197	0.0202	0.0207	0.0212	0.0217	0.0222	0.0228
−1.9	0.0233	0.0239	0.0244	0.0250	0.0256	0.0262	0.0268	0.0274	0.0281	0.0287
−1.8	0.0294	0.0301	0.0307	0.0314	0.0322	0.0329	0.0336	0.0344	0.0351	0.0359
−1.7	0.0367	0.0375	0.0384	0.0392	0.0401	0.0409	0.0418	0.0427	0.0436	0.0446
−1.6	0.0455	0.0465	0.0475	0.0485	0.0495	0.0505	0.0516	0.0526	0.0537	0.0548
−1.5	0.0559	0.0571	0.0582	0.0594	0.0606	0.0618	0.0630	0.0643	0.0655	0.0668
−1.4	0.0681	0.0694	0.0708	0.0721	0.0735	0.0749	0.0764	0.0778	0.0793	0.0808
−1.3	0.0823	0.0838	0.0853	0.0869	0.0885	0.0901	0.0918	0.0934	0.0951	0.0968
−1.2	0.0985	0.1003	0.1020	0.1038	0.1057	0.1075	0.1093	0.1112	0.1131	0.1151
−1.1	0.1170	0.1190	0.1210	0.1230	0.1251	0.1271	0.1292	0.1314	0.1335	0.1357
−1.0	0.1379	0.1401	0.1423	0.1446	0.1469	0.1492	0.1515	0.1539	0.1562	0.1587
−0.9	0.1611	0.1635	0.1660	0.1685	0.1711	0.1736	0.1762	0.1788	0.1814	0.1841
−0.8	0.1867	0.1894	0.1922	0.1949	0.1977	0.2005	0.2033	0.2061	0.2090	0.2119
−0.7	0.2148	0.2177	0.2207	0.2236	0.2266	0.2297	0.2327	0.2358	0.2389	0.2420
−0.6	0.2451	0.2483	0.2514	0.2546	0.2578	0.2611	0.2643	0.2676	0.2709	0.2743
−0.5	0.2776	0.2810	0.2843	0.2877	0.2912	0.2946	0.2981	0.3015	0.3050	0.3085
−0.4	0.3121	0.3156	0.3192	0.3228	0.3264	0.3300	0.3336	0.3372	0.3409	0.3446
−0.3	0.3483	0.3520	0.3557	0.3594	0.3632	0.3669	0.3707	0.3745	0.3783	0.3821
−0.2	0.3859	0.3897	0.3936	0.3974	0.4013	0.4052	0.4090	0.4129	0.4168	0.4207
−0.1	0.4247	0.4286	0.4325	0.4364	0.4404	0.4443	0.4483	0.4522	0.4562	0.4602
−0.0	0.4641	0.4681	0.4721	0.4761	0.4801	0.4840	0.4880	0.4920	0.4960	0.5000

* Adapted from E. L. Grant, *Statistical Quality Control* (2nd Ed.). New York: McGraw-Hill Book Company, Inc. (1952), Table A, pp. 510–511. Reproduced by permission of the publisher.

Table A (*Continued*)

z	0.00	0.01	0.02	0.03	0.04	0.05	0.06	0.07	0.08	0.09
+0.0	0.5000	0.5040	0.5080	0.5120	0.5160	0.5199	0.5239	0.5279	0.5319	0.5359
+0.1	0.5398	0.5438	0.5478	0.5517	0.5557	0.5596	0.5636	0.5675	0.5714	0.5753
+0.2	0.5793	0.5832	0.5871	0.5910	0.5948	0.5987	0.6026	0.6064	0.6103	0.6141
+0.3	0.6179	0.6217	0.6255	0.6293	0.6331	0.6368	0.6406	0.6443	0.6480	0.6517
+0.4	0.6554	0.6591	0.6628	0.6664	0.6700	0.6736	0.6772	0.6808	0.6844	0.6879
+0.5	0.6915	0.6950	0.6985	0.7019	0.7054	0.7088	0.7123	0.7157	0.7190	0.7224
+0.6	0.7257	0.7291	0.7324	0.7357	0.7389	0.7422	0.7454	0.7486	0.7517	0.7549
+0.7	0.7580	0.7611	0.7642	0.7673	0.7704	0.7734	0.7764	0.7794	0.7823	0.7852
+0.8	0.7881	0.7910	0.7939	0.7967	0.7995	0.8023	0.8051	0.8079	0.8106	0.8133
+0.9	0.8159	0.8186	0.8212	0.8238	0.8264	0.8289	0.8315	0.8340	0.8365	0.8389
+1.0	0.8413	0.8438	0.8461	0.8485	0.8508	0.8531	0.8554	0.8577	0.8599	0.8621
+1.1	0.8643	0.8665	0.8686	0.8708	0.8729	0.8749	0.8770	0.8790	0.8810	0.8830
+1.2	0.8849	0.8869	0.8888	0.8907	0.8925	0.8944	0.8962	0.8980	0.8997	0.9015
+1.3	0.9032	0.9049	0.9066	0.9082	0.9099	0.9115	0.9131	0.9147	0.9162	0.9177
+1.4	0.9192	0.9207	0.9222	0.9236	0.9251	0.9265	0.9279	0.9292	0.9306	0.9319
+1.5	0.9332	0.9345	0.9357	0.9370	0.9382	0.9394	0.9406	0.9418	0.9429	0.9441
+1.6	0.9452	0.9463	0.9474	0.9484	0.9495	0.9505	0.9515	0.9525	0.9535	0.9545
+1.7	0.9554	0.9564	0.9573	0.9582	0.9591	0.9599	0.9608	0.9616	0.9625	0.9633
+1.8	0.9641	0.9649	0.9656	0.9664	0.9671	0.9678	0.9686	0.9693	0.9699	0.9706
+1.9	0.9713	0.9719	0.9726	0.9732	0.9738	0.9744	0.9750	0.9756	0.9761	0.9767
+2.0	0.9773	0.9778	0.9783	0.9788	0.9793	0.9798	0.9803	0.9808	0.9812	0.9817
+2.1	0.9821	0.9826	0.9830	0.9834	0.9838	0.9842	0.9846	0.9850	0.9854	0.9857
+2.2	0.9861	0.9864	0.9868	0.9871	0.9875	0.9878	0.9881	0.9884	0.9887	0.9890
+2.3	0.9893	0.9896	0.9898	0.9901	0.9904	0.9906	0.9909	0.9911	0.9913	0.9916
+2.4	0.9918	0.9920	0.9922	0.9925	0.9927	0.9929	0.9931	0.9932	0.9934	0.9936
+2.5	0.9938	0.9940	0.9941	0.9943	0.9945	0.9946	0.9948	0.9949	0.9951	0.9952
+2.6	0.9953	0.9955	0.9956	0.9957	0.9959	0.9960	0.9961	0.9962	0.9963	0.9964
+2.7	0.9965	0.9966	0.9967	0.9968	0.9969	0.9970	0.9971	0.9972	0.9973	0.9974
+2.8	0.9974	0.9975	0.9976	0.9977	0.9977	0.9978	0.9979	0.9979	0.9980	0.9981
+2.9	0.9981	0.9982	0.9982	0.9983	0.9984	0.9984	0.9985	0.9985	0.9986	0.9986
+3.0	0.99865	0.99869	0.99874	0.99878	0.99882	0.99886	0.99889	0.99893	0.99896	0.99900
+3.1	0.99903	0.99906	0.99910	0.99913	0.99915	0.99918	0.99921	0.99924	0.99926	0.99929
+3.2	0.99931	0.99934	0.99936	0.99938	0.99940	0.99942	0.99944	0.99946	0.99948	0.99950
+3.3	0.99952	0.99953	0.99955	0.99957	0.99958	0.99960	0.99961	0.99962	0.99964	0.99965
+3.4	0.99966	0.99967	0.99969	0.99970	0.99971	0.99972	0.99973	0.99974	0.99975	0.99976
+3.5	0.99977	0.99978	0.99978	0.99979	0.99980	0.99981	0.99981	0.99982	0.99983	0.99983

Table B Student's *t* Distribution

df	Percentile point						
	70	80	90	95	97.5	99	99.5
1	.73	1.38	3.08	6.31	12.71	31.82	63.66
2	.62	1.06	1.89	2.92	4.30	6.96	9.92
3	.58	.98	1.64	2.35	3.18	4.54	5.84
4	.57	.94	1.53	2.13	2.78	3.75	4.60
5	.56	.92	1.48	2.01	2.57	3.36	4.03
6	.55	.91	1.44	1.94	2.45	3.14	3.71
7	.55	.90	1.42	1.90	2.36	3.00	3.50
8	.55	.89	1.40	1.86	2.31	2.90	3.36
9	.54	.88	1.38	1.83	2.26	2.82	3.25
10	.54	.88	1.37	1.81	2.23	2.76	3.17
11	.54	.88	1.36	1.80	2.20	2.72	3.11
12	.54	.87	1.36	1.78	2.18	2.68	3.06
13	.54	.87	1.35	1.77	2.16	2.65	3.01
14	.54	.87	1.34	1.76	2.14	2.62	2.98
15	.54	.87	1.34	1.75	2.13	2.60	2.95
16	.54	.86	1.34	1.75	2.12	2.58	2.92
17	.53	.86	1.33	1.74	2.11	2.57	2.90
18	.53	.86	1.33	1.73	2.10	2.55	2.88
19	.53	.86	1.33	1.73	2.09	2.54	2.86
20	.53	.86	1.32	1.72	2.09	2.53	2.84
21	.53	.86	1.32	1.72	2.08	2.52	2.83
22	.53	.86	1.32	1.72	2.07	2.51	2.82
23	.53	.86	1.32	1.71	2.07	2.50	2.81
24	.53	.86	1.32	1.71	2.06	2.49	2.80
25	.53	.86	1.32	1.71	2.06	2.48	2.79
26	.53	.86	1.32	1.71	2.06	2.48	2.78
27	.53	.86	1.31	1.70	2.05	2.47	2.77
28	.53	.86	1.31	1.70	2.05	2.47	2.76
29	.53	.85	1.31	1.70	2.04	2.46	2.76
30	.53	.85	1.31	1.70	2.04	2.46	2.75
40	.53	.85	1.30	1.68	2.02	2.42	2.70
50	.53	.85	1.30	1.67	2.01	2.40	2.68
60	.53	.85	1.30	1.67	2.00	2.39	2.66
80	.53	.85	1.29	1.66	1.99	2.37	2.64
100	.53	.84	1.29	1.66	1.98	2.36	2.63
200	.52	.84	1.29	1.65	1.97	2.34	2.60
500	.52	.84	1.28	1.65	1.96	2.33	2.59
∞	.52	.84	1.28	1.64	1.96	2.33	2.58

Table C Chi Square*

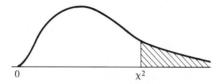

0 χ^2

ν	Probability										
	0.99	0.98	0.95	0.90	0.80	0.20	0.10	0.05	0.02	0.01	0.001
1	0.0³157	0.0³628	0.00393	0.0158	0.0642	1.642	2.706	3.841	5.412	6.635	10.827
2	0.0201	0.0404	0.103	0.211	0.446	3.219	4.605	5.991	7.824	9.210	13.815
3	0.115	0.185	0.352	0.584	1.005	4.642	6.251	7.815	9.837	11.341	16.268
4	0.297	0.429	0.711	1.064	1.649	5.989	7.779	9.488	11.668	13.277	18.465
5	0.554	0.752	1.145	1.610	2.343	7.289	9.236	11.070	13.388	15.086	20.517
6	0.872	1.134	1.635	2.204	3.070	8.558	10.645	12.592	15.033	16.812	22.457
7	1.239	1.564	2.167	2.833	3.822	9.803	12.017	14.067	16.622	18.475	24.322
8	1.646	2.032	2.733	3.490	4.594	11.030	13.362	15.507	18.168	20.090	26.125
9	2.088	2.532	3.325	4.168	5.380	12.242	14.684	16.919	19.679	21.666	27.877
10	2.558	3.059	3.940	4.865	6.179	13.442	15.987	18.307	21.161	23.209	29.588
11	3.053	3.609	4.575	5.578	6.989	14.631	17.275	19.675	22.618	24.725	31.264
12	3.571	4.178	5.226	6.304	7.807	15.812	18.549	21.026	24.054	26.217	32.909
13	4.107	4.765	5.892	7.042	8.634	16.985	19.812	22.362	25.472	27.688	34.528
14	4.660	5.368	6.571	7.790	9.467	18.151	21.064	23.685	26.873	29.141	36.123
15	5.229	5.985	7.261	8.547	10.307	19.311	22.307	24.996	28.259	30.578	37.697
16	5.812	6.614	7.962	9.312	11.152	20.465	23.542	26.296	29.633	32.000	39.252
17	6.408	7.255	8.672	10.085	12.002	21.615	24.769	27.587	30.995	33.409	40.790
18	7.015	7.906	9.390	10.865	12.857	22.760	25.989	28.869	32.346	34.805	42.312
19	7.633	8.567	10.117	11.651	13.716	23.900	27.204	30.144	33.687	36.191	43.820
20	8.260	9.237	10.851	12.443	14.578	25.038	28.412	31.410	35.020	37.566	45.315
21	8.897	9.915	11.591	13.240	15.445	26.171	29.615	32.671	36.343	38.932	46.797
22	9.542	10.600	12.338	14.041	16.314	27.301	30.813	33.924	37.659	40.289	48.268
23	10.196	11.293	13.091	14.848	17.187	28.429	32.007	35.172	38.968	41.638	49.728
24	10.856	11.992	13.848	15.659	18.062	29.553	33.196	36.415	40.270	42.980	51.179
25	11.524	12.697	14.611	16.473	18.940	30.675	34.382	37.652	41.566	44.314	52.620
26	12.198	13.409	15.379	17.292	19.820	31.795	35.563	38.885	42.856	45.642	54.052
27	12.879	14.125	16.151	18.114	20.703	32.912	36.741	40.113	44.140	46.963	55.476
28	13.565	14.847	16.928	18.939	21.588	34.027	37.916	41.337	45.419	48.278	56.893
29	14.256	15.574	17.708	19.768	22.475	35.139	39.087	42.557	46.693	49.588	58.302
30	14.953	16.306	18.493	20.599	23.364	36.250	40.256	43.773	47.962	50.892	59.703

* Abridged from R. A. Fisher and F. Yates, *Statistical Tables for Biological, Agricultural and Medical Research.* Edinburgh and London: Oliver & Boyd, Ltd (1953), Table IV. Reproduced by permission of the authors and publishers. For larger values of v, the expression $\sqrt{2\chi^2} - \sqrt{2v - 1}$ may be used as a normal deviate with unit variance, remembering that the probability for χ^2 corresponds with that of a single tail of a normal curve.

Table D *F* Distribution*

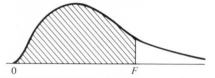

df for denom.	$1 - \alpha$	df for numerator											
		1	2	3	4	5	6	7	8	9	10	11	12
1	.75	5.83	7.50	8.20	8.58	8.82	8.98	9.10	9.19	9.26	9.32	9.36	9.41
	.90	39.9	49.5	53.6	55.8	57.2	58.2	58.9	59.4	59.9	60.2	60.5	60.7
	.95	161	200	216	225	230	234	237	239	241	242	243	244
2	.75	2.57	3.00	3.15	3.23	3.28	3.31	3.34	3.35	3.37	3.38	3.39	3.39
	.90	8.53	9.00	9.16	9.24	9.29	9.33	9.35	9.37	9.38	9.39	9.40	9.41
	.95	18.5	19.0	19.2	19.2	19.3	19.3	19.4	19.4	19.4	19.4	19.4	19.4
	.99	98.5	99.0	99.2	99.2	99.3	99.3	99.4	99.4	99.4	99.4	99.4	99.4
3	.75	2.02	2.28	2.36	2.39	2.41	2.42	2.43	2.44	2.44	2.44	2.45	2.45
	.90	5.54	5.46	5.39	5.34	5.31	5.28	5.27	5.25	5.24	5.23	5.22	5.22
	.95	10.1	9.55	9.28	9.12	9.10	8.94	8.89	8.85	8.81	8.79	8.76	8.74
	.99	34.1	30.8	29.5	28.7	28.2	27.9	27.7	27.5	27.3	27.2	27.1	27.1
4	.75	1.81	2.00	2.05	2.06	2.07	2.08	2.08	2.08	2.08	2.08	2.08	2.08
	.90	4.54	4.32	4.19	4.11	4.05	4.01	3.98	3.95	3.94	3.92	3.91	3.90
	.95	7.71	6.94	6.59	6.39	6.26	6.16	6.09	6.04	6.00	5.96	5.94	5.91
	.99	21.2	18.0	16.7	16.0	15.5	15.2	15.0	14.8	14.7	14.5	14.4	14.4
5	.75	1.69	1.85	1.88	1.89	1.89	1.89	1.89	1.89	1.89	1.89	1.89	1.89
	.90	4.06	3.78	3.62	3.52	3.45	3.40	3.37	3.34	3.32	3.30	3.28	3.27
	.95	6.61	5.79	5.41	5.19	5.05	4.95	4.88	4.82	4.77	4.74	4.71	4.68
	.99	16.3	13.3	12.1	11.4	11.0	10.7	10.5	10.3	10.2	10.1	9.96	9.89
6	.75	1.62	1.76	1.78	1.79	1.79	1.78	1.78	1.77	1.77	1.77	1.77	1.77
	.90	3.78	3.46	3.29	3.18	3.11	3.05	3.01	2.98	2.96	2.94	2.92	2.90
	.95	5.99	5.14	4.76	4.53	4.39	4.28	4.21	4.15	4.10	4.06	4.03	4.00
	.99	13.7	10.9	9.78	9.15	8.75	8.47	8.26	8.10	7.98	7.87	7.79	7.72
7	.75	1.57	1.70	1.72	1.72	1.71	1.71	1.70	1.70	1.69	1.69	1.69	1.68
	.90	3.59	3.26	3.07	2.96	2.88	2.83	2.78	2.75	2.72	2.70	2.68	2.67
	.95	5.59	4.74	4.35	4.12	3.97	3.87	3.79	3.73	3.68	3.64	3.60	3.57
	.99	12.2	9.55	8.45	7.85	7.46	7.19	6.99	6.84	6.72	6.62	6.54	6.47
8	.75	1.54	1.66	1.67	1.66	1.66	1.65	1.64	1.64	1.64	1.63	1.63	1.62
	.90	3.46	3.11	2.92	2.81	2.73	2.67	2.62	2.59	2.56	2.54	2.52	2.50
	95	5.32	4.46	4.07	3.84	3.69	3.58	3.50	3.44	3.39	3.35	3.31	3.28
	.99	11.3	8.65	7.59	7.01	6.63	6.37	6.18	6.03	5.91	5.81	5.73	5.67
9	.75	1.51	1.62	1.63	1.63	1.62	1.61	1.60	1.60	1.59	1.59	1.58	1.58
	.90	3.36	3.01	2.81	2.69	2.61	2.55	2.51	2.47	2.44	2.42	2.40	2.38
	.95	5.12	4.26	3.86	3.63	3.48	3.37	3.29	3.23	3.18	3.14	3.10	3.07
	.99	10.6	8.02	6.99	6.42	6.06	5.80	5.61	5.47	5.35	5.26	5.18	5.11
10	.75	1.49	1.60	1.60	1.59	1.59	1.58	1.57	1.56	1.56	1.55	1.55	1.54
	.90	3.28	2.92	2.73	2.61	2.52	2.46	2.41	2.38	2.35	2.32	2.30	2.28
	.95	4.96	4.10	3.71	3.48	3.33	3.22	3.14	3.07	3.02	2.98	2.94	2.91
	.99	10.0	7.56	6.55	5.99	5.64	5.39	5.20	5.06	4.94	4.85	4.77	4.71
11	.75	1.47	1.58	1.58	1.57	1.56	1.55	1.54	1.53	1.53	1.52	1.52	1.51
	.90	3.23	2.86	2.66	2.54	2.45	2.39	2.34	2.30	2.27	2.25	2.23	2.21
	.95	4.84	3.98	3.59	3.36	3.20	3.09	3.01	2.95	2.90	2.85	2.82	2.79
	.99	9.65	7.21	6.22	5.67	5.32	5.07	4.89	4.74	4.63	4.54	4.46	4.40
12	.75	1.46	1.56	1.56	1.55	1.54	1.53	1.52	1.51	1.51	1.50	1.50	1.49
	.90	3.18	2.81	2.61	2.48	2.39	2.33	2.28	2.24	2.21	2.19	2.17	2.15
	.95	4.75	3.89	3.49	3.26	3.11	3.00	2.91	2.85	2.80	2.75	2.72	2.69
	.99	9.33	6.93	5.95	5.41	5.06	4.82	4.64	4.50	4.39	4.30	4.22	4.16

Table D (*Continued*)

15	20	24	30	40	50	60	100	120	200	500	∞	1 − α	df for denom.
					df for numerator								
9.49	9.58	9.63	9.67	9.71	9.74	9.76	9.78	9.80	9.82	9.84	9.85	.75	
61.2	61.7	62.0	62.3	62.5	62.7	62.8	63.0	63.1	63.2	63.3	63.3	.90	1
246	248	249	250	251	252	252	253	253	254	254	254	.95	
3.41	3.43	3.43	3.44	3.45	3.45	3.46	3.47	3.47	3.48	3.48	3.48	.75	
9.42	9.44	9.45	9.46	9.47	9.47	9.47	9.48	9.48	9.49	9.49	9.49	.90	2
19.4	19.4	19.5	19.5	19.5	19.5	19.5	19.5	19.5	19.5	19.5	19.5	.95	
99.4	99.4	99.5	99.5	99.5	99.5	99.5	99.5	99.5	99.5	99.5	99.5	.99	
2.46	2.46	2.46	2.47	2.47	2.47	2.47	2.47	2.47	2.47	2.47	2.47	.75	
5.20	5.18	5.18	5.17	5.16	5.15	5.15	5.14	5.14	5.14	5.14	5.13	.90	3
8.70	8.66	8.64	8.62	8.59	8.58	8.57	8.55	8.55	8.54	8.53	8.53	.95	
26.9	26.7	26.6	26.5	26.4	26.4	26.3	26.2	26.2	26.2	26.1	26.1	.99	
2.08	2.08	2.08	2.08	2.08	2.08	2.08	2.08	2.08	2.08	2.08	2.08	.75	
3.87	3.84	3.83	3.82	3.80	3.80	3.79	3.78	3.78	3.77	3.76	3.76	.90	
5.86	5.80	5.77	5.75	5.72	5.70	5.69	5.66	5.66	5.65	5.64	5.63	.95	4
14.2	14.0	13.9	13.8	13.7	13.7	13.7	13.6	13.6	13.5	13.5	13.5	.99	
1.89	1.88	1.88	1.88	1.88	1.88	1.87	1.87	1.87	1.87	1.87	1.87	.75	
3.24	3.21	3.19	3.17	3.16	3.15	3.14	3.13	3.12	3.12	3.11	3.10	.90	5
4.62	4.56	4.53	4.50	4.46	4.44	4.43	4.41	4.40	4.39	4.37	4.36	.95	
9.72	9.55	9.47	9.38	9.29	9.24	9.20	9.13	9.11	9.08	9.04	9.02	.99	
1.76	1.76	1.75	1.75	1.75	1.75	1.74	1.74	1.74	1.74	1.74	1.74	.75	
2.87	2.84	2.82	2.80	2.78	2.77	2.76	2.75	2.74	2.73	2.73	2.72	.90	6
3.94	3.87	3.84	3.81	3.77	3.75	3.74	3.71	3.70	3.69	3.68	3.67	.95	
7.56	7.40	7.31	7.23	7.14	7.09	7.06	6.99	6.97	6.93	6.90	6.88	.99	
1.68	1.67	1.67	1.66	1.66	1.66	1.65	1.65	1.65	1.65	1.65	1.65	.75	
2.63	2.59	2.58	2.56	2.54	2.52	2.51	2.50	2.49	2.48	2.48	2.47	.90	7
3.51	3.44	3.41	3.38	3.34	3.32	3.30	3.27	3.27	3.25	3.24	3.23	.95	
6.31	6.16	6.07	5.99	5.91	5.86	5.82	5.75	5.74	5.70	5.67	5.65	.99	
1.62	1.61	1.60	1.60	1.59	1.59	1.59	1.58	1.58	1.58	1.58	1.58	.75	
2.46	2.42	2.40	2.38	2.36	2.35	2.34	2.32	2.32	2.31	2.30	2.29	.90	8
3.22	3.15	3.12	3.08	3.04	3.02	3.01	2.97	2.97	2.95	2.94	2.93	.95	
5.52	5.36	5.28	5.20	5.12	5.07	5.03	4.96	4.95	4.91	4.88	4.86	.99	
1.57	1.56	1.56	1.55	1.55	1.54	1.54	1.53	1.53	1.53	1.53	1.53	.75	
2.34	2.30	2.28	2.25	2.23	2.22	2.21	2.19	2.18	2.17	2.17	2.16	.90	9
3.01	2.94	2.90	2.86	2.83	2.80	2.79	2.76	2.75	2.73	2.72	2.71	.95	
4.96	4.81	4.73	4.65	4.57	4.52	4.48	4.42	4.40	4.36	4.33	4.31	.99	
1.53	1.52	1.52	1.51	1.51	1.50	1.50	1.49	1.49	1.49	1.48	1.48	.75	
2.24	2.20	2.18	2.16	2.13	2.12	2.11	2.09	2.08	2.07	2.06	2.06	.90	10
2.85	2.77	2.74	2.70	2.66	2.64	2.62	2.59	2.58	2.56	2.55	2.54	.95	
4.56	4.41	4.33	4.25	4.17	4.12	4.08	4.01	4.00	3.96	3.93	3.91	.99	
1.50	1.49	1.49	1.48	1.47	1.47	1.47	1.46	1.46	1.46	1.45	1.45	.75	
2.17	2.12	2.10	2.08	2.05	2.04	2.03	2.00	2.00	1.99	1.98	1.97	.90	11
2.72	2.65	2.61	2.57	2.53	2.51	2.49	2.46	2.45	2.43	2.42	2.40	.95	
4.25	4.10	4.02	3.94	3.86	3.81	3.78	3.71	3.69	3.66	3.62	3.60	.99	
1.48	1.47	1.46	1.45	1.45	1.44	1.44	1.43	1.43	1.43	1.42	1.42	.75	
2.10	2.06	2.04	2.01	1.99	1.97	1.96	1.94	1.93	1.92	1.91	1.90	.90	12
2.62	2.54	2.51	2.47	2.43	2.40	2.38	2.35	2.34	2.32	2.31	2.30	.95	
4.01	3.86	3.78	3.70	3.62	3.57	3.54	3.47	3.45	3.41	3.38	3.36	.99	

Table D (*Continued*)

df for denom.	$1-\alpha$	df for numerator											
		1	2	3	4	5	6	7	8	9	10	11	12
13	.75	1.45	1.54	1.54	1.53	1.52	1.51	1.50	1.49	1.49	1.48	1.47	1.47
	.90	3.14	2.76	2.56	2.43	2.35	2.28	2.23	2.20	2.16	2.14	2.12	2.10
	.95	4.67	3.81	3.41	3.18	3.03	2.92	2.83	2.77	2.71	2.67	2.63	2.60
	.99	9.07	6.70	5.74	5.21	4.86	4.62	4.44	4.30	4.19	4.10	4.02	3.96
14	.75	1.44	1.53	1.53	1.52	1.51	1.50	1.48	1.48	1.47	1.46	1.46	1.45
	.90	3.10	2.73	2.52	2.39	2.31	2.24	2.19	2.15	2.12	2.10	2.08	2.05
	.95	4.60	3.74	3.34	3.11	2.96	2.85	2.76	2.70	2.65	2.60	2.57	2.53
	.99	8.86	6.51	5.56	5.04	4.69	4.46	4.28	4.14	4.03	3.94	3.86	3.80
15	.75	1.43	1.52	1.52	1.51	1.49	1.48	1.47	1.46	1.46	1.45	1.44	1.44
	.90	3.07	2.70	2.49	2.36	2.27	2.21	2.16	2.12	2.09	2.06	2.04	2.02
	.95	4.54	3.68	3.29	3.06	2.90	2.79	2.71	2.64	2.59	2.54	2.51	2.48
	.99	8.68	6.36	5.42	4.89	4.56	4.32	4.14	4.00	3.89	3.80	3.73	3.67
16	.75	1.42	1.51	1.51	1.50	1.48	1.48	1.47	1.46	1.45	1.45	1.44	1.44
	.90	3.05	2.67	2.46	2.33	2.24	2.18	2.13	2.09	2.06	2.03	2.01	1.99
	.95	4.49	3.63	3.24	3.01	2.85	2.74	2.66	2.59	2.54	2.49	2.46	2.42
	.99	8.53	6.23	5.29	4.77	4.44	4.20	4.03	3.89	3.78	3.69	3.62	3.55
17	.75	1.42	1.51	1.50	1.49	1.47	1.46	1.45	1.44	1.43	1.43	1.42	1.41
	.90	3.03	2.64	2.44	2.31	2.22	2.15	2.10	2.06	2.03	2.00	1.98	1.96
	.95	4.45	3.59	3.20	2.96	2.81	2.70	2.61	2.55	2.49	2.45	2.41	2.38
	.99	8.40	6.11	5.18	4.67	4.34	4.10	3.93	3.79	3.68	3.59	3.52	3.46
18	.75	1.41	1.50	1.49	1.48	1.46	1.45	1.44	1.43	1.42	1.42	1.41	1.40
	.90	3.01	2.62	2.42	2.29	2.20	2.13	2.08	2.04	2.00	1.98	1.96	1.93
	.95	4.41	3.55	3.16	2.93	2.77	2.66	2.58	2.51	2.46	2.41	2.37	2.34
	.99	8.29	6.01	5.09	4.58	4.25	4.01	3.84	3.71	3.60	3.51	3.43	3.37
19	.75	1.41	1.49	1.49	1.47	1.46	1.44	1.43	1.42	1.41	1.41	1.40	1.40
	.90	2.99	2.61	2.40	2.27	2.18	2.11	2.06	2.02	1.98	1.96	1.94	1.91
	.95	4.38	3.52	3.13	2.90	2.74	2.63	2.54	2.48	2.42	2.38	2.34	2.31
	.99	8.18	5.93	5.01	4.50	4.17	3.94	3.77	3.63	3.52	3.43	3.36	3.30
20	.75	1.40	1.49	1.48	1.46	1.45	1.44	1.42	1.42	1.41	1.40	1.39	1.39
	.90	2.97	2.59	2.38	2.25	2.16	2.09	2.04	2.00	1.96	1.94	1.92	1.89
	.95	4.35	3.49	3.10	2.87	2.71	2.60	2.51	2.45	2.39	2.35	2.31	2.28
	.99	8.10	5.85	4.94	4.43	4.10	3.87	3.70	3.56	3.46	3.37	3.29	3.23
22	.75	1.40	1.48	1.47	1.45	1.44	1.42	1.41	1.40	1.39	1.39	1.38	1.37
	.90	2.95	2.56	2.35	2.22	2.13	2.06	2.01	1.97	1.93	1.90	1.88	1.86
	.95	4.30	3.44	3.05	2.82	2.66	2.55	2.46	2.40	2.34	2.30	2.26	2.23
	.99	7.95	5.72	4.82	4.31	3.99	3.76	3.59	3.45	3.35	3.26	3.18	3.12
24	.75	1.39	1.47	1.46	1.44	1.43	1.41	1.40	1.39	1.38	1.38	1.37	1.36
	.90	2.93	2.54	2.33	2.19	2.10	2.04	1.98	1.94	1.91	1.88	1.85	1.83
	.95	4.26	3.40	3.01	2.78	2.62	2.51	2.42	2.36	2.30	2.25	2.21	2.18
	.99	7.82	5.61	4.72	4.22	3.90	3.67	3.50	3.36	3.26	3.17	3.09	3.03
26	.75	1.38	1.46	1.45	1.44	1.42	1.41	1.40	1.39	1.37	1.37	1.36	1.35
	.90	2.91	2.52	2.31	2.17	2.08	2.01	1.96	1.92	1.88	1.86	1.84	1.81
	.95	4.23	3.37	2.98	2.74	2.59	2.47	2.39	2.32	2.27	2.22	2.18	2.15
	.99	7.72	5.53	4.64	4.14	3.82	3.59	3.42	3.29	3.18	3.09	3.02	2.96
28	.75	1.38	1.46	1.45	1.43	1.41	1.40	1.39	1.38	1.37	1.36	1.35	1.34
	.90	2.89	2.50	2.29	2.16	2.06	2.00	1.94	1.90	1.87	1.84	1.81	1.79
	.95	4.20	3.34	2.95	2.71	2.56	2.45	2.36	2.29	2.24	2.19	2.15	2.12
	.99	7.64	5.45	4.57	4.07	3.75	3.53	3.36	3.23	3.12	3.03	2.96	2.90

Table D (*Continued*)

df for numerator													$1-\alpha$	df for denom.
15	20	24	30	40	50	60	100	120	200	500	∞			
1.46	1.45	1.44	1.43	1.42	1.42	1.42	1.41	1.41	1.40	1.40	1.40	.75	
2.05	2.01	1.98	1.96	1.93	1.92	1.90	1.88	1.88	1.86	1.85	1.85	.90	13
2.53	2.46	2.42	2.38	2.34	2.31	2.30	2.26	2.25	2.23	2.22	2.21	.95	
3.82	3.66	3.59	3.51	3.43	3.38	3.34	3.27	3.25	3.22	3.19	3.17	.99	
1.44	1.43	1.42	1.41	1.41	1.40	1.40	1.39	1.39	1.39	1.38	1.38	.75	
2.01	1.96	1.94	1.91	1.89	1.87	1.86	1.83	1.83	1.82	1.80	1.80	.90	
2.46	2.39	2.35	2.31	2.27	2.24	2.22	2.19	2.18	2.16	2.14	2.13	.95	14
3.66	3.51	3.43	3.35	3.27	3.22	3.18	3.11	3.09	3.06	3.03	3.00	.99	
1.43	1.41	1.41	1.40	1.39	1.39	1.38	1.38	1.37	1.37	1.36	1.36	.75	
1.97	1.92	1.90	1.87	1.85	1.83	1.82	1.79	1.79	1.77	1.76	1.76	.90	15
2.40	2.33	2.29	2.25	2.20	2.18	2.16	2.12	2.11	2.10	2.08	2.07	.95	
3.52	3.37	3.29	3.21	3.13	3.08	3.05	2.98	2.96	2.92	2.89	2.87	.99	
1.41	1.40	1.39	1.38	1.37	1.37	1.36	1.36	1.35	1.35	1.34	1.34	.75	
1.94	1.89	1.87	1.84	1.81	1.79	1.78	1.76	1.75	1.74	1.73	1.72	.90	16
2.35	2.28	2.24	2.19	2.15	2.12	2.11	2.07	2.06	2.04	2.02	2.01	.95	
3.41	3.26	3.18	3.10	3.02	2.97	2.93	2.86	2.84	2.81	2.78	2.75	.99	
1.40	1.39	1.38	1.37	1.36	1.35	1.35	1.34	1.34	1.34	1.33	1.33	.75	
1.91	1.86	1.84	1.81	1.78	1.76	1.75	1.73	1.72	1.71	1.69	1.69	.90	17
2.31	2.23	2.19	2.15	2.10	2.08	2.06	2.02	2.01	1.99	1.97	1.96	.95	
3.31	3.16	3.08	3.00	2.92	2.87	2.83	2.76	2.75	2.71	2.68	2.65	.99	
1.39	1.38	1.37	1.36	1.35	1.34	1.34	1.33	1.33	1.32	1.32	1.32	.75	
1.89	1.84	1.81	1.78	1.75	1.74	1.72	1.70	1.69	1.68	1.67	1.66	.90	18
2.27	2.19	2.15	2.11	2.06	2.04	2.02	1.98	1.97	1.95	1.93	1.92	.95	
3.23	3.08	3.00	2.92	2.84	2.78	2.75	2.68	2.66	2.62	2.59	2.57	.99	
1.38	1.37	1.36	1.35	1.34	1.33	1.33	1.32	1.32	1.31	1.31	1.30	.75	
1.86	1.81	1.79	1.76	1.73	1.71	1.70	1.67	1.67	1.65	1.64	1.63	.90	19
2.23	2.16	2.11	2.07	2.03	2.00	1.98	1.94	1.93	1.91	1.89	1.88	.95	
3.15	3.00	2.92	2.84	2.76	2.71	2.67	2.60	2.58	2.55	2.51	2.49	.99	
1.37	1.36	1.35	1.34	1.33	1.33	1.32	1.31	1.31	1.30	1.30	1.29	.75	
1.84	1.79	1.77	1.74	1.71	1.69	1.68	1.65	1.64	1.63	1.62	1.61	.90	20
2.20	2.12	2.08	2.04	1.99	1.97	1.95	1.91	1.90	1.88	1.86	1.84	.95	
3.09	2.94	2.86	2.78	2.69	2.64	2.61	2.54	2.52	2.48	2.44	2.42	.99	
1.36	1.34	1.33	1.32	1.31	1.31	1.30	1.30	1.30	1.29	1.29	1.28	.75	
1.81	1.76	1.73	1.70	1.67	1.65	1.64	1.61	1.60	1.59	1.58	1.57	.90	22
2.15	2.07	2.03	1.98	1.94	1.91	1.89	1.85	1.84	1.82	1.80	1.78	.95	
2.98	2.83	2.75	2.67	2.58	2.53	2.50	2.42	2.40	2.36	2.33	2.31	.99	
1.35	1.33	1.32	1.31	1.30	1.29	1.29	1.28	1.28	1.27	1.27	1.26	.75	
1.78	1.73	1.70	1.67	1.64	1.62	1.61	1.58	1.57	1.56	1.54	1.53	.90	24
2.11	2.03	1.98	1.94	1.89	1.86	1.84	1.80	1.79	1.77	1.75	1.73	.95	
2.89	2.74	2.66	2.58	2.49	2.44	2.40	2.33	2.31	2.27	2.24	2.21	.99	
1.34	1.32	1.31	1.30	1.29	1.28	1.28	1.26	1.26	1.26	1.25	1.25	.75	
1.76	1.71	1.68	1.65	1.61	1.59	1.58	1.55	1.54	1.53	1.51	1.50	.90	26
2.07	1.99	1.95	1.90	1.85	1.82	1.80	1.76	1.75	1.73	1.71	1.69	.95	
2.81	2.66	2.58	2.50	2.42	2.36	2.33	2.25	2.23	2.19	2.16	2.13	.99	
1.33	1.31	1.30	1.29	1.28	1.27	1.27	1.26	1.25	1.25	1.24	1.24	.75	
1.74	1.69	1.66	1.63	1.59	1.57	1.56	1.53	1.52	1.50	1.49	1.48	.90	28
2.04	1.96	1.91	1.87	1.82	1.79	1.77	1.73	1.71	1.69	1.67	1.65	.95	
2.75	2.60	2.52	2.44	2.35	2.30	2.26	2.19	2.17	2.13	2.09	2.06	.99	

Table D (*Continued*)

df for denom.	1 − α	df for numerator											
		1	2	3	4	5	6	7	8	9	10	11	12
30	.75	1.38	1.45	1.44	1.42	1.41	1.39	1.38	1.37	1.36	1.35	1.35	1.34
	.90	2.88	2.49	2.28	2.14	2.05	1.98	1.93	1.88	1.85	1.82	1.79	1.77
	.95	4.17	3.32	2.92	2.69	2.53	2.42	2.33	2.27	2.21	2.16	2.13	2.09
	.99	7.56	5.39	4.51	4.02	3.70	3.47	3.30	3.17	3.07	2.98	2.91	2.84
40	.75	1.36	1.44	1.42	1.40	1.39	1.37	1.36	1.35	1.34	1.33	1.32	1.31
	.90	2.84	2.44	2.23	2.09	2.00	1.93	1.87	1.83	1.79	1.76	1.73	1.71
	.95	4.08	3.23	2.84	2.61	2.45	2.34	2.25	2.18	2.12	2.08	2.04	2.00
	.99	7.31	5.18	4.31	3.83	3.51	3.29	3.12	2.99	2.89	2.80	2.73	2.66
60	.75	1.35	1.42	1.41	1.38	1.37	1.35	1.33	1.32	1.31	1.30	1.29	1.29
	.90	2.79	2.39	2.18	2.04	1.95	1.87	1.82	1.77	1.74	1.71	1.68	1.66
	.95	4.00	3.15	2.76	2.53	2.37	2.25	2.17	2.10	2.04	1.99	1.95	1.92
	.99	7.08	4.98	4.13	3.65	3.34	3.12	2.95	2.82	2.72	2.63	2.56	2.50
120	.75	1.34	1.40	1.39	1.37	1.35	1.33	1.31	1.30	1.29	1.28	1.27	1.26
	.90	2.75	2.35	2.13	1.99	1.90	1.82	1.77	1.72	1.68	1.65	1.62	1.60
	.95	3.92	3.07	2.68	2.45	2.29	2.17	2.09	2.02	1.96	1.91	1.87	1.83
	.99	6.85	4.79	3.95	3.48	3.17	2.96	2.79	2.66	2.56	2.47	2.40	2.34
200	.75	1.33	1.39	1.38	1.36	1.34	1.32	1.31	1.29	1.28	1.27	1.26	1.25
	.90	2.73	2.33	2.11	1.97	1.88	1.80	1.75	1.70	1.66	1.63	1.60	1.57
	.95	3.89	3.04	2.65	2.42	2.26	2.14	2.06	1.98	1.93	1.88	1.84	1.80
	.99	6.76	4.71	3.88	3.41	3.11	2.89	2.73	2.60	2.50	2.41	2.34	2.27
∞	.75	1.32	1.39	1.37	1.35	1.33	1.31	1.29	1.28	1.27	1.25	1.24	1.24
	.90	2.71	2.30	2.08	1.94	1.85	1.77	1.72	1.67	1.63	1.60	1.57	1.55
	.95	3.84	3.00	2.60	2.37	2.21	2.10	2.01	1.94	1.88	1.83	1.79	1.75
	.99	6.63	4.61	3.78	3.32	3.02	2.80	2.64	2.51	2.41	2.32	2.25	2.18

Table D* (*Continued*)

15	20	24	30	40	50	60	100	120	200	500	∞	$1 - \alpha$	df for denom.
\multicolumn{12}{c}{df for numerator}		
1.32	1.30	1.29	1.28	1.27	1.26	1.26	1.25	1.24	1.24	1.23	1.23	.75	
1.72	1.67	1.64	1.61	1.57	1.55	1.54	1.51	1.50	1.48	1.47	1.46	.90	30
2.01	1.93	1.89	1.84	1.79	1.76	1.74	1.70	1.68	1.66	1.64	1.62	.95	
2.70	2.55	2.47	2.39	2.30	2.25	2.21	2.13	2.11	2.07	2.03	2.01	.99	
1.30	1.28	1.26	1.25	1.24	1.23	1.22	1.21	1.21	1.20	1.19	1.19	.75	
1.66	1.61	1.57	1.54	1.51	1.48	1.47	1.43	1.42	1.41	1.39	1.38	.90	40
1.92	1.84	1.79	1.74	1.69	1.66	1.64	1.59	1.58	1.55	1.53	1.51	.95	
2.52	2.37	2.29	2.20	2.11	2.06	2.02	1.94	1.92	1.87	1.83	1.80	.99	
1.27	1.25	1.24	1.22	1.21	1.20	1.19	1.17	1.17	1.16	1.15	1.15	.75	
1.60	1.54	1.51	1.48	1.44	1.41	1.40	1.36	1.35	1.33	1.31	1.29	.90	60
1.84	1.75	1.70	1.65	1.59	1.56	1.53	1.48	1.47	1.44	1.41	1.39	.95	
2.35	2.20	2.12	2.03	1.94	1.88	1.84	1.75	1.73	1.68	1.63	1.60	.99	
1.24	1.22	1.21	1.19	1.18	1.17	1.16	1.14	1.13	1.12	1.11	1.10	.75	
1.55	1.48	1.45	1.41	1.37	1.34	1.32	1.27	1.26	1.24	1.21	1.19	.90	120
1.75	1.66	1.61	1.55	1.50	1.46	1.43	1.37	1.35	1.32	1.28	1.25	.95	
2.19	2.03	1.95	1.86	1.76	1.70	1.66	1.56	1.53	1.48	1.42	1.38	.99	
1.23	1.21	1.20	1.18	1.16	1.14	1.12	1.11	1.10	1.09	1.08	1.06	.75	
1.52	1.46	1.42	1.38	1.34	1.31	1.28	1.24	1.22	1.20	1.17	1.14	.90	200
1.72	1.62	1.57	1.52	1.46	1.41	1.39	1.32	1.29	1.26	1.22	1.19	.95	
2.13	1.97	1.89	1.79	1.69	1.63	1.58	1.48	1.44	1.39	1.33	1.28	.99	
1.22	1.19	1.18	1.16	1.14	1.13	1.12	1.09	1.08	1.07	1.04	1.00	.75	
1.49	1.42	1.38	1.34	1.30	1.26	1.24	1.18	1.17	1.13	1.08	1.00	.90	∞
1.67	1.57	1.52	1.46	1.39	1.35	1.32	1.24	1.22	1.17	1.11	1.00	.95	
2.04	1.88	1.79	1.70	1.59	1.52	1.47	1.36	1.32	1.25	1.15	1.00	.99	

Table E.1 Upper 5 Percent Points of Studentized Range q*

n_2										p**									
	2	3	4	5	6	7	8	9	10	11	12	13	14	15	16	17	18	19	20
1	18.0	26.7	32.8	37.2	40.5	43.1	45.4	47.3	49.1	50.6	51.9	53.2	54.3	55.4	56.3	57.2	58.0	58.8	59.6
2	6.09	8.28	9.80	10.89	11.73	12.43	13.03	13.54	13.99	14.39	14.75	15.08	15.38	15.65	15.91	16.14	16.36	16.57	16.77
3	4.50	5.88	6.83	7.51	8.04	8.47	8.85	9.18	9.46	9.72	9.95	10.16	10.35	10.52	10.69	10.84	10.98	11.12	11.24
4	3.93	5.00	5.76	6.31	6.73	7.06	7.35	7.60	7.83	8.03	8.21	8.37	8.52	8.67	8.80	8.92	9.03	9.14	9.24
5	3.61	4.54	5.18	5.64	5.99	6.28	6.52	6.74	6.93	7.10	7.25	7.39	7.52	7.64	7.75	7.86	7.95	8.04	8.13
6	3.46	4.34	4.90	5.31	5.63	5.89	6.12	6.32	6.49	6.65	6.79	6.92	7.04	7.14	7.24	7.34	7.43	7.51	7.59
7	3.34	4.16	4.68	5.06	5.35	5.59	5.80	5.99	6.15	6.29	6.42	6.54	6.65	6.75	6.84	6.93	7.01	7.08	7.16
8	3.26	4.04	4.53	4.89	5.17	5.40	5.60	5.77	5.92	6.05	6.18	6.29	6.39	6.48	6.57	6.65	6.73	6.80	6.87
9	3.20	3.95	4.42	4.76	5.02	5.24	5.43	5.60	5.74	5.87	5.98	6.09	6.19	6.28	6.36	6.44	6.51	6.58	6.65
10	3.15	3.88	4.33	4.66	4.91	5.12	5.30	5.46	5.60	5.72	5.83	5.93	6.03	6.12	6.20	6.27	6.34	6.41	6.47
11	3.11	3.82	4.26	4.58	4.82	5.03	5.20	5.35	5.49	5.61	5.71	5.81	5.90	5.98	6.06	6.14	6.20	6.27	6.33
12	3.08	3.77	4.20	4.51	4.75	4.95	5.12	5.27	5.40	5.51	5.61	5.71	5.80	5.88	5.95	6.02	6.09	6.15	6.21
13	3.06	3.73	4.15	4.46	4.69	4.88	5.05	5.19	5.32	5.43	5.53	5.63	5.71	5.79	5.86	5.93	6.00	6.06	6.11
14	3.03	3.70	4.11	4.41	4.64	4.83	4.99	5.13	5.25	5.36	5.46	5.56	5.64	5.72	5.79	5.86	5.92	5.98	6.03
15	3.01	3.67	4.08	4.37	4.59	4.78	4.94	5.08	5.20	5.31	5.40	5.49	5.57	5.65	5.72	5.79	5.85	5.91	5.96
16	3.00	3.65	4.05	4.34	4.56	4.74	4.90	5.03	5.15	5.26	5.35	5.44	5.52	5.59	5.66	5.73	5.79	5.84	5.90
17	2.98	3.62	4.02	4.31	4.52	4.70	4.86	4.99	5.11	5.21	5.31	5.39	5.47	5.55	5.61	5.68	5.74	5.79	5.84
18	2.97	3.61	4.00	4.28	4.49	4.67	4.83	4.96	5.07	5.17	5.27	5.35	5.43	5.50	5.57	5.63	5.69	5.74	5.79
19	2.96	3.59	3.98	4.26	4.47	4.64	4.79	4.92	5.04	5.14	5.23	5.32	5.39	5.46	5.53	5.59	5.65	5.70	5.75
20	2.95	3.58	3.96	4.24	4.45	4.62	4.77	4.90	5.01	5.11	5.20	5.28	5.36	5.43	5.50	5.56	5.61	5.66	5.71
24	2.92	3.53	3.90	4.17	4.37	4.54	4.68	4.81	4.92	5.01	5.10	5.18	5.25	5.32	5.38	5.44	5.50	5.55	5.59
30	2.89	3.48	3.84	4.11	4.30	4.46	4.60	4.72	4.83	4.92	5.00	5.08	5.15	5.21	5.27	5.33	5.38	5.43	5.48
40	2.86	3.44	3.79	4.04	4.23	4.39	4.52	4.63	4.74	4.82	4.90	4.98	5.05	5.11	5.17	5.22	5.27	5.32	5.36
60	2.83	3.40	3.74	3.98	4.16	4.31	4.44	4.55	4.65	4.73	4.81	4.88	4.94	5.00	5.06	5.11	5.15	5.20	5.24
120	2.80	3.36	3.69	3.92	4.10	4.24	4.36	4.47	4.56	4.64	4.71	4.78	4.84	4.90	4.95	5.00	5.04	5.09	5.13
∞	2.77	3.32	3.63	3.86	4.03	4.17	4.29	4.39	4.47	4.55	4.62	4.68	4.74	4.80	4.84	4.89	4.93	4.97	5.01

* From J. M. May, "Extended and Corrected Tables of the Upper Percentage Points of the Studentized Range," *Biometrika*, vol. 39 (1952), pp. 192–193. Reproduced by permission of the trustees of *Biometrika*.

** p is the number of quantities (for example, means) whose range is involved. n_2 is the degrees of freedom in the error estimate.

Table E.2 Upper 1 Percent Points of Studentized Range q

n_2^*	p^* 2	3	4	5	6	7	8	9	10	11	12	13	14	15	16	17	18	19	20
1	90.0	135	164	186	202	216	227	237	246	253	260	266	272	277	282	286	290	294	298
2	14.0	19.0	22.3	24.7	26.6	28.2	29.5	30.7	31.7	32.6	33.4	34.1	34.8	35.4	36.0	36.5	37.0	37.5	37.9
3	8.26	10.6	12.2	13.3	14.2	15.0	15.6	16.2	16.7	17.1	17.5	17.9	18.2	18.5	18.8	19.1	19.3	19.5	19.8
4	6.51	8.12	9.17	9.96	10.6	11.1	11.5	11.9	12.3	12.6	12.8	13.1	13.3	13.5	13.7	13.9	14.1	14.2	14.4
5	5.70	6.97	7.80	8.42	8.91	9.32	9.67	9.97	10.24	10.48	10.70	10.89	11.08	11.24	11.40	11.55	11.68	11.81	11.93
6	5.24	6.33	7.03	7.56	7.97	8.32	8.61	8.87	9.10	9.30	9.49	9.65	9.81	9.95	10.08	10.21	10.32	10.43	10.54
7	4.95	5.92	6.54	7.01	7.37	7.68	7.94	8.17	8.37	8.55	8.71	8.86	9.00	9.12	9.24	9.35	9.46	9.55	9.65
8	4.74	5.63	6.20	6.63	6.96	7.24	7.47	7.68	7.87	8.03	8.18	8.31	8.44	8.55	8.66	8.76	8.85	8.94	9.03
9	4.60	5.43	5.96	6.35	6.66	6.91	7.13	7.32	7.49	7.65	7.78	7.91	8.03	8.13	8.23	8.32	8.41	8.49	8.57
10	4.48	5.27	5.77	6.14	6.43	6.67	6.87	7.05	7.21	7.36	7.48	7.60	7.71	7.81	7.91	7.99	8.07	8.15	8.22
11	4.39	5.14	5.62	5.97	6.25	6.48	6.67	6.84	6.99	7.13	7.25	7.36	7.46	7.56	7.65	7.73	7.81	7.88	7.95
12	4.32	5.04	5.50	5.84	6.10	6.32	6.51	6.67	6.81	6.94	7.06	7.17	7.26	7.36	7.44	7.52	7.59	7.66	7.73
13	4.26	4.96	5.40	5.73	5.98	6.19	6.37	6.53	6.67	6.79	6.90	7.01	7.10	7.19	7.27	7.34	7.42	7.48	7.55
14	4.21	4.89	5.32	5.63	5.88	6.08	6.26	6.41	6.54	6.66	6.77	6.87	6.96	7.05	7.12	7.20	7.27	7.33	7.39
15	4.17	4.83	5.25	5.56	5.80	5.99	6.16	6.31	6.44	6.55	6.66	6.76	6.84	6.93	7.00	7.07	7.14	7.20	7.26
16	4.13	4.78	5.19	5.49	5.72	5.92	6.08	6.22	6.35	6.46	6.56	6.66	6.74	6.82	6.90	6.97	7.03	7.09	7.15
17	4.10	4.74	5.14	5.43	5.66	5.85	6.01	6.15	6.27	6.38	6.48	6.57	6.66	6.73	6.80	6.87	6.94	7.00	7.05
18	4.07	4.70	5.09	5.38	5.60	5.79	5.94	6.08	6.20	6.31	6.41	6.50	6.58	6.65	6.72	6.79	6.85	6.91	6.96
19	4.05	4.67	5.05	5.33	5.55	5.73	5.89	6.02	6.14	6.25	6.34	6.43	6.51	6.58	6.65	6.72	6.78	6.84	6.89
20	4.02	4.64	5.02	5.29	5.51	5.69	5.84	5.97	6.09	6.19	6.29	6.37	6.45	6.52	6.59	6.65	6.71	6.76	6.82
24	3.96	4.54	4.91	5.17	5.37	5.54	5.69	5.81	5.92	6.02	6.11	6.19	6.26	6.33	6.39	6.45	6.51	6.56	6.61
30	3.89	4.45	4.80	5.05	5.24	5.40	5.54	5.65	5.76	5.85	5.93	6.01	6.08	6.14	6.20	6.26	6.31	6.36	6.41
40	3.82	4.37	4.70	4.93	5.11	5.27	5.39	5.50	5.60	5.69	5.77	5.84	5.90	5.96	6.02	6.07	6.12	6.17	6.21
60	3.76	4.28	4.60	4.82	4.99	5.13	5.25	5.36	5.45	5.53	5.60	5.67	5.73	5.79	5.84	5.89	5.93	5.98	6.02
120	3.70	4.20	4.50	4.71	4.87	5.01	5.12	5.21	5.30	5.38	5.44	5.51	5.56	5.61	5.66	5.71	5.75	5.79	5.83
∞	3.64	4.12	4.40	4.60	4.76	4.88	4.99	5.08	5.16	5.23	5.29	5.35	5.40	5.45	5.49	5.54	5.57	5.61	5.65

* p is the number of quantities (for example, means) whose range is involved. n_2 is the degrees of freedom in the error estimate.

Table F Coefficients of Orthogonal Polynomials

k	Polynomial	1	2	3	4	5	6	7	8	9	10	$\Sigma \xi_j^2$	λ
3	Linear	−1	0	1								2	1
	Quadratic	1	−2	1								6	3
	Linear	−3	−1	1	3							20	2
4	Quadratic	1	−1	−1	1							4	1
	Cubic	−1	3	−3	1							20	$\frac{10}{3}$
	Linear	−2	−1	0	1	2						10	1
5	Quadratic	2	−1	−2	−1	2						14	1
	Cubic	−1	2	0	−2	1						10	$\frac{5}{6}$
	Quartic	1	−4	6	−4	1						70	$\frac{35}{12}$
	Linear	−5	−3	−1	1	3	5					70	2
6	Quadratic	5	−1	−4	−4	−1	5					84	$\frac{3}{2}$
	Cubic	−5	7	4	−4	−7	5					180	$\frac{5}{3}$
	Quartic	1	−3	2	2	−3	1					28	$\frac{7}{12}$
	Linear	−3	−2	−1	0	1	2	3				28	1
7	Quadratic	5	0	−3	−4	−3	0	5				84	1
	Cubic	−1	1	1	0	−1	−1	1				6	$\frac{1}{6}$
	Quartic	3	−7	1	6	1	−7	3				154	$\frac{7}{12}$
	Linear	−7	−5	−3	−1	1	3	5	7			168	2
	Quadratic	7	1	−3	−5	−5	−3	1	7			168	1
8	Cubic	−7	5	7	3	−3	−7	−5	7			264	$\frac{2}{3}$
	Quartic	7	−13	−3	9	9	−3	−13	7			616	$\frac{7}{12}$
	Quintic	−7	23	−17	−15	15	17	−23	7			2184	$\frac{7}{10}$
	Linear	−4	−3	−2	−1	0	1	2	3	4		60	1
	Quadratic	28	7	−8	−17	−20	−17	−8	7	28		2772	3
9	Cubic	−14	7	13	9	0	−9	−13	−7	14		990	$\frac{5}{6}$
	Quartic	14	−21	−11	9	18	9	−11	−21	14		2002	$\frac{7}{12}$
	Quintic	−4	11	−4	−9	0	9	4	−11	4		468	$\frac{3}{20}$
	Linear	−9	−7	−5	−3	−1	1	3	5	7	9	330	2
	Quadratic	6	2	−1	−3	−4	−4	−3	−1	2	6	132	$\frac{1}{2}$
10	Cubic	−42	14	35	31	12	−12	−31	−35	−14	42	8580	$\frac{5}{3}$
	Quartic	18	−22	−17	3	18	18	3	−17	−22	18	2860	$\frac{5}{12}$
	Quintic	−6	14	−1	−11	−6	6	11	1	−14	6	780	$\frac{1}{10}$

Answers to Odd-Numbered Problems

CHAPTER 2

2.1 $\bar{Y} = 18{,}472.9$, $s^2 = 41.7$.

2.3 For a two-sided alternative, some points are
$\mu = 18{,}473;\ 18{,}472;\ 18{,}471;\ 18{,}470;\ 18{,}469;\ 18{,}468;\ 18{,}467$
$P_a = \beta = 0.08, 0.39, 0.80, 0.95, 0.80, 0.39, 0.08$.

2.5 $n = 7$.

2.7 Do not reject hypothesis, since $\chi^2 = 17.05$ with 11 df.

2.9 Do not reject hypothesis, since $F = 1.8$.

2.11 Reject hypothesis at 1 percent level, since $|t| = 13.4$.

2.13 1. Reject hypothesis, since $z = 4.0$.
2. Do not reject hypothesis, since $z = 1.71$.
3. Do not reject hypothesis z, since $z < 1$.

2.15 For $\quad\quad\mu = 13\quad 15\quad 17\quad 19$
Power $(1 - \beta) = 0.04\quad 0.27\quad 0.70\quad 0.95$.

2.17 Reject hypothesis at $\alpha = 0.05$, since $t = 8.32$.

2.19 Reject hypothesis of equal variances at $\alpha = 0.05$, since $F = 17.4$. Do not reject hypothesis of equal means at $\alpha = 0.05$, since t' is < 1.

2.21 Reject hypothesis at $\alpha = 0.05$, since $t = 2.68$.

2.23 Do not reject hypothesis at $\alpha = 0.05$, since $t = 1.62$.

2.25 Reject hypothesis at $\alpha = 0.05$, since $z = -1.96$.

CHAPTER 3

3.1

Source	df	SS	MS
Between A levels	4	253.04	63.26
Error	20	76.80	3.84
Totals	24	329.84	

Significant at the 1 percent level.

3.3 Two such contrasts might be

	SS
$C_1 = T_A - T_C = 31$	60.06
$C_2 = T_A - 2T_B + T_C = -79$	130.02
	190.08

Neither is significant at the 5 percent level.

3.5

	B	A	C
$\bar{Y}_{.j} =$	25.4	22.4	18.5

None significantly different.

3.7 Two sets and their sums of squares might be

SS

Set 1:
$$C_1 = \ \ 2T_1 \qquad\qquad\qquad\qquad - \ \ 2T_5 = \ -28 \quad 49.0$$
$$C_2 = \qquad\qquad 4T_2 \qquad - \ \ 6T_4 \qquad = -108 \quad 48.6$$
$$C_3 = 11T_1 + 11T_2 - 14T_3 + 11T_4 + 11T_5 = \quad 71 \quad 1.3$$
$$C_4 = 10T_1 - \ \ 4T_2 \qquad\qquad - \ 4T_4 + 10T_5 = \quad 8 \quad 0.1$$
$$ 99.0$$

SS

Set 2:
$$C_1 = \ \ 6T_1 - \ \ 2T_2 \qquad\qquad\qquad\qquad = \ -18 \quad 3.4$$
$$C_2 = 11T_1 + 11T_2 - \ \ 8T_3 \qquad\qquad = -237 \quad 33.6$$
$$C_3 = \ \ 4T_1 + \ \ 4T_2 + \ \ 4T_3 - 19T_4 \qquad = -252 \quad 36.3$$
$$C_4 = \ \ 2T_1 + \ \ 2T_2 + \ \ 2T_3 + \ \ 2T_4 - 23T_5 = -172 \quad 25.7$$
$$ 99.0$$

3.9 75.6 percent due to levels of A, 24.4 percent due to error.

3.11 Two might be $C_1 = T_{.1} - T_{.2}$ and $C_2 = 2T_{.1} - T_{.2} - T_{.3}$ and $C_1 = -24$, which is less than $As_{C_1} = 60.65$, so is not significant at 5 percent, and $C_2 = 7$, which is less than $As_{C_2} = 105.07$, so is not significant at 5 percent.

3.13

Source	df	SS	MS	F
Between temperatures	4	1268.54	317.14	70.3***
Error	25	112.83	4.51	

*** Significant at the .1 percent level.

3.15

Source	df	SS	MS	F
Between coatings	3	1135.0	378.3	29.8***
Error	16	203.2	12.7	

3.17 Newman–Keuls tests show:

Coating	IV	III	II	I
Means	42.0	43.6	57.2	58.4

3.19 One case based on $n - 1$ df and the other on Newman–Keuls. The latter is better if variances are homogeneous.

3.21 One set of orthogonal contrasts might be

$$C_1 = 4T_{.1} - T_{.2} - T_{.3} - T_{.4} - T_{.5}$$
$$C_2 = \qquad\quad T_{.2} - T_{.3}$$
$$C_3 = \qquad\qquad\qquad\qquad T_{.4} - T_{.5}$$
$$C_4 = \qquad\quad T_{.2} + T_{.3} - T_{.4} - T_{.5}$$

3.23 For C_1 in Problem 3.21, $s_C = 24.5$.

3.25 Litter to litter differences account for 28.6 percent of the total variance.

CHAPTER 4

4.1

Source	df	SS	MS
Between coater types	3	1.53	0.51*
Between days	2	0.21	0.10
Error	6	0.54	0.09
Totals	11	2.28	

* Significant at the 5 percent level.

4.3

	K	A	L	M
$\bar{Y}_{.j}$:	5.27	4.87	4.73	4.27

K and M are significantly different at the 5 percent level.

4.5 Four contrasts might be

$C_1 = T_M - T_A$ Nonsignificant
$C_2 = T_M - T_K$ Significant at 5 percent by Scheffé
$C_3 = T_M - T_L$ Nonsignificant
$C_4 = T_A - T_K$ Nonsignificant

4.7

Source	df	SS	MS
Electrodes	4	4.2966	1.0742**
Positions	4	0.5696	0.1424
Strips	4	0.3136	0.0784
Error	12	0.8058	0.0672

** Significant at the 1 percent level.

4.9 Proof.

4.11 Error df $= 0$.

4.13 Three groups (treatments) and six lessons (blocks) ANOVA:

Source	df
Groups	2
Lessons	5
Error	10

4.15 Newman–Keuls on means:

Group	C	B	A
Means	12.33	19.17	22.17

4.17 a. $Y_{ijk} = \mu + S_i + t_j + D_k + \varepsilon_{ijk}$ and ANOVA:

Source	df
Scales	4
Times	4
Days	4
Error	12

b. ANOVA:

Source	df
Scales	4
Days	4
Error	16

4.19 $H_0: F_j = 0$ for all j. Model: $Y_{ij} = \mu + F_j + \beta_i + \varepsilon_{ij}$. $F = 8.75$, which is significant at the 1 percent level.

4.21 From -0.38 to 0.88.

4.23 For example,

$$C_1 = T_{.1} - T_{.2}$$
$$C_2 = T_{.1} + T_{.2} - 2T_{.3}$$
$$C_3 = \qquad\qquad T_{.3} - T_{.4}$$

4.25

Source	df	SS	MS	F
Vendors	2	32.89	16.44	3.02
Scales	2	0.22	0.11	<1
Inspectors	2	0.22	0.11	<1
Error	2	10.89	5.44	

Nothing significant at the 5 percent level.

CHAPTER 5

5.1

Source	df	SS	MS	F
Phosphors	2	1,244.44	622.22	8.96**
Glass	1	13,338.89	13,338.89	192.09***
$P \times G$ interaction	2	44.45	22.22	<1
Error	12	833.33	69.44	

5.3 G_2 better than G_1—lower current. For phosphors, Newman–Keuls:

Phosphor: C   A   B
Means: 253.33   260.00   273.33

so A or C is better than B.

5.5

Source	df	SS	MS
Humidity	2	9.07	4.53
Temperature	2	8.66	4.33
$T \times H$ interaction	4	6.07	1.52
Error	27	28.50	1.06
Totals	35	52.30	

5.7 At the 5 percent significance level reject H_2 and H_3 but not H_1.

5.9 Plot of cell totals versus feed for the two material lines are not parallel (interaction), and both material and feed effect are obvious.

5.11

Source	df	SS	MS
Temperature	2	600.09	300.04**
Mix	2	1.59	0.80*
$T \times M$	4	0.98	0.25*
Laboratory	3	3.85	1.28**
$T \times L$	6	1.54	0.26*
$M \times L$	6	0.76	0.13*
$T \times M \times L$	12	0.90	0.07*
Error	36	0.67	0.02
Totals	71	610.38	

5.13

Source	df	SS	MS
Soil	4	41.47	10.37
Fertilizer	2	68.87	34.44
Interaction	8	16.13	2.02
Error	15	52.50	3.50

5.15 Newman–Keuls on fertilizers:

Types: 3   2   1
Means: 3.6 5.2 7.3

Hence type 1 gives best yield with any of the five soil types.

5.17 a. No.   b. Yes.   c. Yes.

5.19 Two-dimensional stimulus more variable than three dimensional.

J by A means:

1.04   1.14   1.17   1.20   1.24   1.42   1.58   1.59   1.70   1.70

5.21

Source	df	SS	MS
Thickness A	1	846.81	846.81**
Temperature B	1	5041.00	5041.00**
AB	1	509.63	509.63**
Drying condition C	1	5.88	5.88
AC	1	1.44	1.44
BC	1	15.21	15.21*
ABC	1	0.14	0.14
Length of wash L	3	69.76	23.25*
AL	3	15.79	5.26
BL	3	3.04	1.01
ABL	3	11.45	3.82
CL	3	9.65	3.22
ACL	3	8.07	2.69
BCL	3	6.43	2.14
$ABCL$	3	5.66	1.89
Error	32	87.21	2.73
Totals	63	6637.17	

5.23 a.

Source	df
Waxes	3
Times	2
$W \times T$ interaction	6
Error	12

b. Changes in gloss index due to the four waxes are different for the three polishing times.

5.25 a. Significants are A, B, C, and the ABC interaction.
   b. Run Newman–Keuls on 24 cell means
   c. $4 \times 3 \times 2$ layout with five observations per treatment.
   d. 7.0 to 12.2.

CHAPTER 6

6.1

Source	df	SS	MS
Factor A	1	0.08	0.08
Factor B	1	70.08	70.08**
$A \times B$ interaction	1	24.08	24.08
Error	8	36.68	4.58
Totals	11	130.92	

6.3

Source	df	SS	MS
A	1	2704.00	2704.00
B	1	26,732.25	26,732.25
AB	1	7744.00	7744.00
C	1	24,025.00	24,025.00
AC	1	16,256.25	16,256.25
BC	1	64,516.00	64,516.00
ABC	1	420.25	420.25
Error	8	246,284.00	30,785.50
Totals	15	388,681.75	

Nothing significant at the 5-percent level.

6.5 No plots, since there are no significant effects at the 5 percent level.

6.7

Source	df	SS	MS
A	1	13,736.5	
B	1	6188.3	
AB	1	22,102.5	Same
C	1	22.7	as
AC	1	22,525.0	SS
BC	1	12,051.3	
ABC	1	20,757.0	
D	1	81,103.8	
AD	1	145,665.0*	
BD	1	9214.3	
ABD	1	126,630.3*	
CD	1	148.8	
ACD	1	6757.0	
BCD	1	294.0	
ABCD	1	19,453.8	
Error	16	431,599.4	26,975.0
Totals	31	918,249.7	

6.9 a. (1), a, b, ab, c, ac, bc, abc, d, ad, bd, abd, cd, acd, bcd, $abcd$.

b.

Source	df
Main effects	1 each for 4
Two-way interaction	1 each for 6
Three-way interaction	1 each for 4 ⎫ Use as error with 5 df
Four-way interaction	1 each for 1 ⎭
Total	15

c. Find A effect by plus signs where a is present in treatment combination and minus signs where absent. Same for D and then multiply signs to get AD signs.

6.11 a. 32.

b.

Source	df
Main effects	1 each for 5
Two-way interaction	1 each for 10
Three-way interaction	1 each for 10
Four-way interaction	1 each for 5 ⎫ Use as error.
Five-way interaction	1 each for 1 ⎭
Total	31

c. Get signs for A, C, and E and multiply.

d. True if no interactions are present. Otherwise use Newman–Keuls on treatment means.

6.13 Feed rate, 0.015; tool condition, new; tool type, precision.

6.15 2^6 factorial. Probably use four-, five-, and six-way interactions as error term with 22 df.

6.17

Source	df	SS and MS
Drop-out temperature (A)	1	3.00
Back-zone temperature (B)	1	4.06*
$A \times B$ interaction	1	1.20
Atmosphere (C)	1	3.00
$A \times C$ interaction	1	1.71
$B \times C$ interaction	1	4.65*
$A \times B \times C$ interaction	1	15.96**

Significant based on MS for error of 0.37 with 3 df.

6.19

Source	df	SS and MS
A	1	45.60*
B	1	1.36
AB	1	53.56*
C	1	0.01
AC	1	0.00
BC	1	0.01
ABC	1	0.04

A and AB significant based on MS for error of 4.41 with 3 df.

Newman–Keuls on A, B means: 0.00   4.20   4.35   9.95.

No significant gaps as error is too large.

6.21 Plot the following: $(1) = 0$, $a = -3$, $b = -2$, $ab = 5$, $c = -7$, $ac = 0$, $bc = 1$, and $abc = -2$.

CHAPTER 7

7.1 $Y'_X = 6.57 + 0.50X_j$.

7.3 $\eta^2 = 0.5981$, $r^2 = 0.5023$, $s_{Y.X} = 0.756$.

7.5   Scattergram indicates quadratic.

7.7 $Y'_u = 144.65 + 3.53u + 1.23u^2 - 0.46u^3 - 1.05u^4$.

7.9 a. Second degree.

b. $Y'_u = 96.19 + 2.69u + 1.13u^2$.

7.11 $r^2 = 0.22, \eta^2 = 0.27$.

7.13

Source	df	SS	
Between ages	4	82.28	
Linear	1		72.25**
Quadratic	1		9.78*
Cubic	1		0.25
Quartic	1		0.00
Error	45	17.50	

$MS_{error} = 0.39$

Curve is $Y'_u = 3.91 + 0.85u - 0.264u^2$ and $R^2 = 0.82$.

7.15

Source	df	SS	MS
Gate	2	10.58	5.29**
Resin	1	0.01	0.01
$G \times R$ interaction	2	0.00	0.00
Error	6	0.07	0.01

7.17 Linear only on gate setting giving: $Y'_u = 2.517 + 1.15u$ or $Y'_X = 0.217 + 0.575X_j$.

7.19

Source	df		Source	df	
R	1		$G \times W$	4	
G	2		$G_L \times W_L$		1
G_L		1	$G_L \times W_Q$		1
G_Q		1	$G_Q \times W_L$		1
$R \times G$	2		$G_Q \times W_Q$		1
$R \times G_L$		1	$R \times G \times W$	4	
$R \times G_Q$		1	$R \times G_L \times W_L$		1
W	2		$R \times G_L \times W_Q$		1
W_L		1	$R \times G_Q \times W_L$		1
W_Q		1	$R \times G_Q \times W_Q$		1
$R \times W$	2		Error	18	
$R \times W_L$		1			
$R \times W_Q$		1	Total	35	

7.21 Graph.

7.23

Source	df		Source	df
Radius (R)	3		$R_L \times F_L$	1
R_L	1		$R_L \times F_Q$	1
R_Q	1		$R_L \times F_C$	1
R_C	1		$R_Q \times F_L$	1
Force (F)	3		$R_Q \times F_Q$	1
F_L	1		$R_Q \times F_C$	1
F_Q	1		$R_C \times F_L$	1
F_C	1		$R_C \times F_Q$	1
$R \times F$ interaction	9		$R_C \times F_C$	1
			Error	32
			Total	47

7.25 Plot.

CHAPTER 8

8.1 a. $\bar{Y} = 9.2875$, $s_Y = 3.5292$; $\bar{X}_1 = 1.5250$, $s_1 = 0.2252$; $\bar{X}_2 = 10.0125$, $s_2 = 4.3066$; $r_{Y1} = -0.7940$, $r_{Y2} = 0.5012$, $r_{12} = -0.1904$.
   b. $Y' = 28.2641 - 12.4431 X_1$.

8.3 $Y' = 68.78 + 1.92 X_1$ adequate. $r_{Y1}^2 = 0.34$.
   For both x's: $Y' = 64.69 + 1.92 X_1 + 0.0127 X_2$.
   Note: $r_{12} = 0$ as X_1 and X_2 are taken orthogonal.

8.5 a. One with largest correlation with Y.
   b. One with largest partial correlation with Y after first X is removed.
   c. When no more partials are significant.
   d. $F_{2,20} = 3.33 < 3.49$ at 5 percent, so stop here.

8.7 a. $R_{Y.1234} = 0.71$.
   b. X_2 and X_3 only.
   c. ± 12.89 (two standard errors).

8.9 Adequate equation: $Y' = -0.66 + 0.29 X_1 + 0.45 X_3$, where X_1 = ethnic group and X_3 = interview score. $R_{Y.13}^2 = 0.65$.

8.11 a. X_{14}.
   b. $X'_5 = 16.18 + 0.89 X_{14}$.
   c. $r_{5.13}^2 = 0.89$.
   d. $s_{5.14} = 443.02$.

8.13 Stop now and use equation

$$X'_5 = 260.26 + 0.71 X_{14} + 81.74 X_6 - 27.22 X_8 + 0.43 X_{402}$$
$$- 0.80 X_9 - 79.01 X_{15}.$$
$$R^2 = 0.93.$$

8.15 a. Scattergram looks like quadratic.
   b. $Y' = 52.40 + 9.03X - 0.22X^2$ and $R^2_{Y.XX^2} = 0.91$.
   c. Linear: $Y' = 132.77 + 0.41X$ and $r^2_{YX} = 0.12$.

CHAPTER 9

9.1

Source	df	SS	MS
Temperature	2	348.44	174.22
Humidity	2	333.77	166.88
$T \times H$ interaction	4	358.23	89.56
Error	9	680.00	75.66
Totals	17	1720.44	

None significant at the 5 percent level.

9.3 $AB = 320.11$
    $AB^2 = 38.11$ } 358.22

9.5

Source	df	SS	MS
V	2	2.79	1.40*
T	2	9.01	4.50**
$V \times T$	4	1.99	0.50
Error	9	2.15	

9.7 $VT = 0.94$ and $VT^2 = 1.04$, each with 2 df.

9.9 Any combination except $T_H V_H$.

9.11 After coding data by subtracting 2.6 and multiplying by 10, we have

Source	df	SS	MS
Surface thickness A	2	2544.71	1272.35
Base thickness B	2	4787.37	2393.68
$A \times B$ interaction	4	185.18	46.29
Subbase thickness C	2	4165.15	2082.57
$A \times C$ interaction	4	189.72	47.48
$B \times C$ interaction	4	27.74	6.93
$A \times B \times C$ interaction	8	100.06	12.51
Error	27	71.50	2.65

All except $B \times C$ are significant at the 5 percent level, but the main effects predominate.

9.13

Source	df	SS	
AB	2	66.93	\} 185.19
AB^2	2	118.26	
AC	2	58.93	\} 189.75
AC^2	2	130.82	
BC	2	5.82	\} 27.75
BC^2	2	21.93	
ABC	2	17.60	
ABC^2	2	8.26	\} 100.05
AB^2C	2	53.48	
AB^2C^2	2	20.71	

9.15 Plots show strong linear effects of A, B, and C. They show AB and AC interaction—but it is slight compared with the main effects. A slight quadratic trend in A is also noted.

9.17

Source	df		SS	MS
Resin type R	1		0.050	0.050*
Weight fraction S	2		0.540	
S_L		1	0.540	0.540**
S_Q		1	0.000	0.000
Gate setting G	2		30.300	
G_L		1	30.150	30.150**
G_Q		1	0.150	0.150*
RS	2		0.040	0.020
RG	2		0.060	0.030*
SG	4		0.343	
$S_L G_L$		1	0.280	0.280**
$S_L G_Q$		1	0.002	0.002
$S_Q G_L$		1	0.010	0.010
$S_Q G_Q$		1	0.051	0.051*
RSG	4		0.050	0.013
Error	18		0.150	0.0083
Totals	35		31.530	

9.19 $Y'_x = 3.20$, $Y'_x = 3.02$.

CHAPTER 10

10.1

Source	EMS
O_i	$\sigma_\varepsilon^2 + 10\sigma_O^2$
A_j	$\sigma_\varepsilon^2 + 2\sigma_{OA}^2 + 6\phi_A$
OA_{ij}	$\sigma_\varepsilon^2 + 2\sigma_{OA}^2$
$\varepsilon_{k(ij)}$	σ_ε^2

Tests are indicated by arrows. None significant at the 5 percent level.

10.3

Source	EMS
A_i	$\sigma_\varepsilon^2 + nc\sigma_{AB}^2 + nbc\sigma_A^2$
B_j	$\sigma_\varepsilon^2 + nc\sigma_{AB}^2 + nac\sigma_B^2$
AB_{ij}	$\sigma_\varepsilon^2 + nc\sigma_{AB}^2$
C_k	$\sigma_\varepsilon^2 + n\sigma_{ABC}^2 + na\sigma_{BC}^2 + nb\sigma_{AC}^2 + nab\phi_C$
AC_{ik}	$\sigma_\varepsilon^2 + n\sigma_{ABC}^2 + nb\sigma_{AC}^2$
BC_{jk}	$\sigma_\varepsilon^2 + n\sigma_{ABC}^2 + na\sigma_{BC}^2$
ABC_{ijk}	$\sigma_\varepsilon^2 + n\sigma_{ABC}^2$
$\varepsilon_{m(ijk)}$	σ_ε^2

Tests are obvious.
No direct test on C.

10.5

Source	EMS
A_i	$\sigma_\varepsilon^2 + nb\sigma_{ACD}^2 + nbc\sigma_{AD}^2 + nbd\sigma_{AC}^2 + nbcd\phi_A$
B_j	$\sigma_\varepsilon^2 + na\sigma_{BCD}^2 + nac\sigma_{BD}^2 + nad\sigma_{BC}^2 + nacd\phi_B$
AB_{ij}	$\sigma_\varepsilon^2 + n\sigma_{ABCD}^2 + nc\sigma_{ABD}^2 + nd\sigma_{ABC}^2 + ncd\phi_{AB}$
C_k	$\sigma_\varepsilon^2 + nab\sigma_{CD}^2 + nabd\sigma_C^2$
AC_{ik}	$\sigma_\varepsilon^2 + nb\sigma_{ACD}^2 + nbd\sigma_{AC}^2$
BC_{jk}	$\sigma_\varepsilon^2 + na\sigma_{BCD}^2 + nad\sigma_{BC}^2$
ABC_{ijk}	$\sigma_\varepsilon^2 + n\sigma_{ABCD}^2 + nd\sigma_{ABC}^2$
D_m	$\sigma_\varepsilon^2 + nab\sigma_{CD}^2 + nabc\sigma_D^2$
AD_{im}	$\sigma_\varepsilon^2 + nb\sigma_{ACD}^2 + nbc\sigma_{AD}^2$
BD_{jm}	$\sigma_\varepsilon^2 + na\sigma_{BCD}^2 + nac\sigma_{BD}^2$
ABD_{ijm}	$\sigma_\varepsilon^2 + n\sigma_{ABCD}^2 + nc\sigma_{ABD}^2$
CD_{km}	$\sigma_\varepsilon^2 + nab\sigma_{CD}^2$
ACD_{ikm}	$\sigma_\varepsilon^2 + nb\sigma_{ACD}^2$
BCD_{jkm}	$\sigma_\varepsilon^2 + na\sigma_{BCD}^2$
$ABCD_{ijkm}$	$\sigma_\varepsilon^2 + n\sigma_{ABCD}^2$
$\varepsilon_{q(ijkm)}$	σ_ε^2

No direct test on A, B, or AB.

10.7   a. $Y_{ijk} = \mu + T_i + S_j + TS_{ij} + \varepsilon_{k(ij)}$.
       b.

Source	df	EMS
T_i	3	$\sigma_\varepsilon^2 + 2\sigma_{TS}^2 + 10\phi_T$
S_j	4	$\sigma_\varepsilon^2 + 8\sigma_S^2$
TS_{ij}	12	$\sigma_\varepsilon^2 + 2\sigma_{TS}^2$
$\varepsilon_{k(ij)}$	20	σ_ε^2

c. See arrows above.

10.9

Source	EMS
O_i	$\sigma_\varepsilon^2 + 2\sigma_{OI}^2 + 8\sigma_O^2$
I_j	$\sigma_\varepsilon^2 + 2\sigma_{OI}^2 + 8\sigma_I^2$
OI_{ij}	$\sigma_\varepsilon^2 + 2\sigma_{OI}^2$
$\varepsilon_{k(ij)}$	σ_ε^2

Instruments:	40 percent
Operators:	32 percent
Operator/	
instrument interaction:	15 percent
Error:	13 percent

10.11 $n_0 = 4.44$, $s_\varepsilon^2 = 1.16$, $s_A^2 = 5.32$.

10.15   To test A, $MS = MS_{AB} + MS_{AC} - MS_{ABC}$
       B, $MS = MS_{AB} + MS_{BC} - MS_{ABC}$
       C, $MS = MS_{AC} + MS_{BC} - MS_{ABC}$

CHAPTER 11

11.1

Source	df	SS	MS	EMS
L_i	2	27.42	13.71	$\sigma_\varepsilon^2 + 3\sigma_R^2 + 12\phi_L$
$R_{j(i)}$	9	36.38	4.04	$\sigma_\varepsilon^2 + 3\sigma_R^2$
$\varepsilon_{k(ij)}$	24	21.80	0.91	σ_ε^2
Totals	35	85.60		

Difference between rolls within lots is significant at the 1 percent level.

11.3

Source	EMS
A_i	$\sigma_\varepsilon^2 + 2\sigma_C^2 + 6\sigma_B^2 + 24\phi_A$
$B_{j(i)}$	$\sigma_\varepsilon^2 + 2\sigma_C^2 + 6\sigma_B^2$
$C_{k(ij)}$	$\sigma_\varepsilon^2 + 2\sigma_C^2$
$\varepsilon_{m(ijk)}$	σ_ε^2

Tests are obvious.

11.5

Source	df	SS	MS	EMS
T_i	1	5,489,354	5,489,354	$\sigma_\varepsilon^2 + 6\sigma_M^2 + 12\phi_T$
$M_{j(i)}$	2	408,250	204,125	$\sigma_\varepsilon^2 + 6\sigma_M^2$
S_k	1	8971	8971	$\sigma_\varepsilon^2 + 3\sigma_{MS}^2 + 12\phi_S$
TS_{ik}	1	37,445	37,445	$\sigma_\varepsilon^2 + 3\sigma_{MS}^2 + 6\phi_{TS}$
$MS_{kj(i)}$	2	31,520	15,760	$\sigma_\varepsilon^2 + 3\sigma_{MS}^2$
$\varepsilon_{m(ijk)}$	16	316,930	19,808	σ_ε^2

11.7

Source	EMS
T_i	$\sigma_\varepsilon^2 + 18\sigma_m^2 + 36\phi_t$
$M_{j(i)}$	$\sigma_\varepsilon^2 + 18\sigma_m^2$
S_k	$\sigma_\varepsilon^2 + 9\sigma_{ms}^2 + 36\phi_s$
TS_{ik}	$\sigma_\varepsilon^2 + 9\sigma_{ms}^2 + 18\phi_{ts}$
$MS_{kj(i)}$	$\sigma_\varepsilon^2 + 9\sigma_{ms}^2$
P_m	$\sigma_\varepsilon^2 + 6\sigma_{mp}^2 + 24\phi_p$
TP_{im}	$\sigma_\varepsilon^2 + 6\sigma_{mp}^2 + 12\phi_{tp}$
$MP_{mj(i)}$	$\sigma_\varepsilon^2 + 6\sigma_{mp}^2$
SP_{km}	$\sigma_\varepsilon^2 + 3\sigma_{msp}^2 + 12\phi_{sp}$
TSP_{ikm}	$\sigma_\varepsilon^2 + 3\sigma_{msp}^2 + 6\phi_{tsp}$
$MSP_{mkj(i)}$	$\sigma_\varepsilon^2 + 3\sigma_{msp}^2$
$\varepsilon_{q(ijkm)}$	σ_ε^2

11.9 a, b.

Source	df	EMS
C_i	1	$\sigma_\varepsilon^2 + 2\sigma_F^2 + 8\sigma_C^2$
$F_{j(i)}$	6	$\sigma_\varepsilon^2 + 2\sigma_F^2$
V_k	1	$\sigma_\varepsilon^2 + \sigma_{FV}^2 + 4\sigma_{CV}^2 + 8\phi_V$
CV_{ik}	1	$\sigma_\varepsilon^2 + \sigma_{FV}^2 + 4\sigma_{CV}^2$
$FV_{kj(i)}$	6	$\sigma_\varepsilon^2 + \sigma_{FV}^2$

c. 75 percent.

11.11 Newman–Keuls shows experimental posttest group means better than control and one experimental group better than the other.

11.13 Newman–Keuls gives: 54.66   54.96   57.00   65.66

So posttest experimental is best.

11.15 Discuss based on EMS and results.

11.17 $Y_{ijkm} = \mu + G_i + C_j + GC_{ij} + S_{k(ij)} + T_m + GT_{im} + CT_{jm}$
$+ GCT_{ijm} + TS_{mk(ij)}$

11.19

Source	df	EMS
D_i	2	$2\sigma_S^2 + 22\sigma_D^2$
G_j	1	$\sigma_{SG}^2 + 11\sigma_{DG}^2 + 33\phi_G$
DG_{ij}	2	$\sigma_{SG}^2 + 11\sigma_{DG}^2$
$S_{k(i)}$	30	$2\sigma_S^2$
$SG_{jk(i)}$	30	σ_{SG}^2

11.21

Source	df	EMS
R_i	1	$\sigma_\varepsilon^2 + 6\sigma_H^2 + 36\sigma_R^2$
$H_{j(i)}$	10	$\sigma_\varepsilon^2 + 6\sigma_H^2$
P_k	2	$\sigma_\varepsilon^2 + 2\sigma_{PH}^2 + 12\sigma_{RP}^2 + 24\phi_P$
RP_{ik}	2	$\sigma_\varepsilon^2 + 2\sigma_{PH}^2 + 12\sigma_{RP}^2$
$PH_{kj(i)}$	20	$\sigma_\varepsilon^2 + 2\sigma_{PH}^2$
$\varepsilon_{m(ijk)}$	36	σ_ε^2

11.23

Source	df	EMS
F_i	4	$\sigma_\varepsilon^2 + 9\sigma_S^2 + 54\phi_F$
C_j	1	$\sigma_\varepsilon^2 + 9\sigma_S^2 + 135\phi_C$
FC_{ij}	4	$\sigma_\varepsilon^2 + 9\sigma_S^2 + 27\phi_{FC}$
$S_{k(ij)}$	20	$\sigma_\varepsilon^2 + 9\sigma_S^2$
P_m	2	$\sigma_\varepsilon^2 + 3\sigma_{SP}^2 + 90\phi_P$
FP_{im}	8	$\sigma_\varepsilon^2 + 3\sigma_{SP}^2 + 18\phi_{FP}$
CP_{jm}	2	$\sigma_\varepsilon^2 + 3\sigma_{SP}^2 + 45\phi_{CP}$
FCP_{ijm}	8	$\sigma_\varepsilon^2 + 3\sigma_{SP}^2 + 9\phi_{FCP}$
$SP_{mk(ij)}$	40	$\sigma_\varepsilon^2 + 3\sigma_{SP}^2$
$\varepsilon_{q(ijkm)}$	180	σ_ε^2

11.25 Recommend 1:1 concentration and any of the fillers—iron filings, iron oxide, or copper.

CHAPTER 12

12.1

Source	df	SS	MS	EMS
R_i	2	337.15	168.58	$\sigma_\varepsilon^2 + 18\sigma_R^2$
S_j	1	16.66	16.66	$\sigma_\varepsilon^2 + 9\sigma_{RS}^2 + 27\phi_S$
RS_{ij}	2	126.78	63.39	$\sigma_\varepsilon^2 + 9\sigma_{RS}^2$
H_k	2	15.59	7.80	$\sigma_\varepsilon^2 + 6\sigma_{RH}^2 + 18\phi_H$
HS_{jk}	2	80.12	40.06	$\sigma_\varepsilon^2 + 3\sigma_{SRH}^2 + 9\phi_{HS}$
RH_{ik}	4	62.85	15.71	$\sigma_\varepsilon^2 + 6\sigma_{RH}^2$
SRH_{ijk}	4	229.44	57.36	$\sigma_\varepsilon^2 + 3\sigma_{SRH}^2$
$\varepsilon_{m(ijk)}$	36	200.00	5.56	σ_ε^2
Totals	53	1068.59		

Replications (blocks) and replications by all other factors are significant at the 5 percent level; however, these are not the factors of chief interest.

12.3

Source	df	SS	EMS
R_i	1	As before	$\sigma_\varepsilon^2 + 3\sigma_{RM}^2 + 12\sigma_R^2$
T_j	1	with R in	$\sigma_\varepsilon^2 + 3\sigma_{RM}^2 + 6\sigma_M^2 + 6\sigma_{RT}^2 + 12\phi_T$
RT_{ij}	1	place of S	$\sigma_\varepsilon^2 + 3\sigma_{RM}^2 + 6\sigma_{RT}^2$
$M_{k(j)}$	2		$\sigma_\varepsilon^2 + 3\sigma_{RM}^2 + 6\sigma_M^2$
$RM_{ik(j)}$	2		$\sigma_\varepsilon^2 + 3\sigma_{RM}^2$
$\varepsilon_{m(ijk)}$	16		σ_ε^2
Total	23		

No direct test on types, but results look highly significant. No other significant effects.

12.5

Source	df	EMS	
R_i	2	$10\sigma_R^2$	No test
M_j	4	$2\sigma_{RM}^2 + 6\phi_M$	
RM_{ij}	8	$2\sigma_{RM}^2$	No test
S_k	1	$5\sigma_{RS}^2 + 15\phi_S$	Poor test
RS_{ik}	2	$5\sigma_{RS}^2$	No test
MS_{jk}	4	$\sigma_{RMS}^2 + 3\phi_{MS}$	
RMS_{ijk}	8	σ_{RMS}^2	No test

12.7

Source	df	EMS
R_i	2	$30\sigma_R^2$
V_j	4	$6\sigma_{RV}^2 + 18\phi_V$
RV_{ij}	8	$6\sigma_{RV}^2$
S_k	5	$5\sigma_{RS}^2 + 15\phi_S$
RS_{ik}	10	$5\sigma_{RS}^2$
VS_{jk}	20	$\sigma_{RVS}^2 + 3\phi_{VS}$
RVS_{ijk}	40	σ_{RVS}^2

12.9

Source	df	EMS	
R_i	1	$\sigma_C^2 + 36\sigma_R^2$	
P_j	2	$\sigma_C^2 + 12\sigma_{RP}^2 + 24\phi_P$	Poor test
RP_{ij}	2	$\sigma_C^2 + 12\sigma_{RP}^2$	
H_k	5	$\sigma_C^2 + 6\sigma_{RH}^2 + 12\phi_H$	
RH_{ik}	5	$\sigma_C^2 + 6\sigma_{RH}^2$	
PH_{jk}	10	$\sigma_C^2 + 2\sigma_{RPH}^2 + 4\phi_{PH}$	
RPH_{ijk}	10	$\sigma_C^2 + 2\sigma_{RPH}^2$	
$C_{m(ijk)}$	36	σ_C^2	

12.11 Compare models.

12.13 Complete: $Y_{ijkm} = \mu + R_k + M_i + RM_{ik} + H_{j(i)} + RH_{kj(i)} + \varepsilon_{m(ijk)}$.
Reduced: $Y_{ijkm} = \mu + R_k + M_i + H_{j(i)} + \varepsilon_{m(ijk)}$.
Take fewer readings per head.

12.15 Complete: $Y_{ijkmq} = \mu + R_q + M_i + RM_{iq} + G_j + RG_{jq} + MG_{ij} + RMG_{ijq}$
$+ T_{k(j)} + RT_{qk(j)} + MT_{ik(j)} + RMT_{iqk(j)} + \varepsilon_{m(ijkq)}$

Reduced: Omit all interactions with R.

CHAPTER 13

13.1

	Source	df	EMS
Whole plot	R_i	1	$\sigma_\varepsilon^2 + 6\sigma_{RS}^2 + 18\sigma_R^2$
	H_j	2	$\sigma_\varepsilon^2 + 2\sigma_{RHS}^2 + 4\sigma_{HS}^2 + 6\sigma_{RH}^2 + 12\phi_H$
	RH_{ij}	2	$\sigma_\varepsilon^2 + 2\sigma_{RHS}^2 + 6\sigma_{RH}^2$
Split- plot	S_k	2	$\sigma_\varepsilon^2 + 6\sigma_{RS}^2 + 12\sigma_S^2$
	RS_{ik}	2	$\sigma_\varepsilon^2 + 6\sigma_{RS}^2$
	HS_{jk}	4	$\sigma_\varepsilon^2 + 2\sigma_{RHS}^2 + 4\sigma_{HS}^2$
	RHS_{ijk}	4	$\sigma_\varepsilon^2 + 2\sigma_{RHS}^2$
	$\varepsilon_{m(ijk)}$	18	σ_ε^2
	Total 35		

13.3   a. Fill nine spaces in random order.

b. Choose one of the orifice sizes at random, randomize flow rate through this orifice, then move to a second orifice size, and so on.

c.

Source	df
R_i	2
O_j	2
RO_{ij}	4
F_k	2
RF_{ik}	4
OF_{jk}	4
ROF_{ijk}	8

d. Take more replications.

13.5

	Source	df	EMS
Whole	R_i	3	$\sigma_\varepsilon^2 + 8\sigma_{RA}^2 + 40\sigma_R^2$
plot	S_j	1	$\sigma_\varepsilon^2 + 4\sigma_{RSA}^2 + 16\sigma_{SA}^2 + 20\sigma_{RS}^2 + 80\phi_S$
	RS_{ij}	3	$\sigma_\varepsilon^2 + 4\sigma_{RSA}^2 + 20\sigma_{RS}^2$
	J_k	1	$\sigma_\varepsilon^2 + 4\sigma_{RJA}^2 + 16\sigma_{JA}^2 + 20\sigma_{RJ}^2 + 80\phi_J$
	RJ_{ik}	3	$\sigma_\varepsilon^2 + 4\sigma_{RJA}^2 + 20\sigma_{RJ}^2$
	SJ_{jk}	1	$\sigma_\varepsilon^2 + 2\sigma_{RSJA}^2 + 10\sigma_{RSJ}^2 + 8\sigma_{SJA}^2 + 40\phi_{SJ}$
	RSJ_{ijk}	3	$\sigma_\varepsilon^2 + 2\sigma_{RSJA}^2 + 10\sigma_{RSJ}^2$
Split-	A_m	4	$\sigma_\varepsilon^2 + 8\sigma_{RA}^2 + 32\sigma_A^2$
plot	RA_{im}	12	$\sigma_\varepsilon^2 + 8\sigma_{RA}^2$
	SA_{jm}	4	$\sigma_\varepsilon^2 + 4\sigma_{RSA}^2 + 16\sigma_{SA}^2$
	RSA_{ijm}	12	$\sigma_\varepsilon^2 + 4\sigma_{RSA}^2$
	JA_{km}	4	$\sigma_\varepsilon^2 + 4\sigma_{RJA}^2 + 16\sigma_{JA}^2$
	RJA_{ikm}	12	$\sigma_\varepsilon^2 + 4\sigma_{RJA}^2$
	SJA_{jkm}	4	$\sigma_\varepsilon^2 + 2\sigma_{RSJA}^2 + 8\sigma_{SJA}^2$
	$RSJA_{ijkm}$	12	$\sigma_\varepsilon^2 + 2\sigma_{RSJA}^2$
	$\varepsilon_{q(ijkm)}$	80	σ_ε^2

13.7

	Source	df	MS	EMS
Whole	R_i	2	60.46	$\sigma_\varepsilon^2 + 5\sigma_{RO}^2 + 20\sigma_R^2$
plot	D_j	4	276.19	$\sigma_\varepsilon^2 + \sigma_{RDO}^2 + 3\sigma_{DO}^2 + 4\sigma_{RD}^2 + 12\phi_D$
	RD_{ij}	8	40.63	$\sigma_\varepsilon^2 + \sigma_{RDO}^2 + 4\sigma_{RD}^2$
Split-	O_k	3	16.73	$\sigma_\varepsilon^2 + 5\sigma_{RO}^2 + 15\sigma_O^2$
plot	RO_{ik}	6	57.44	$\sigma_\varepsilon^2 + 5\sigma_{RO}^2$
	DO_{jk}	12	12.78	$\sigma_\varepsilon^2 + \sigma_{RDO}^2 + 3\sigma_{DO}^2$
	RDO_{ijk}	24	46.83	$\sigma_\varepsilon^2 + \sigma_{RDO}^2$
	Total	59		

Days are significant by an F' test.

13.9   In a nested experiment the levels of B, say, are different within each level of A. In a split-plot experiment the same levels of B (in the split) are used at each level of A.

13.11 Laboratory means: 1   3   2

         11.22   11.25   12.53

         Mix means: B A C

         10.29   11.58   13.13

Laboratory $\times$ mix means all different in decreasing order: 165, 155, 145 with B, A, and C.

13.13

Source	EMS
R_i	$\sigma_\varepsilon^2 + 9\sigma_{RL}^2 + 27\sigma_R^2$
L_j	$\sigma_\varepsilon^2 + 9\sigma_{RL}^2 + 36\sigma_L^2$
RL_{ij}	$\sigma_\varepsilon^2 + 9\sigma_{RL}^2$
T_k	$\sigma_\varepsilon^2 + 3\sigma_{RLT}^2 + 12\sigma_{LT}^2 + 9\sigma_{RT}^2 + 36\phi_T$
RT_{ik}	$\sigma_\varepsilon^2 + 3\sigma_{RLT}^2 + 9\sigma_{RT}^2$
LT_{jk}	$\sigma_\varepsilon^2 + 3\sigma_{RLT}^2 + 12\sigma_{LT}^2$
RLT_{ijk}	$\sigma_\varepsilon^2 + 3\sigma_{RLT}^2$
M_m	$\sigma_\varepsilon^2 + 3\sigma_{RLM}^2 + 12\sigma_{LM}^2 + 9\sigma_{RM}^2 + 36\phi_M$
RM_{im}	$\sigma_\varepsilon^2 + 3\sigma_{RLM}^2 + 9\sigma_{RM}^2$
LM_{jm}	$\sigma_\varepsilon^2 + 3\sigma_{RLM}^2 + 12\sigma_{LM}^2$
RLM_{ijm}	$\sigma_\varepsilon^2 + 3\sigma_{RLM}^2$
TM_{km}	$\sigma_\varepsilon^2 + \sigma_{RLTM}^2 + 4\sigma_{LTM}^2 + 3\sigma_{RTM}^2 + 12\phi_{TM}$
RTM_{ikm}	$\sigma_\varepsilon^2 + \sigma_{RLTM}^2 + 3\sigma_{RTM}^2$
LTM_{jkm}	$\sigma_\varepsilon^2 + \sigma_{RLTM}^2 + 4\sigma_{LTM}^2$
$RTLM_{ijkm}$	σ_ε^2

13.15 a. Split-plot.

b. Source	df	EMS
R_i	$r - 1$	$\sigma_\varepsilon^2 + 15\sigma_R^2$
M_j	2	$\sigma_\varepsilon^2 + 5\sigma_{RM}^2 + 10\phi_M$
RM_{ij}	$2(r - 1)$	$\sigma_\varepsilon^2 + 5\sigma_{RM}^2$
T_k	4	$\sigma_\varepsilon^2 + 3\sigma_{RT}^2 + 6\phi_T$
RT_{ik}	$4(r - 1)$	$\sigma_\varepsilon^2 + 3\sigma_{RT}^2$
MT_{jk}	8	$\sigma_\varepsilon^2 + \sigma_{RMT}^2 + 2\phi_{MT}$
RMT_{ijk}	$8(r - 1)$	$\sigma_\varepsilon^2 + \sigma_{RMT}^2$

$c \cdot r \geqq 4$.

13.17

Source	df	EMS
R_i	1	$\sigma_\varepsilon^2 + 18\sigma_R^2$
M_j	2	$\sigma_\varepsilon^2 + 6\sigma_{RM}^2 + 12\phi_M$
RM_{ij}	2	$\sigma_\varepsilon^2 + 6\sigma_{RM}^2$
T_k	5	$\sigma_\varepsilon^2 + 3\sigma_{RT}^2 + 6\phi_T$
RT_{ik}	5	$\sigma_\varepsilon^2 + 3\sigma_{RT}^2$
MT_{jk}	10	$\sigma_\varepsilon^2 + \sigma_{RMT}^2 + 2\phi_{MT}$
RMT_{ijk}	10	$\sigma_\varepsilon^2 + \sigma_{RMT}^2$

13.19 Take another replication.

CHAPTER 14

14.1 Show. $SS_{blocks} = SS_{AC} = 10.125$.

14.3

replication I confound AB		replication II confound AC		replication III confound BC	
(1)	a	a	(1)	(1)	b
ab	b	c	ac	bc	c
c	ac	ab	b	a	ab
abc	bc	bc	abc	abc	ac

14.5 BC.

14.7 $SS_{blocks} = 29{,}153.5 = SS_{AB} + SS_{ACD} + SS_{BCD}$.

14.9 The principal block contains: (1), $ab, ad, bd, c, abc, acd, bcd$.

14.11 One scheme is to confound ABC, CDE, and $ABDE$. The principal block contains (1), $ab, de, abde, ace, bce, acd, bcd$.

Source	df
Main effects	5
Two-way interaction	10
Three-way interaction	8
Four-way interaction	4
Five-way interaction	1
Blocks (or ABC, CDE, $ABDE$)	3
Total	31

14.13 One scheme is to confound $ABCD^2$. The principal block contains 0000, 1110, 1102, 1012, 2111, 2122, 0221, 0210, 0120, 0101, 1211, 1200, 2220, 0022, 0011, 0202, 0112, 1121, 1020, 1001, 1222, 2010, 2201, 2100, 2002, 2021, 2212.

14.15 Verify.

14.17 Confounded *ABC, CDE, ADF, BEF, ABDE, BCDF, ACEF* gives a principal block of (1), *acd, bce, abf, abde, bcdf, acef, def*.

Source	df
Main effects	1 each for 6
Two-way interaction	1 each for 15
Three-way interaction (except confounds)	1 each for 16
Four-way interaction (except confounds)	1 each for 12
Five-way interaction	1 each for 6
Six-way interaction	1 each for 1
Blocks (or confounds)	7
Total	63

14.19 Yates with seven columns—pool to get temperature and its interactions.

14.21 Confound *TABC* with blocks (mod 2), $L = X_1 + X_2 + X_3 + X_4$.

14.23 Principal blocks are

Replication I   (1), *ab, de, abde, bcd, acd, bce, ace*
Replication II (1), *be, cd, bcde, ace, abc, ade, abd*
Replication III (1), *bc, abe, ace, cde, bde, abcd, ad*

Source	df	
Replication	2	
Blocks in replication	9	11
Main effects	1 each for 5	
Two-way interaction	1 each for 10	
Three-way interaction	1 each for 10	
Four-way interaction	1 each for 5	
Five-way interaction	1 each for 1	
Replications × all	53	84
Total	95	

CHAPTER 15

15.1 Aliases:

Source	df	SS	
$A = C$	1	49	
$B = ABC$	1	0	No tests.
$AB = BC$	1	1	

15.3   Aliases are

$$A = BC = AB^2C^2$$
$$B = AB^2C = AC$$
$$C = ABC^2 = AB$$
$$AB^2 = AC^2 = BC^2$$

15.5 $A = BD, B = AD, C = ABCD, D = AB, AC = BCD, BC = ACD, CD = ABC.$

15.7 $A = AB^2C^2D = BCD^2$
         $B = AB^2CD^2 = ACD^2$
         $C = ABC^2D^2 = ABD^2$
         $D = ABC = ABCD$
        $AB = ABC^2D = CD^2$
       $AB^2 = AC^2D = BC^2D$
        $AC = AB^2CD = BD^2$
       $AC^2 = AB^2D = BC^2D^2$
        $AD = AB^2C^2 = BCD$
        $BC = AB^2C^2D^2 = AD^2$
       $BC^2 = AB^2D^2 = AC^2D^2$
        $BD = AB^2C = ACD$
        $CD = ABC^2 = ABD$

15.9

Source	df
Main effects with 5- or 6-way	7
2-way interaction with 4- or 6-way	21
3-way interaction, and so on as error	35
Total	63

15.11

Source	df	SS	MS
Heats (or 2-way)	6	51.92	8.65*
Treatments (or part of H)	1	12.04	12.04*
Positions (or 3-way)	2	0.58	0.29
PH (or 3-way)	12 ⎱	29.83 ⎱ 33.42	2.39
PT (or 2-way)	2 ⎰	3.59 ⎰	
Totals	23	97.96	

15.13 Combining three- and four-way interactions as error with 5 df gives

Source	MS
A	12.44
B	53.40*
AB	0.00
C	33.44
AC	7.06
BC	0.00
D	31.22
AD	42.48*
BD	1.62
CD	0.07
Error	5.73

B and AD significant at 5 percent level.

15.15 a. Replication I: (1), ab, de, abde, bcd, acd, bce, ace.
  b. BC = A = BDE = ACDE.
  c. Replications × three-way = 34 df.

15.17 a. ACD.

b.

Source	df	MS
A (or CD)	1	5729.85*
B (or ABCD)	1	91.80
C (or AD)	1	4199.86*
D (or AC)	1	1217.71
AB, BC, BD, and their aliases	3	281.88

Some bad aliases, better to confound ABCD.

15.19 BCDEF confounded.

Source	df
Main (or four-way)	1 each for 5
Two-way or three-way	1 each for 10

No good error term unless one assumes no interactions.

CHAPTER 16

16.1 Before covariance: $F_{6,100} = 50$ and significant.
  After covariance: $F_{6,99} = 16.5$ and still significant.

16.3

Source	df	$\sum x^2$	$\sum xy$	$\sum y^2$	Adjusted $\sum y^2$	df	MS
Between lots	3	500	− 620	790			
Error	16	1182	428	406	251.02	15	16.73
Totals	19	1682	− 192	1196	1174.08	18	
					923.06	3	307.69*

16.5

End of cycle	Estimate of s	Significance
2, block 1	0.60	None
2, block 2	0.64	All except AB
3, block 1	0.60	
3, block 2	0.61	All

16.7

Source	df	MS
Machines M	1	0.0136
Lumber grade L	1	0.0003
ML	1	0.0066
Replications R	1	0.0001
MR	1	0.0028
LR	1	0.0021
MLR	1	0.0015

16.9

Source	df	SS	MS
Days	3	34,292	11,431
Treatments (adjusted)	3	508	169
Error	5	5025	1005

Treatments not significant at 5 percent.

16.11 If seven were used,

Source	df
Treatments	1
Blocks (men)	6
Error	6
Total	13

Index

α, probability of type I error, 20
AB^2 and AB components of A $\times$ B interaction, 195
Aliases, 304–306
Analysis, 2, 6
Analysis of variance (ANOVA), 37, 351
  covariance analysis and, 319–321
  factorial experiments rationale, 98–103
  fundamental equation, 40, 73
  nested experiments rationale, 232–233
  one-way, 37–46, 50, 53, 54, 56, 57, 59, 61
  randomized block design rationale, 72–73
  rationale, 38–46
  two-way, 68, 73
Anderson, 270
Attribute data, 318, 351–360
  philosophy, 351

β, probability of type II error, 20–23
Barnett, E. H., 341
Bartlett-Box F test, 60
Batson, H. C., 351, 355
Behrens-Fisher problem, 28
Bennett, C. A., 210, 222
Binomial variable, 7
Blocks, 68
  See also Randomized block design
Box, G.E.P., 339, 341

Canonical form, 340
Central composite design, 340

Change in mean effect (CIM), 341, 345
Chi square
  factorial, 351–354
  table, 382
Cochran's C test, 60
Coefficient of determination r^2, 135
Completely randomized design, 11, 67–76, 266, 319
  distinguished from split-plot design, 270
Components of variance, 55–59
Computer programs
  factorial experiments, 96–97
  homogeneity of variance, 60–61
  multiple regression, 167, 175–176
  for nested or nested-factorial experiments, 238–239
  Newman-Keuls test (SNK), 53–54
  for one-way ANOVA, 43–44, 50
  problems with quantitative levels of a factor, 144, 145
Concomitant variable X, 318
Confidence
  interval, 19
  limits, 19, 25, 26, 54
Confounding in blocks, 281–299
  complete, 290–291
  confounding systems, 282–285
  partial, 291–294
  in 2^f factorials, 285–290
  in 3^f factorials, 294–299
Consistent statistic, 18
Contrasts, 46–53, 114

Correlation, 2
  coefficient (r), 135, 168, 177
  ratio R and R^2, 139
Covariance analysis, 318–328
  philosophy, 318–319
Critical region (of test statistic), 20, 21
Curvilinear regression, 137–139

Davies, O. L., 25, 297
Decision rule, 20, 21, 25
Defining contrast, 283
Degrees of freedom, 73, 75, 77, 78
Dependent variable, 3
Design
  completely randomized, 11, 67–76,
    266, 270, 319
  general, 4–6
  Graeco-Latin square, 79–80
  incomplete block, 80, 355–360
  randomized block, 67–76
  split-plot, 265–276
  split-split plot, 273–276
  Youden square, 80, 360–362
Developmental research, 2

Effect of a factor, defined, 114
Error
  restriction, 270
  split-plot, 268
  type I, 20
  type II, 20–23
  whole-plot, 268
Estimate
  interval, 17, 19
  point, 17
Estimation, 17–19
Eta squared, 135
Evolutionary operation (EVOP), 318,
    341–351
  form rationale, 346–351
  philosophy, 341
Expected mean square (EMS), 56–59,
    212–223
  derivations, 218–220
  pseudo-F test, 221–223
  rules, 214–218

Expected value, 17–19, 58, 59, 218–220
Experiment
  defined, 1
  factorial, 3, 86–104, 112–124, 187–
    206, 252–259, 265–276, 281–299
  general, 2–4
  nested, 227–233, 238–244, 259
  nested-factorial, 233–244, 259
  single-factor, 36–61, 67–80, 103–104,
    211–212, 224, 312–313, 319
Experimental research, 1, 164
Ex-post-facto research, 2, 164

F distribution, 26–27, 38
  table of, 383–388
F'' test, 222–223
  pseudo-F, 221–223
Factorial chi square, 351–354
  table, 382
Factorial experiments, 3, 86–104, 112–
    124
  advantages, 88
  analysis, 91–94
  ANOVA rationale, 98–103
  computer program, 96–97
  confounding in blocks, 281–299
  defined, 88–89
  interpretations, 94–96
  in Latin square design, 258–259
  in randomized block design, 252–258
  in split-plot design, 265–276
  2^2, 112–119
  2^3, 10, 120–123
  3^3, 187–198
  3^3, 198–206
Factors. See Levels of factors
Fisher, R. A., 77, 140, 355
Fixed model, 210–214, 220
Fractional factorial, 304–314
Fractional replication, 303–314
  aliases, 304–306
  2^f, 306–310
  3^f, 310–314
Franklin, N. L., 210, 222
Fundamental equation of analysis of
    variance, 40, 73

Graeco-Latin square design, 79–80, 259

Hierarchical experiments, 3, 227
Historical research, 2
Homogeneity of variances tests, 60–61
Hypotheses
    statistical, 19
    tests of, 19–21, 25–27, 206

I and J components of interaction, 194
Incomplete block design, 80, 355–360
Independent variables, ways of handling,
    5–6
Inference. *See* Statistical inference
Interaction
    A × B, AB components of, 195
    factorial experiments, 88–90, 93–95,
        97, 99, 100, 102–104
    factorial experiments in randomized
        block design, 253, 256, 258
    high-order, increase in, 303
    I and J components of, 194
    nested-factorial experiments, 234
    with quantitative and qualitative fac-
        tors, 149–151, 153, 154, 156
    2^2 factorial experiments, 113–116
    2^3 factorial experiments, 120, 121
    3^2 factorial experiments, 189–198
    3^3 factorial experiments, 198–206
    Intercorrelation matrix, 168
    Interval estimate, 17, 19

J and I interaction, 194

Kempthorne, O., 283, 294
Keuls. *See* Newman-Keuls

Latin square design, 76–80, 328, 360
    factorial experiments in, 253, 258–259
    fractional replication and, 312, 313
    Graeco-Latin squares, 79–80
    nested or nested-factorial experiments
        in, 259
    Youden square, 80, 360–362
Least squares method, 132, 164
Least squares normal equations, 132
Levels of factors

    qualitative and quantitative, 3, 5, 6,
        144–158
    random and fixed, 36–37, 210–224
Linear regression, 129, 130–137

McLean, 270
Manipulation, 1, 2
Mathematical model, 5, 6
Max-min-con principle, 6
Mean of population, 11, 16
    of sample, 17
Mean square (MS), 18, 41
Means
    confidence limits on, 54
    tests on, 27–30, 47, 51
Minimum variance, 18
Minimum-variance estimate, 18
Missing values, 73–76
Mixed model, 210, 211, 213, 214, 220,
    223
Model
    fixed, 210–214, 220
    mathematical, 5, 6
    mixed, 210, 211, 213, 214, 220, 223
    random, 55, 210–214, 220, 223
Modulus of 3, 195
Multiple regression, 129, 164–177, 331
    computer programs, 167, 175–176

Nested experiments, 3, 227–233
    computer programs, 238–239
    in randomized blocks or Latin square
        design, 259
    repeated-measures designs as, 239–244
Nested-factorial experiments, 233–238
    computer programs, 238–239
    in randomized blocks or Latin square
        designs, 259
    repeated-measures designs as, 239–244
Newman-Keuls range test, 51–54, 78,
    94–95, 198, 231, 238
NID, 36, 37
Normal curve, table of, 379–380
Null hypothesis form, 5

Operating characteristic (OC) curve, 22–
    23

Orthogonal contrasts, 47–51, 359
Orthogonal polynomials, 129, 139–144, 159
  coefficients of, table, 391
Ostle, B., 319, 328
Owen, D. B., 25

Parameters, 16
Path of steepest ascent method, 330
Pearson product-moment correlation coefficient (r's), 135, 168, 177
Plots, 266
Point estimates, 17
Population standard deviation, 16
Population variance, 16
Power curve, 23
Power of a test, 23
Principal block, 284
Probability, 15
Proportions, tests on, 30–32
Pseudo-F test, 221–223

Quadratic regression, 137–139
Qualitative/Quantitative factors, 129, 144–159
  two factors, both quantitative, 152–158
  two factors, one qualitative, one quantitative, 145–152

Random model, 55, 210–214, 220, 223
Random order, 4–5
Random samples, 16
Random variable, average or expected value of, 16
Randomization, 1, 2
  of order of experimentation, 4–5
Randomized block design
  complete, 67–76
  factorial experiments in, 252–258
  incomplete, 80, 355–360
  nested or nested-factorial design in, 259
Randomized blocks, 328

Regression, 2, 319, 321
  covariance analysis and, 319, 321, 324, 326, 328
  curvilinear, 137–139
  linear, 129–137
  multiple, 129, 164–177, 331
  orthogonal polynomials, 139–144
Repeated-measures designs, 71
  as nested and nested-factorial experiments, 239–244
Research
  defined, 1
  types, 2
Research hypothesis, 5
Response-surface experimentation, 318, 328–340, 341
  more complex surfaces, 338–340
  philosophy, 328–329
  two-fold problem, 329–338
Response variable Y, 3
Restriction error, 270
Risks of making incorrect decisions, 20
Rotatable design, 339

Sample, random, 16
Sample size, determining, 4, 23–25, 224
Sample standard deviation, 17
Sample statistics, 16, 19
Satterthwaite, 222
Scheffé's test, 51, 52–54
Single-factor experiments, 36–61, 312–313
  compared with factorial experiments with one observation per cell, 103–104
  completely randomized, 36–59
  covariance analysis and, 319
  Latin square designs, 76–80
  levels of treatment, 211–212
  randomized block design, 67–76
  sample size, 224
Snedecor, G. W., 319
SNK, 53
Split-plot design, 265–276

Split-plot error, 268
Split-split-plot design, 273–276
SPSS (Statistical Package for the Social Sciences), 43, 44, 53
Standard deviation
  of population, 16
  of sample, 17
Standard error of estimate, 135
Statistical hypothesis, 19
Statistical inference, 16–32
  application to tests on means, 27–30
  application to tests on proportions, 30–32
  application to tests on variances, 25–27
  estimation, 16–19
  operating characteristic curve, 21–23
  sample size, determining, 23–25
  tests of hypotheses, 19–21
Statistics
  defined, 15
  sample, 16
Stepwise Regression program, 168, 173, 176
Studentized ranges, table of, 389–390
Student's t distribution, 54, 381
Sum of squares (SS), 17, 18, 41–43, 56
  degrees of freedom, 73
Survey research, 2

Test statistics, 19
Treatment effect, 36, 211
Type I error, 20
Type II error, 20–23

Unbiased statistic, 17
Uncertainty, 15

Variance
  components of, 55–59
  of population, 16
  of sample, 16, 17
Variance ratio, 26
Variance tests, 25–27
  homogeneity, 60–61
  maximum/minimum, 60
Variances, homogeneous, tests for, 60–61

Whole-plot error, 268
Winer, 243

Yates, F., 77, 118, 140, 355
Yates method, 118, 119, 121–123, 288, 308, 309, 352, 353
Youden square, 80, 360–362